Prediction of Protein Structures, Functions, and Interactions

Prediction of Protein Structures, Functions, and Interactions

Edited by

JANUSZ M. BUJNICKI

Laboratory of Bioinformatics and Protein Engineering, International Institute of Molecular and Cell Biology, Warsaw, Poland

and

Laboratory of Bioinformatics, Institute of Molecular Biology and Biotechnology, Faculty of Biology, Adam Mickiewicz University, Poznan, Poland

A John Wiley and Sons, Ltd., Publication

This edition first published 2009
© 2009 John Wiley & Sons Ltd

Registered Office
John Wiley & Sons Ltd, The Atrium, Southern Gate, Chichester, West Sussex, PO19 8SQ, United Kingdom

For details of our global editorial offices, for customer services and for information about how to apply for permission to reuse the copyright material in this book please see our website at www.wiley.com.

The right of the author to be identified as the author of this work has been asserted in accordance with the Copyright, Designs and Patents Act 1988.

All rights reserved. No part of this publication may be reproduced, stored in a retrieval system, or transmitted, in any form or by any means, electronic, mechanical, photocopying, recording or otherwise, except as permitted by the UK Copyright, Designs and Patents Act 1988, without the prior permission of the publisher.

Wiley also publishes its books in a variety of electronic formats. Some content that appears in print may not be available in electronic books.

Designations used by companies to distinguish their products are often claimed as trademarks. All brand names and product names used in this book are trade names, service marks, trademarks or registered trademarks of their respective owners. The publisher is not associated with any product or vendor mentioned in this book. This publication is designed to provide accurate and authoritative information in regard to the subject matter covered. It is sold on the understanding that the publisher is not engaged in rendering professional services. If professional advice or other expert assistance is required, the services of a competent professional should be sought.

The Publisher and the Author make no representations or warranties with respect to the accuracy or completeness of the contents of this work and specifically disclaim all warranties, including without limitation any implied warranties of fitness for a particular purpose. This work is sold with the understanding that the Publisher is not engaged in rendering professional services. The advice and strategies contained herein may not be suitable for every situation. In view of ongoing research, equipment modifications, changes in governmental regulations, and the constant flow of information relating to the use of experimental reagents, equipment, and devices, the reader is urged to review and evaluate the information provided in the package insert or instructions for each chemical, piece of equipment, reagent, or device for, among other things, any changes in the instructions or indication of usage and for added warnings and precautions. The fact that an organization or Website is referred to in this work as a citation and/or a potential source of further information does not mean that the Author or the Publisher endorses the information the organization or Website may provide or recommendations it may make. Further, readers should be aware that Internet Websites listed in this work may have changed or disappeared between when this work was written and when it is read. No warranty may be created or extended by any promotional statements for this work. Neither the Publisher nor the Author shall be liable for any damages arising herefrom.

Library of Congress Cataloging-in-Publication Data

Prediction of protein structures, functions, and interactions / [edited by] Janusz M. Bujnicki.
       p. ; cm.
    Includes bibliographical references and index.
    ISBN 978-0-470-51767-3
    1. Proteins–Structures.   2. Amino acid sequence.   3. Protein-protein interactions.   I. Bujnicki, Janusz M.
    [DNLM: 1. Sequence Analysis, Protein–methods.   2. Amino Acid
Sequence–physiology.   3. Models, Molecular.   4. Protein Folding. QU 450 P923 2008]
    QP551.P656   2008
    572'.633–dc22   2008040271

A catalogue record for this book is available from the British Library.

ISBN 978-0-470-51767-3 (H/B)

Set in 10/12pt Times by Aptara Inc., New Delhi, India
Printed and bound in Great Britain by CPI Antony Rowe, Chippenham, Wiltshire

I dedicate this book with love to my parents.
Thank you for everything.

Contents

List of Contributors

Alexandre M.J.J. Bonvin, Bijvoet Center for Biomolecular Research, Science Faculty, Utrecht University, Padualaan 8, 3584CH, Utrecht, The Netherlands

Janusz M. Bujnicki, Laboratory of Bioinformatics and Protein Engineering, International Institute of Molecular and Cell Biology, ul. Ks. Trojdena 4, 02-109 Warsaw, Poland, and Laboratory of Bioinformatics, Institute of Molecular Biology and Biotechnology, Faculty of Biology, Adam Mickiewicz University, ul. Umultowska 89, 61-614 Poznan, Poland

Meghana Chitale, Department of Computer Science, College of Science, Purdue University, West Lafayette, IN, 47907, USA

Sjoerd J. de Vries, Bijvoet Center for Biomolecular Research, Science Faculty, Utrecht University, Padualaan 8, 3584CH, Utrecht, The Netherlands

Arne Elofsson, Center for Biomembrane Research and Stockholm Bioinformatics Center, Stockholm University, SE-106 91 Stockholm, Sweden

Dominik Gront, Faculty of Chemistry, University of Warsaw, Pateura 1, 02-093 Warsaw, Poland

Troy Hawkins, Department of Biological Sciences, College of Science, Purdue University, West Lafayette, IN, 47907, USA

Marcin Jąkalski, Laboratory of Bioinformatics, Institute of Molecular Biology and Biotechnology, Faculty of Biology, Adam Mickiewicz University, ul. Umultowska 89, 61-614 Poznan, Poland

Katarzyna H. Kaminska, Laboratory of Bioinformatics and Protein Engineering, International Institute of Molecular and Cell Biology, ul. Ks. Trojdena 4, 02-109 Warsaw, Poland, and Laboratory of Bioinformatics, Institute of Molecular Biology and Biotechnology, Faculty of Biology, Adam Mickiewicz University, ul. Umultowska 89, 61-614 Poznan, Poland

Joanna M. Kasprzak, Laboratory of Bioinformatics, Institute of Molecular Biology and Biotechnology, Faculty of Biology, Adam Mickiewicz University, ul. Umultowska 89, 61-614 Poznan, Poland

Daisuke Kihara, Department of Biological Sciences/Computer Science, Markey Center for Structural Biology, College of Science, Purdue University, West Lafayette, IN, 47907, USA

Kengo Kinoshita, Human Genome Center, Institute of Medical Science, University of Tokyo, 4-6-1 Shirokanedai, Minato-ku, Tokyo 108-8639, Japan

Andrzej Kolinski, Faculty of Chemistry, University of Warsaw, Pateura 1, 02-093 Warsaw, Poland

Hidetoshi Kono, Computational Biology, Quantum Beam Science Directorate, Japan Atomic Energy Agency, 8-1 Umemidai, Kizugawa, Kyoto 619-0215, Japan

Jan Kosiński, Laboratory of Bioinformatics and Protein Engineering, International Institute of Molecular and Cell Biology, ul. Ks. Trojdena 4, 02-109 Warsaw, Poland

Łukasz Kozłowski, Laboratory of Bioinformatics and Protein Engineering, International Institute of Molecular and Cell Biology, ul. Ks. Trojdena 4, 02-109 Warsaw, Poland

Mateusz Kurcinski, Faculty of Chemistry, University of Warsaw, Pateura 1, 02-093 Warsaw, Poland

Dorota Latek, Faculty of Chemistry, University of Warsaw, Pateura 1, 02-093 Warsaw, Poland

Gonzalo Lopez, Structural Biology and Biocomputing Programme, Spanish National Cancer Research Centre (CNIO), C. Melchor Fernandez Almagro, 3, 28029 Madrid, Spain

Karolina Majorek, Laboratory of Bioinformatics, Institute of Molecular Biology and Biotechnology, Faculty of Biology, Adam Mickiewicz University, ul. Umultowska 89, 61-614 Poznan, Poland

Kaja Milanowska, Laboratory of Bioinformatics, Institute of Molecular Biology and Biotechnology, Faculty of Biology, Adam Mickiewicz University, ul. Umultowska 89, 61-614 Poznan, Poland

Elena Nabieva, Department of Computer Science & Lewis-Sigler Institute for Integrative Genomics, Carl Icahn Laboratory, Princeton University, Princeton, NJ 08544, USA

Mona Singh, Department of Computer Science & Lewis-Sigler Institute for Integrative Genomics, Carl Icahn Laboratory, Princeton University, Princeton, NJ 08544, USA

Janet M. Thornton, European Bioinformatics Institute, Wellcome Trust Genome Campus, Hinxton, Cambridge, CB10 1SD, UK

Karolina L. Tkaczuk, Laboratory of Bioinformatics and Protein Engineering, International Institute of Molecular and Cell Biology, ul. Ks. Trojdena 4, 02-109 Warsaw, Poland

James W. Torrance, European Bioinformatics Institute, Wellcome Trust Genome Campus, Hinxton, Cambridge, CB10 1SD, UK

Michael Tress, Structural Biology and Biocomputing Programme, Spanish National Cancer Research Centre (CNIO), C. Melchor Fernandez Almagro, 3, 28029 Madrid, Spain

Alfonso Valencia, Structural Biology and Biocomputing Programme, Spanish National Cancer Research Centre (CNIO), C. Melchor Fernandez Almagro, 3, 28029 Madrid, Spain

Marc van Dijk, Bijvoet Center for Biomolecular Research, Science Faculty, Utrecht University, Padualaan 8, 3584CH, Utrecht, The Netherlands

Björn Wallner, Center for Biomembrane Research and Stockholm Bioinformatics Center, Stockholm University, SE-106 91 Stockholm, Sweden

Kei Yura, Center for Informational Biology, Ochanomizu University, 2-1-1, Otsuka, Bunkyo, Tokyo 112-8610, Japan

Preface

Nucleotide sequences of nucleic acids (DNA or RNA) and amino acid sequences of proteins are the most widely used type of biological information. Nucleic acid sequences are relatively easy to determine experimentally, and can be used to predict the sequences of proteins they encode. Thanks to advances in high-throughput DNA sequencing, the number of sequences in public databases grows exponentially. Thereby, sequence databases became one of the largest data resources in biology, providing researchers with a wealth of information, and challenging computer scientists to develop tools that would allow for rapid discrimination between relevant and irrelevant information for various types of analyses. While DNA sequence is mostly a carrier of genetic information, genes typically exert their function through the RNA and protein molecules they encode. A growing body of evidence points toward the importance of various RNA molecules that do not encode proteins, but fulfill various regulatory and/or catalytic roles. However, proteins that are produced by translation of coding RNAs remain the most important type of macromolecules required for the vital functions of the cell. Therefore, this book focuses on computational analyses of proteins, although many of these analyses are relevant also to nucleic acids (e.g. sequence alignment and prediction of protein-nucleic acid complexes).

Proteins perform most essential structural, enzymatic, transport, and regulatory functions in the cell. Protein functions are strictly determined by their structures, which can be organized into four levels of hierarchies with increasing complexity. These levels are: primary, secondary, tertiary, and quaternary structure. Above the structure of individual proteins and complexes there is another level of complexity, namely networks of interactions. The description of this hierarchy extends beyond the area of traditional structural bioinformatics and enters the realm of systems biology. The chapters in this book are meant to provide summaries of the concepts underlying protein sequence-structure-function relationships at these different levels of organization, and to serve as a comprehensive resource of state-of-the-art methods (as of 2008) for the corresponding bioinformatics analyses. The book is aimed at advanced graduate students, postdocs and faculty members. Although individual chapters have been written by several different authors and each can stand on its own, the book has been designed as a cohesive whole rather than a collection of independent reviews.

The primary structure corresponds to a linear sequence of amino acid residues linked together by peptide bonds into a polypeptide chain. The C-α atoms and peptide bond atoms form the main chain backbone, and other atoms protrude away as side chains. Functionally and evolutionarily relevant elements of primary structure are sequence motifs, domains and modules. In this volume, the bioinformatic tools for protein sequence analysis, including detection of motifs and domains, are reviewed in Chapter 1 by Kaminska *et al.*, while tools for protein function prediction from sequence are reviewed in Chapter 3 by Chitale *et al.* The level above primary structure is the secondary structure, defined as the local conformation of a polypeptide chain. It is characterized by the presence of recurring elements stabilized by hydrogen bonds between the main chain carboxyl and amino groups of different residues. Two very common types of secondary structures are corkscrew-like helices and extended strands; these regular elements are connected by loops that exhibit a plethora of different conformations. Further, some regions of polypeptide chain may exhibit no stable conformation and appear intrinsically disordered. Bioinformatics methods for predicting secondary structure and regions of disorder are reviewed in Chapter 2 by Majorek *et al.* The next level up is tertiary structure, which describes the three-dimensional arrangement of secondary structural elements and connecting regions within individual protein domains. Two different approaches for tertiary structure prediction, by 'template-based' and 'template-free' methods, have been reviewed in Chapter 4 by Kosinski *et al.* and in Chapter 5 by Gront and Kolinski, respectively. Methods for quality assessment of protein models (i.e. MQAPs) have been reviewed by Wallner and Elofsson in Chapter 6. Methods for predicting possible sites of interactions with other molecules and potential enzymatic functions from protein structure are reviewed in Chapter 7 by Kinoshita and coworkers, and Chapter 8 by Torrance and Thornton, respectively. Beyond the tertiary structure is the quaternary structure, which usually refers to the non-covalent association of two or more polypeptide chains into a protein complex. Here, this term will be also used to describe non-covalent interactions between domains within a single polypeptide. The so-called docking methods for prediction of the structure of protein-protein and protein-nucleic acid complexes from the structure of the isolated subunits are reviewed in Chapter 9 by de Vries and coworkers. The relationship of protein networks to protein function is reviewed in Chapter 10 by Nabieva and Singh, while Chapter 11 by Tress and coworkers integrates the whole spectrum of studies, from sequences to structures and networks, and places them in the context of systems biology.

The successful publication of this book would not have been possible without much appreciated support of many people. First, I owe an enormous debt to the various authors who contributed to this volume. Their hard work, dedication, expertise have made this book something we can all be very proud of. Thank you! Sincere thanks are extended to Paul Deards and Richard Davies, along with everyone else at Wiley who helped to manage this publication project. Thanks are also owed to members of my laboratory, both in Warsaw and in Poznań, for tolerating my prolonged absence during the major writing and editing phase. I must thank Ichizo Kobayashi and the Univeristy of Tokyo for hosting and supporting me while this work has been carried out, and all members of the Kobayashi laboratory for their hospitality (and nourishing me with sushi, sashimi, tempura, okonomiyaki, fugu, soba, udon, dried octopus, and yes of course, sake and shochu).

Janusz M. Bujnicki
Warsaw-Poznań-Tokyo
2007–08

1

The Basics of Protein Sequence Analysis

Katarzyna H. Kaminska, Kaja Milanowska and Janusz M. Bujnicki

1.1 Introduction

Genes and proteins are products of evolution. Over the course of evolution, the nucleotide sequences of genes undergo numerous changes. First, duplications (or deletions) may lead to creation of additional copies (or removal) of genes or gene fragments. Second, local mutations: substitutions, insertions and deletions within genes may result in changes to the amino acid sequence of proteins they encode. Thus, the initially identical copies of duplicated genes over time accumulate divergent mutations that make their sequences progressively dissimilar. Not all positions of protein-encoding genes are equally susceptible to mutation, as some amino acid residues may be very important for protein function, stability, or folding and may thus be more constrained in the residue types allowed. Therefore, although mutations are random, in nature we observe only such protein variants, in which sequence changes have been 'accepted' by the evolutionary pressure. Proteins with mutations that cause detrimental changes in structure and/or function are usually eliminated. If the protein is important to the integrity of the organism, the organism that bears the mutant gene dies, and the structurally/functionally compromised variant ceases to exist; or if it is not important, then the inactivated gene may be eventually 'purged' from the genome by random deletions. On the other hand, if the mutant variant brings an additional and beneficial new function to the organism, it is likely to be retained and further 'optimized' towards the activity favored by the selective pressure.

The above-mentioned evolutionary mechanisms have given rise to families of evolutionarily related proteins (homologs), which share a common ancestor. Duplicated proteins are

Prediction of Protein Structures, Functions, and Interactions   Edited by Janusz Bujnicki
© 2009 John Wiley & Sons, Ltd

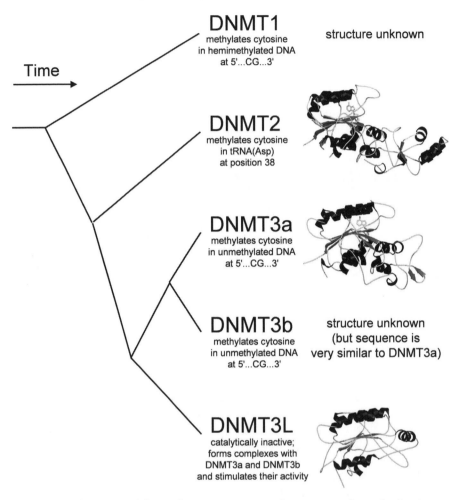

Figure 1.1   *In the course of the evolution, protein-encoding genes undergo duplications, and the resulting copies accumulate differentiating mutations (substitutions, insertions, deletions). As long as a small subset of residues important for internal stability and interactions with key partner molecules is preserved, the overall structure and mode of action of diverging homologous proteins is likely to remain similar. As a result, we observe that extant homologous proteins retain similar tertiary structure, while sequence similarity becomes less and less evident. Mutations may cause the protein or one of its paralogous copies to lose its function (and be eliminated), or to develop a new function, usually by somehow modifying the previous function. Example: a family of cytosine-C5 methyltransferases. Most members methylate cytosine in DNA; however, DNMT2 has apparently changed its specificity and acts on tRNA, and DNMT3L has lost the original catalytic activity, but instead gained a new regulatory activity*

described as paralogs, and in these relatives the sequences and functions can diverge considerably from the original variant (see Figure 1.1). A general function of paralogs (such as the ability to bind a certain type of molecule or to catalyze a certain type of chemical reaction) often remains conserved, but they tend to specialize in different specific roles (e.g. catalysis of a similar reaction on different substrates, different mode of regulation, or being directed to different cellular compartments etc.). It has been found that new activities and entire biochemical pathways evolve by recruitment and 'tinkering' of enzymes that are already capable of performing the desired chemistry, rather than by developing new functions from scratch.[1] Thus, paralogous enzymes are often found to carry out similar reactions in different pathways. Examples of large groups of paralogous proteins include: different kinases or different helix-turn-helix transcription factors encoded in the human genome. On the other hand, proteins from different organisms that have diverged from the ancestral gene present in the last common ancestor of these organisms (i.e. copies of 'the same protein in different organisms') are called orthologs. They tend to retain very similar functions and their sequences usually show higher conservation than between paralogs (for a detailed review on orthology and paralogy and discussion of several caveats, see ref.[2]). Thus, members of a protein family exhibit divergence, but usually share a specific biological function despite high sequence diversity.

As the number of known protein structures solved by X-ray crystallography and NMR techniques increased, it became clear that protein structure is much more highly conserved throughout evolution than protein's sequence.[3] While in many families sequence identity between members can drop below 5% identical residues, they tend to retain most of their common structural scaffold, mainly in the core of the protein. Structure is also more conserved than function; remote paralogs that retain common fold but replace functionally important residues may fulfill completely different roles in the cell (examples include a non-enzymatic heme-binding protein nitrophorin of a bedbug, which is related to an enzyme inositol polyphosphate 5-phosphatase[4]). Counterexamples may be found: proteins that exhibit high sequence similarity but different functions and/or structures (for a review see ref.[5]), however they are relatively rare. This suggests that structure comparison is the best method to detect remote evolutionary relationship.[6] Unfortunately, protein structure determination is considerably more costly and time consuming than gene sequencing, therefore the sequence databases have always been several orders of magnitude larger than the structure databases. There has been an exponential increase in the sizes of both types of data since the early 1970s but the largest sequence database GenBank[7] doubles in size roughly every 18 months, while the number of protein structures deposited in the Protein Data Bank[8] doubles roughly every three years, hence the gap keeps growing and is unlikely to be closed in the near future.

Not only have the structures lagged behind sequences, but also functional characterization. With the current pace of data generation by high-throughput sequencing projects, it is an impossible task to study all proteins by experiment. Thus, it is imperative to develop methods that use sequence information to identify evolutionary relationships and/or predict common structures and functions (or at least some aspect thereof). In this chapter, we discuss bioinformatic approaches for analyzing protein sequences, in particular aiming at identification and basic characterization of evolutionary relationships. The following chapters in this volume focus on direct prediction of functional properties from sequence (Chitale *et al.*), prediction of local conformation (Majorek *et al.*), and construction of three-dimensional structural models based on sequence analyses (Kosinski *et al.*). Here, we first define the primary functional units in protein sequence (domains and motifs) and describe how domains are duplicated and combined in various ways to give different protein families. We then briefly describe the major classifications and databases of protein families, domains, and motifs. In the main part of the chapter we review algorithms for protein sequence analyses, with the particular focus on their implementations that have been made freely available as web servers or downloadable computer software. We concentrate on methods for database searches and identification of sequence similarities, clustering of sequences into homologous families, multiple sequence alignment, and inference of evolutionary relationships. Finally, we consider an iterative procedure utilizing these methods for identification of domains and motifs in the protein sequence and their functional characterization.

1.2 Domains: Primary Functional Units in Protein Sequence

Proteins are modular, containing discrete regions that perform different roles. The primary modular unit is called a domain. Regrettably, there is no standard definition of what a domain really is. Structural biologists put emphasis on structural autonomy, biochemists and geneticists refer to regions with autonomous function detectable in their experimental assays, while evolutionary biologists focus on regions that are conserved throughout the evolution. Here, we adapt a definition based mostly on structural and evolutionary criteria.

The structural domain has been first defined in the 1960s with the advent of the first structures of water-soluble globular proteins determined by X-ray crystallography (for review see ref. [9]). Globular domains are characterized by ellipsoidal or spherical shape, and a relatively stable internal structure, which is defined by the amino acid sequence. In structural domains the backbone of a polypeptide chain exhibits elements of regular secondary structure (α-helices and/or β-strands) that forms a unique three-dimensional arrangement called a 'fold', which serves as a scaffold for functionally important side chains of amino acid residues. Some amino acid residues form a hydrophobic core, from which water molecules are excluded, while others are exposed at the hydrophilic surface, where they form sites of interactions with other molecules. In order to satisfy these requirements, protein domains are typically formed by amino acid sequences of high informational complexity. Globular domains typically range from 50 to 300 residues with a few larger and smaller exceptions (review: [6]). Domains located within biological membranes exhibit similar structures, with a few exceptions: they are usually barrel-shaped, with a hydrophobic 'belt' on the outside that ensures a seamless fit to the hydrocarbon tails of the lipid bilayer. One type of transmembrane (TM) proteins is composed exclusively of α-helices, while the other contains only β-strands; the latter type of structures form pores, and contain an internal hydrophilic channel instead of the hydrophobic core (review: [10]).

Since 1970 it emerged that structural domains may recur in different structural contexts or in multiple copies in the same polypeptide chain. More recent comparative analyses of large numbers of protein sequences and structures confirmed that a structural domain is also a fundamental unit in evolution (reviews: [6,11]). The same domains can be found in different proteins in all three forms of life, Archaea, Bacteria and Eukaryota, as well as in viruses that infect them. Examples of frequently recurring domains include: a helix-turn-helix domain often found in DNA-binding proteins (20~100 residues), a TIM-barrel domain present in many enzymes (~200 residues), or a transmembrane domain found in G-protein coupled receptors (~250 residues).

Gene fragments encoding domains may undergo duplication (e.g. leading to proteins with tandem copies of the same domain), fusion with other genes or gene fragments (leading to multi-domain proteins). Protein families are usually defined based on a presence of one common domain, which does not exclude the possible presence of additional domains. For example in enzyme families, the common homologous domain is usually responsible for performing catalysis, while the auxiliary domains may be responsible for recognition of various substrates. (e.g. in enzymes acting on DNA, they may recognize different specific DNA sequences). These auxiliary domains may formally belong to different families or even exhibit different folds. Thus, it is important to remember that proteins may comprise multiple domains, of which some may be homologous, while other may be non-homologous. Certain combinations of domains that are found recurring in diverse proteins are often referred to as modules or supradomains. They duplicate and are selected as one evolutionary unit either because it is functionally beneficial to have both activities present in one polypeptide or because the functional site is created between the domains.[12] Examples of such modules can be found e.g. in nucleic acid polymerases, which often have the polymerization domain fused to an exonuclease proof-reading domain or in proteins involved in signal transduction, which have a nuclear receptor ligand-binding domain fused to a DNA-binding domain.

It is important to remember that a conserved 3D structure in the different context does not guarantee the same amino-acid sequence or function; in fact these features may differ substantially for remotely related proteins and domains. An insertion of one domain into another may cause the latter domain to become discontinuous in sequence, even though its original three-dimensional fold is preserved, with distant sequence elements brought together to form a stable structure. Another example of a complex rearrangement is circular permutation (review: [13]), when a sequence fragment from one terminus is transferred to the other terminus, thereby changing the order of sequence motifs within the domain. A circularly permuted sequence may still form the same three-dimensional fold, albeit with a different connectivity of the polypeptide chain (N- and C-termini of a protein appear in a different position in the structure). Sequence rearrangements that do not preserve the order of primary sequence make detection of structurally conserved domains a very difficult task (see the final section of this chapter).

In addition to stably folded domains, many proteins possess segments that are non-globular in the sense that they lack a tightly packed hydrophobic core. They are often formed by compositionally biased sequences that are poor in hydrophobic residues and enriched in charged residues, and exhibit different types of 'low complexity' regions, e.g. short-period repeats, near-homopolymeric residue clusters, or aperiodic mosaics of only a few residue types.[14,15] Such segments may form fibrous or filamentous structures (e.g. in collagen or keratins) or exhibit conformational heterogeneity, so called 'intrinsic disorder' (see refs. [16,17]). Some of these regions form linkers that permit the correct spacing between globular domains, but others play more specific roles, in particular harbor sites for interactions with other molecules, including proteins and nucleic acids. The review of the variety of structures assumed by non-globular regions is beyond the scope of this chapter; here, we will discuss only those of their features that are directly related to sequence–function relationships. For recent reviews on structure–function relationships of fibrous and intrinsically disordered proteins (IDPs) proteins the reader should consult ref. [18] and refs [19–21], respectively. Bioinformatics methodology for prediction of regions of disorder is reviewed in detail in the chapter by Majorek *et al.* in this volume.

1.3 Sequence Motifs

While essentially all protein sequences can be subdivided into globular domains and non-globular segments, the most basic functional unit in protein sequence is called a motif. Motifs usually correspond to short sequence fragments (a few, typically up to 10 amino acids) that reflect some vital biological role in terms of structure or function (e.g. are responsible for stabilizing interactions or promote a particular conformation within a protein molecule or take part in binding of another molecule). Motifs occur frequently both in globular and non-globular sequence segments, but depending on the structural context, they fulfill different roles. Structured motifs (SMs) are fingerprints of globular domains. They are conserved in the evolution because of critical involvement in activity, for which the entire domain is selected, e.g. binding of the ligand that serves as a cofactor in the enzymatic reaction catalyzed by the enzyme. They are usually conformationally rigid (or at least their fragments are, while some parts may show mobility required for function). Examples of SMs include Walker A GXXGXGK(T/S) and Walker B '(R/K)X(6-7)Lh(4)D' motifs involved in ATP-binding in a large group of ATP-utilizing enzymes.[22] Other SMs may be required for structural stability, e.g. contain Zn-binding Cys and His residues in e.g. C_2H_2–type Zn-finger domains: 'CX(2-4)C. . .HX(2-4)H'.[23] The presence of a common SM in a particular set of domains suggests the presence of a similar well-defined structure required for binding of a ligand, but may or may not indicate homology. In particular, motifs involved in binding of widespread ligands are found in several protein families that are unrelated to each other. Thus, caution must be exerted when identification of a single common SM is used to infer evolutionary relationship, and it should be accompanied by analysis of global sequence similarity (see below) and preferably, also global structural similarity.

Linear motifs (LMs) are a different group of functionally heterogeneous sites. They mediate interactions of proteins with other molecules, are responsible for cell compartment targeting, or represent the sites of post-translational modification, such as phosphorylation, glycosylation, fucosylation, methylation etc. (review: [24]). Motifs of this kind are typically embedded in locally unstructured regions, but possess a few specificity-determining residues favoring disorder-order transition upon binding. LMs have a unique amino acid composition, dissimilar to either globular domains or non-globular segments; they are enriched in Pro, hydrophobic residues Trp, Leu, Phe, and Tyr, as well as charged residues Arg and Asp.[25] Examples of LMs include the PXXP motif for binding to SH3 domains, the NPXY motif for the interaction with PTB domains, the WXXW C-mannosylation site, and the WXXX(Y/F) peroxisomal targeting signal. LMs rarely occur in 'conventional' globular domains, but if they do, these domains almost invariably undergo posttranslational modifications. LMs also show completely different conservation patterns than SMs. SMs are evolutionarily constrained by many interactions within the globular domains and/or stable binding to high-affinity ligands, therefore they are often conserved in entire protein families or superfamilies. LMs are typically involved in transient interactions, rely on a very few specific interactions and their structure is loosely constrained, therefore they may be easily created as well as removed due to few accidental mutations. If they appear in locations that confer selective advantage, e.g. due to introducing a regulatory switch, they may be preserved in the course of the evolution. However, due to the relative redundancy of LMs, removal of a single site (e.g. one of many phosphorylation sites within a regulatory region of a particular protein) rarely has as drastic effects as removal of an individual SM (e.g. a catalytic motif in the enzyme). As a result LMs tend to be conserved only among very close homologs, and are frequently in non-homologous proteins that nonetheless share the same functionality (e.g. the ability to be phosphorylated by the same protein kinase). Thus, Nature appears to use LMs as evolutionary interaction switches.[24]

1.4   Databases of Protein Families, Domains, and Motifs

The importance of domains as structural building blocks, basic elements of biochemical function, and elements of evolution, has brought about many automated methods for their identification and classification in proteins of known structure. However, as mentioned before there is no standard definition of what a domain really is, therefore assigning domain boundaries even for proteins with known structures is not a trivial task. While human experts disagree for approximately 10% of structures, automatic methods for domain assignment show much larger discrepancy even for structures that the human experts agree on.[26] Expectedly, assignment of domains for proteins in the absence of structural information varies enormously; hence prediction of domains from sequence remains a challenging problem. However, before we describe bioinformatic methods that approach this problem, we will describe databases of protein families and domains, and tools for database searches and multiple sequence alignments.

A number of databases have been created to facilitate classification and identification of domains and motifs, and using them for protein function prediction. They usually classify proteins based on the presence of conserved domains (defined according to many different criteria) and/or motifs and group them according to sequence or structural similarity or based on predicted evolutionary relationships, such as orthology. Table 1.1 lists some of the most comprehensive and well-established databases of families, domains, and motifs, whose entries have been created and are curated at least partially by protein experts.

The most popular databases that classify protein domains based on structural comparisons are SCOP [27] and CATH.[28] Domain definitions used by these databases are based on very similar geometric criteria and therefore usually coincide with each other. Both databases are organized hierarchically, with the top level corresponding to structural class of a domain, i.e. the proportion of residues adopting α-helical or β-strand conformation (see the chapter by Majorek *et al.* for the discussion on secondary structure assignment). Within each class, domains are classified into folds, which group together proteins exhibiting significant structural similarity, both in terms of the arrangement of structures in three dimensions, and connectivity between them (as a result, circularly permuted variants that differ in connectivity should fall into different folds, hence this criterion is sometimes relaxed). Further, proteins with the same fold and evidence for evolutionary relationships are classified into homologous superfamilies. Within superfamilies proteins with clear sequence similarity are grouped into families. SCOP is maintained by mostly manual analysis for recognizing relationships to generate superfamilies, while CATH uses a combination of automatic and manual analysis.

Table 1.1 Databases of domains and protein families

Method	URL (http://)	Data*	Description
			Structural
SCOP[27]	scop.mrc-lmb.cam.ac.uk/scop	D	Hierarchical classification of domain structures, with four levels: Class, Fold, Superfamily, Family. Linked to the SUPERFAMILY database[48], which maps protein sequences from fully sequenced genomes onto SCOP superfamilies and represents the resulting sequence superfamilies as HMMs.
CATH[28]	www.biochem.ucl.ac.uk/bsm/cath	D	Hierarchical classification of domain structures, with four levels: Class, Architecture, Topology, and Homologous superfamily. Linked to the Gene3D sequence database[49], which assigns proteins into HMMs based on CATH domain families.
			Sequence
InterPro[29]	www.ebi.ac.uk/interpro/	F, D, SM, LM	An integrative 'meta-database' that collects annotations at the level of families, domains, and motifs.
Pfam[31]	pfam.sanger.ac.uk/	F, D	Collects MSAs and HMMs covering protein families and domains. Provides information about protein domain architectures, species distributions, and known protein structures.
PANTHER[50]	www.pantherdb.org/	F	Classifies genes and proteins into families and subfamilies by their functions, using published experimental evidence and predictions based on evolutionary relationships. Families and subfamilies are then categorized by molecular function and biological process ontology terms. For some entries pathway information is also provided.
TIGRFAMs[51]	www.tigr.org/TIGRFAMs/	F	Collection of protein families encoded as HMMs.
iProClass[52]	http://pir.georgetown.edu/iproclass/	F	Classifies proteins according to PIR superfamilies and annotates them with PROSITE signatures.
ProDom[53]	prodom.prabi.fr/prodom/current/html/home.php	D	Contains domain families automatically generated from SWISS-PROT and TrEMBL sequence databases, and information about protein domain architectures
SMART[54]	smart.embl-heidelberg.de/	D	Stores information about protein domain architectures and protein-protein interactions.

Table 1.1 *(continued)*

PRINTS[55]	http://www.bioinf.manchester.ac.uk/dbbrowser/PRINTS/	SM	Database of protein family/domain fingerprints
PROSITE[33]	www.expasy.ch/prosite/	SM, LM, F, D,	Consists of documentation entries describing protein domains, families and functional sites, associated patterns and profiles to identify them. Complemented by ProRule, a collection of rules based on profiles and patterns, which increases the discriminatory power of profiles and patterns by providing additional information about functionally and/or structurally critical amino acids.
CDD[35]	www.ncbi.nlm.nih.gov/Structure/cdd/cdd.shtml	F, D	Collection of MSAs for domains and full-length proteins, contains its own curated domains and un-curated entries imported from other databases. Allows searching for proteins with similar sequences (CD-search) and similar domain architectures (CDART)
COG[37]	www.ncbi.nih.gov/COG/	F, D	Collection of MSAs and phyletic profiles for groups of orthologs or close paralogs from at least 3 fully sequenced genomes.

*D, F, SM, and LM indicate domains, families, structured motifs and linear motifs.

Among protein family/domain databases that classify protein sequences, there are two comprehensive meta-databases developed at the EBI in the UK and at the NCBI in the USA EBI's INTERPRO[29,30] is a major resource for protein families, domains and functional sites, which integrates the protein sequence database UniProt (which itself is a meta-database of Swiss-Prot, TrEMBL, and PIR) with databases of protein structure (MSD, SCOP, and CATH) and databases of families, domains, and patterns: Pfam, PROSITE, PRINTS, ProDom, SMART, TIGRFAMs, PIRSF, SUPERFAMILY, Gene3D, and PAN-THER (see Table 1.1). Among the latter class of databases, particularly important are Pfam,[31,32] currently the most comprehensive primary protein family/domain resource based on sequence data, and PROSITE,[33,34] which focuses on motifs. Pfam-A entries are high quality, manually curated families, which are further grouped into higher order clans (based on sequence or structure similarity). Pfam-B entries are additional entries generated automatically by referring to the ProDom database.

Conserved Domain Database (CDD) is a sequence database meta-resource within NCBI's Entrez database system.[35,36] The CDD collection contains MSAs of protein families and domains imported from Pfam, SMART and COG databases, as well as additional domains curated at NCBI. CDD-specific domains are organized into evolutionary hierarchies. The Clusters of Orthologous Groups (COG/KOG) database[37] groups together families of entire proteins and evolutionarily conserved modules from completely sequenced genomes, which are predicted to form orthologous clusters. The database is split into a two components: COGs group together proteins encoded by numerous bacterial and archaeal genomes and two yeast genomes, while KOGs group together a relatively smaller number of eukaryotic genomes (including yeasts). COG and Pfam definitions of families are the most commonly referred to in the scientific literature to describe yet uncharacterized proteins or domains.

In addition to databases of protein families curated by experts, a number of studies have reported databases resulting from fully automatic clustering of protein sequences, where 'families' indicate groups of proteins classified according to certain numerical value of sequence similarity. Examples among recently created or updated databases include: CluSTr (http://www.ebi.ac.uk/clustr/),[38] ProtoNet (http://www.protonet.cs.huji.ac.il/),[39] SYSTERS (http://systers.molgen.mpg.de/),[40] eggNOG (http://eggnog.embl.de),[41] InPara-noid (http://InParanoid.sbc.su.se/),[42] OrthoDB (http://cegg.unige.ch/orthodb),[43] SIMAP (http://mips.gsf.de/simap/).[44]

While protein databases contain thousands of domain families and associated SMs, known LMs are limited in number. There are also only a few general LM databases such as ELM[45] or Scansite.[46] A number of programs specialize in cataloging and predicting motifs with narrowly defined function and distribution, e.g. sites of different posttranslational modification, often restricted to particular taxonomic groups (see ref. [47] for review). Table 1.2 lists some of the databases and predictive servers; however a comprehensive review of such databases is beyond the scope of this chapter.

1.5 Database Searches and Pairwise Alignments

The key step in analyzing our sequence of interest (hereafter referred to as 'query' or 'target') is to determine whether it shows any similarity to other protein sequences. The

Table 1.2 Databases of motifs and software for motif finding

Name	URL	Description
SCANSITE[46]	scansite.mit.edu	A database of LMs that are recognized by globular domains (phosphorylation and protein-binding sites: contains LM-domain pairs) and a search utility. Allows searches for combinations of motifs
ELM[45]	elm.eu.org	Catalogues LMs in eukaryotic proteins, searches for defined motifs in a query sequence. Employs a set of filters
Phospho.ELM	phospho.elm.eu.org	ELM version specialized in phosphorylation sites
Minimotif Miner[56]	mmm.engr.uconn.edu	A database of known motifs (partially compiled from other databases) and a search utility. Scores motifs with several methods
CBS prediction servers[47,57]	www.cbs.dtu.dk/services/	A number of independent methods for identification of known posttranslational modification sites, targeting sites, and peptide cleavage sites
MEME/MAST[58]	meme.sdsc.edu/meme/	Searches for user-defined motifs (MAST) or performs de novo motif identification.(MEME)
Gibbs Sampler[59]	bayesweb.wadsworth.org/gibbs/gibbs.html	Searches for user-defined motifs or performs de novo motif identification for a set of up to 1000 sequences. Allows sampling by different strategies: Site, Motif, Recursive, or Centroid
EasyGibbs	www.cbs.dtu.dk/biotools/EasyGibbs/	Requires submission of training examples and evaluation examples to train a motif prediction method
IBM's Bioinformatics and Pattern Discovery[60]	cbcsrv.watson.ibm.com/Tspd.html	A number of tools for sequence pattern discovery
DILIMOT server[61]	dilimot.embl.de	De novo LM finder for a set of unaligned sequences, employs a set of filters
SLiMDisc[62]	bioware.ucd.ie/~slimdisc/	De novo motif finder. Uses TEIRESIAS algorithm to find patterns in a set of unaligned sequences, down-weights motifs found in groups of proteins found to be mutually related
NestedMICA[63]	www.sanger.ac.uk/Software/analysis/nmica	De novo motif finder for a set of sequences
DEME[64]	bioinformatics.org.au/deme/	De novo discriminative motif finder, searches only for patterns that can differentiate the two sets of sequences. Uses an informative Bayesian prior on protein motif columns, allowing it to incorporate prior knowledge of residue characteristics
IBM[65]	www.research.ibm.com/bioinformatics	De novo identification of over-represented motifs

determination of sequence similarity, from which functional similarity and/or homology is inferred, may be carried out in two independent (and complementary) ways, namely searches for patterns of characters or applying statistical models such as profiles of Hidden Markov Models (HMMs). Typically, searches against a database of motifs and full sequences are employed in parallel to see whether the query protein exhibits known LMs, SMs and domains.

Motifs can be represented as strings of characters from a specific alphabet, which discriminates between invariant residues, alternative conserved residues, unspecified residues, excluded residues, repetitions, and other features. A motif can be written as a regular expression such as 'Y.A(4){C}[DE]$', which can be interpreted as Y followed by any residue, followed by four As, followed by a non-C residue, followed by D or E, followed by C-terminus. With this representation, identification of exact matches between the sequence and a database of motifs is fairly simple, as the regular expression either is present in the sequence or not. However, this way of searching is likely to miss relevant motif variants that exhibit slight variations. Allowing for approximate matches allows for detection of more variants, but inevitably causes appearance of false positives. The major limitation of regular expressions is that they do not take into account the information about the relative frequency of residues at different positions. Statistical models such as profiles (also called positional weight matrices, PWMs) give the probability of observing each amino acid in each position. They allow for partial matches and in general have stronger predictive power, i.e. enable detection of diverged but genuine motifs. Some popular software tools for detection of known motifs and '*de novo*' discovery of previously unknown motifs in functionally related sequences are summarized in Table 1.2. Once sequences sharing a common motif are identified and the motif variants are aligned with each other (see below for explanation of alignment techniques), they can be represented as Sequence Logos[66] for visual inspection (e.g. using the WebLogo server[67] at http://weblogo.berkeley.edu/logo.cgi).

Recognition of very short motifs (e.g. most of LMs) remains problematic, as they are often presented in many sequences solely to the sequence composition of the proteome. Thus database searches with most method yield many false positives that have to be filtered out by considering additional information, e.g. presence of globular domains, which usually contain SMs but are depleted in functionally relevant LMs. On the one hand, presence of non-globular, e.g. disordered regions, can be exploited to detect certain LMs, such as phosphorylation sites; this rule has been implemented in the DisPhos server[68] (http://www.ist.temple.edu/DISPHOS/). On the other hand, homologous globular domains often contain conserved sets of SMs, e.g. in spatially adjacent regions involved in formation of binding sites. Typically, the order of SMs is preserved and a pattern of motifs may be exploited to build a diagnostic tool for detection of new members of a protein family. Nonetheless, because of problems with assessment of statistical significance of short motifs, it is recommended that homology predicted via motif searches is confirmed by one of the tools that provides a more global estimate of sequence similarity, e.g. sequence alignment.

Sequence alignments usually assume (or search for) evolutionary conservation, as opposed to similarity of short motifs that may result from convergent evolution. The statistical significance of alignment can be established by estimating the likelihood that the similarity between two sequences is due to their divergence from a common ancestor, rather than pure accident. First, the query sequence and the potentially homologous sequence are searched for a series of similar amino acid residues or residue patterns that are in the same order. Then, gaps are inserted between the residues and sequence fragments are shifted so that residues with identical or similar characters in both sequences are aligned in successive columns. If two sequences are indeed homologous (i.e. they diverged from a common ancestor), matches in the alignment represent residues that have been conserved in the evolution, while mismatches can be interpreted as point mutations and gaps as indels (that is, insertion or deletion mutations) introduced in one or both lineages in the time since they diverged from one another. The biological relevance of sequence alignment is usually assessed by comparison with a structure-based alignment, in which residues are considered homologous if they are spatially superimposable. Structural alignments are considered a 'gold standard' in bioinformatics (review: [69]). Since only a small fraction of protein sequences have known structures, the accuracy of sequence alignment measured on the references is merely an estimation of how well a given algorithm reproduces a structurally correct alignment for a collection of standard datasets.

There are two types of algorithms for sequence alignment based on dynamic programming: global Needleman-Wunsch[70] and local Smith-Waterman.[71] In global alignment, an attempt is made to align the entire sequence, using as many matching amino acid residues as possible, up to both ends of each sequence. Thus, best candidates for global alignment are sequences that are approximately the same length. In local alignment, stretches of sequence with the highest density of matches are aligned, thus generating one or more 'islands' of matches or subalignments in the aligned sequences. Local alignments are more suitable for aligning sequences that are similar along some of their lengths but dissimilar in others and/or sequences that differ in length. Local alignment is particularly useful for identification of regions of homology between proteins composed of different domains, i.e. sequences that are only partially homologous. Such multidomain proteins are very common in Eukaryota, in contrast to Prokaryota (Bacteria and Archaea), which are more frequently composed of single domains and exhibit 'global' homology.

The above methods of establishing sequence relationships have been utilized in database similarity searches. In the initial step the query sequence is compared to every sequence in the selected database, and similar sequences are identified. Pairwise alignments between the target sequence and the best-matching database entries are constructed, typically using dynamic programming algorithms, and scored. Although percent identity of amino acid residues between two sequences is intuitive and easy to calculate, it is a poor measure of protein similarity, especially for more diverged sequences. Protein alignments are typically aligned and scored using substitution matrices that reflect statistical probabilities of one residue being substituted by another. PAM[72] (and its newer versions Gonnet[73] or Jones-Taylor-Thornton/JTT[74]) and BLOSUM[75] are the two most commonly used types of matrices, with PAM being based on an evolutionary model and extrapolation of probabilities calculates for closely related sequences and BLOSUM based on alignments of more remotely related sequences. Different matrices allow for detecting sequences with varying levels of divergence. A scoring function includes also penalties for the introduction of gaps corresponding to insertion or deletion (indel) mutations. Finally, statistical methods are used to determine the likelihood of a particular alignment between sequences or sequence regions arising by chance, given the size and composition of the database being searched. Alignments that have a low probability of occurrence by chance are interpreted as likely to indicate homology. However, the likelihood of finding a given alignment by chance can vary significantly depending on the size and composition of the database. For the search for homologs to be effective and the score to be accurately estimated, the database must contain many unrelated sequences. It is important to remember that pairwise similarities (especially if confined to very short regions) can also reflect convergent evolution or simply coincidental resemblance. Thus, repetitive sequences in the database or query can distort both the search results and the assessment of statistical significance.

The most popular methods for sequence database searches (Table 1.3) are FASTA[76] and BLAST.[77] They identify a series of short non-overlapping subsequences in the query sequence that are then matched to candidate database sequences. Query-database matches are subsequently extended and combined into a local pairwise alignment using a variation of the Smith-Waterman algorithm. Both FASTA and BLAST employ extreme value distributions to estimate the distribution of the scores between the query and the database entries and a probability of a random match.[78,79] The result of a database search is a list of pairwise alignments ranked according to the expectation value (E) that represents a number of sequences that are not related to the query sequence and are predicted to produce as good an alignment score as the query sequence. As a rule of thumb, alignments that exhibit small E value (<0.001 for large databases), presence of long stretches of aligned regions without gaps, and absence of low-complexity regions are likely to indicate homology. Nonetheless, homologous sequences can be so diverged that their pairwise similarity scores are in the range of random noise.

Detection of more remote relationships requires taking into account not only individual sequence pairs, but also analyzing similarities in the context of entire families of homologous proteins. For instance, PSI-BLAST (Position-Specific Iterated BLAST) allows for finding very distant relatives of a protein by first invoking regular BLAST and retrieving statistically significant alignments, calculating a 'sequence profile', or a position-specific score matrix (PSSM) that describes the frequency of amino acids found at each position in aligned sequences, and then searching the database using this matrix.[80] Alternatively to PSSMs, the set of query-database alignments can be used to create a Hidden Markov Model (HMM), which also can be iteratively compared with the database to identify new statistically significant matches (as implemented in methods such as HMMER[81]). The list of detected statistically similar (and presumably homologous) sequences aligned to the query can be then updated with new sequences and searches can be carried out in an iterative fashion until no new sequences are reported with the similarity score above the threshold of statistical significance. It must be emphasized that in rounds >1 the similarity scores are calculated with respect to the whole group of aligned sequences (represented by PSSM or a HMM) rather than to the single query sequence, therefore erroneous addition of unrelated sequences at an early stage of the search can lead to further degeneration of the result and inclusion of many false positives. Thus, e.g. for PSI-BLAST it is recommended to initialize searches with a stringent E-value threshold for inclusion of database sequences in the query PSSM (e.g. 10^{-20}-10^{-3} for typical protein families), and progressive relaxation of the threshold (to e.g. 10^{-3}) in subsequent iterations, depending on the number of reported sequences and their similarity to the query.

The 'intermediate sequence search' (ISS) strategy[82,83] is an alternative to profile-based methods. It employs a series of database searches initiated with the query and then continued in a pairwise manner with its homologs. Saturated BLAST is a freely available software package that performs ISS with BLAST in an automated manner.[84] Since all

Table 1.3 Elected representative methods for sequence database searches

Method	Search strategy	URL (http://)	Description
FASTA[76]	query sequence vs. database	www.ebi.ac.uk/fasta/index.html	Searches for matching sequence patterns or words, rescans matched regions using scoring matrices, and trims the ends of the region to include only sequence contributing to the highest score. It uses a Smith-Waterman algorithm to calculate an optimal score for a local alignment
BLAST[77]	query sequence vs. database	www.ncbi.nlm.nih.gov/blast/Blast.cgi	Uses a heuristic approach to search for exact matches of a small fixed length between the query and sequences in the database, tries to extend the match in both directions, and performs a gapped alignment between the query sequence and the database sequence using a variation of the Smith-Waterman algorithm. Faster than FASTA
PSI-BLAST[80]	iteratated profile search	www.ncbi.nlm.nih.gov/blast/Blast.cgi	A BLAST search is performed and an alignment from the best local hits is built. This alignment is then used as a query for the next round of search. After each round the search alignment is updated
RPS-BLAST[88]	iteratated profile search	www.ncbi.nlm.nih.gov/blast/Blast.cgi	RPS-BLAST (Reverse PSI-BLAST) searches a query sequence against a database of profiles
SENSER[85]	profile-sequence	available from the authors	Performs a PSI-BLAST search, in addition to significant matches extracts candidates for remote homologs from alignments reported with scores below the level of statistical significance. Candidates are then validated by reciprocal PSI-BLAST searches. Aligned homologs are used to build a HMM that is used as a query in subsequent database searches. The procedure may be iterated
PROF_SIM[89]	profile-profile	available from the authors	Compares two input profiles (like those that are generated by PSI-BLAST) and assigns a similarity score to assess their statistical similarity
COMPASS[90]	profile-profile	prodata.swmed.edu/compass/compass.php	Derives numerical profiles from given multiple sequence alignments, constructs local profile-profile alignments and analytically estimates E-values for the detected similarities
HHsearch[91]	profile-profile	toolkit.tuebingen.mpg.de/hhpred	Builds a profile-HMM from a query sequence and compares it with a database of HMMs representing annotated protein families (e.g. PFAM, COGs,) or domains with known structure (PDB, SCOP)
HHsenser[92]	profile-profile	toolkit.tuebingen.mpg.de/hhsenser	Similar to SENSER, but involves 'profile-HMM to profile-HMM' instead of 'profile-HMM to sequence' comparisons to search for remote similarities between whole (super)families

homologs are used as search targets, this strategy is computationally demanding, but it can identify links to remotely related outliers, which may be missed by MSA-based profile or HMM searches that preferentially detect typical sequences. A variant of ISS strategy that includes profile-sequence searches with PSI-BLAST and attempts to extract remote homologs from alignments reported with scores below the level of statistical significance, has been implemented in the method SENSER.[85]

The introduction of profile-based methods, in particular PSI-BLAST, has truly revolutionarized the field of evolutionary bioinformatics, resulting in characterization of numerous conserved domains and detection of remote homologies between many sequences and sequence families that were undetectable in pairwise searches.[86,87] It has also prompted development of several databases of protein families or protein domains (see below), accompanied by the appearance of special bioinformatics tools for searching of these databases. One example is RPS-BLAST (Reverse Position-Specific BLAST) implemented in the IMPALA package,[88] which, as its name implies, reverses the PSI-BLAST approach by comparing a single query sequence against a collection of PSSMs pre-calculated for a number of previously characterized protein families, to determine whether the query sequence is likely to belong to one of these families. Currently the most widely used algorithms for sequence database searches (apart from still extremely popular PSI-BLAST) belong to the newer generation of methods that carry out profile-profile comparisons and allow for detection of even more remote relationships than profile-sequence comparisons. These tools are typically available as web servers; they parse the query sequence provided by the user, automatically run PSI-BLAST to retrieve a profile corresponding to reliably identified candidate homologs (i.e. the query family), and compare it with profiles pre-calculated for a large number of protein families. Examples include PROF_SIM,[89] COMPASS,[90] and HHsearch.[91] Profile-profile search methods have been also adapted to assist in template-based protein structure prediction (described in more detail in chapter by Kosinski *et al.*). The last generation of methods for automated database searches is represented by HHsenser, which combines SENSER-like exhaustive intermediate profile-sequence searches with HHsearch-like pairwise comparison of HMMs.[92]

Once an initial search for homologs of the query sequence is performed, the detected sequences are extracted from the database. Database searches are usually carried out with local alignment programs and extraction of sequences results in retrieval of full length entries from database. Outside the homologous region that has been detected by a local search these sequences may contain regions that are non-homologous to the query, or regions that are homologous to the query but local alignment methods failed to detect them. As mentioned earlier, database searches may result in retrieval of false positives, i.e. sequences that exhibit similarity score above the threshold (e.g. due to biased sequence composition), but nonetheless are not true homologs of the query. Besides, all major databases contain redundant multiple copies of the same protein that differ by only a few residues (e.g. variants with alternative translation codons or results of different sequencing experiments) or exhibit various errors (e.g. terminal truncations or indels caused by incorrect prediction of gene boundaries or exon/intron structure). Such incorrect or redundant sequence variants have to be removed from the preliminary sequence dataset (or corrected, if need be) prior to any advanced analyses. Identification of erroneous sequences is best done at the level of global multiple sequence alignment, which facilitates visualization of missing or redundant regions corresponding to erroneous deletions and insertions. Although there exist a number of fully automated methods for multiple sequence alignment (MSA, see below), thus far no method allows for automated 'purging' of the alignment of all incorrect sequences and this stage has to be done manually, with the aid of methods for graphical representation and editing of alignments. Such analysis becomes very difficult when the number of sequences to be analyzed is significantly larger than 100, and the workstation's screen becomes too small to display them all. On the other hand, identification of redundant sequences is best done by clustering analysis which may or may not require prior calculation of the MSA, and is capable of processing large number of sequences. In our experience, the most useful procedure is to carry out general clustering first to identify major subgroups (potential families) that are possible to handle by alignment editors, followed by calculation and editing of MSA for each subgroup, followed by merging of all edited sequence groups and repeating MSA and carrying out final quality checks. Below, we describe in more detail methods for MSA-independent sequence clustering, MSA construction, and for MSA-based calculation of phylogenetic trees.

1.6   Sequence Clustering

It is well known that protein families can be classified into subfamilies using phylogenetic analysis to calculate a hierarchy of relationships. The traditional representation of this hierarchy is a treelike dendrogram, with individual elements ('leaves') at one end and a single cluster containing every element ('root') at the other. Phylogenetic analysis requires, however, the availability of MSA and intensive calculations to obtain evolutionary distances and generate an accurate treelike representation of mutual relationships within the protein family. There have been many attempts to circumvent this problem, in particular by using various 'surrogate' measures of pairwise sequence similarity, rather than evolutionary distances, and by applying various hierarchical clustering techniques to build treelike representations.

An important step in clustering is to select a distance measure, which will determine how the similarity of two elements is calculated. Sequence clustering algorithms typically employ the value of pairwise sequence similarity, e.g. calculated by BLAST or the Smith-Waterman algorithm (see above) and aim at identifying groups of sequences that are more similar to each other than to other members of the input set. Typically, the aim of protein sequence clustering is to identify groups of homologs exhibiting statistically significant similarity, thus the threshold value for cutting the tree should correspond to the desired evolutionary distance (e.g. to split a superfamily into families and then into subfamilies). An appropriate cutoff should also separate true homologs from non-homologs, which can be used to purge the initial dataset from potential false positives. Clustering can also be used to split a group of functionally similar but not necessarily evolutionarily related proteins into subgroups of homologs that are further analyzed independently from each other. The presence of well-characterized proteins within a family can then allow one to reliably assign functions to other family members whose functions are not known or not well understood. Finding proteins with different functions within the same family may suggest caution in extrapolating functional information. On the other hand, finding families with only uncharacterized members may prompt them as sources of interesting candidates for experimental analyses.

Single linkage (SL) clustering is a simple and intuitive algorithm, in which the distance between two clusters is computed as the distance between the two closest elements in these clusters. It has been implemented e.g. in the BLASTCLUST method from the popular BLAST package[80] (ftp://ftp.ncbi.nih.gov/blast/, also available via a third-party web server http://toolkit.tuebingen.mpg.de/blastclust/). The SL algorithm is known to produce accurate clustering when different subgroups show similar level of internal similarity and an appropriate threshold is given to separate families from each other. A drawback of this method is that clusters may be forced together due to single elements being similar to each other, even though other elements in each cluster may be dissimilar to each other. Thus, the SL analysis is not appropriate for analyzing sets of largely non-homologous multidomain proteins, which may be falsely chained to each other (e.g. a cluster of many proteins comprising domain A and one protein with domains A and B may be chained to a cluster composed of proteins with domain B and one protein with domains B and C, then chained to a cluster of domains C and so on). In particular, many proteins possess small, widespread protein domains (e.g. SH2, WD40, and DnaJ) that are known to have very different functions. The presence of such a common domain within a group of proteins does not necessarily imply that these proteins perform the same function. Ideally, these types of proteins should be classified into a single cluster only if they exhibit highly similar domain architectures. Another drawback is that in many protein superfamilies the degree of similarity within different families varies greatly, and e.g. subfamilies within one family may be more diverged from each other than two other families. Therefore, application of only one average threshold may produce many too small clusters and a few too large clusters.

Due to the fact that SL method has difficulty in detecting an appropriate threshold for identification of clusters, modern protein clustering applications employ other algorithms. In particular graph theory allows the classification of objects into groups based on a global treatment of all relationships in similarity space simultaneously. Thus, proteins and their similarities may be represented as vertices and edges of a graph, respectively, and the initial partition produced, e.g. by SL clustering, may be post-processed by a graph partitioning algorithm (see Chapter 10 by Nabieva and Singh in this volume for a detailed discussion of different clustering algorithms, in the context of graphs representing networks of protein–protein interactions). CLANS (CLuster ANalysis of Sequences)[93] (ftp://ftp.tuebingen.mpg.de/pub/protevo/CLANS) is a freely available Java application, which runs all-against-all BLAST searches for all sequences in the input set, and then applies the Fruchterman–Reingold graph layout algorithm to visualize pairwise sequence similarities based on BLAST P-values in either two-dimensional or three-dimensional space. CLANS allows the user to select different thresholds and parameters for calculation of distances and to carry out clustering using several different algorithms, including single and multiple linkage, network-based, and convex clustering. LGL[94] is a similar clustering algorithm with a Java front end for visualization, however it requires pre-computed similarity values as an input. ProClust (http://pig-pbil.ibcp.fr/magos)[95] is another graph-based clustering algorithm, which scales similarity values based on the length of the protein sequences compared, and takes into account the significance of alignment scores to filter for spurious links. Post-processing to merge clusters is based on comparison of clusters with each other using profile-HMMs (see further sections in this chapter for review of methodology for profile-profile comparisons). MCL[96] relies on the Markov cluster (MCL) algorithm, which finds clusters by calculating the probabilities associated with a transition from one protein to another within the graph and passing the matrix of probabilities through iterative rounds of 'multiplication' and 'inflation' until convergence. The 'inflation' value parameter is used to control the 'tightness' of final clusters. The MCL algorithm is relatively insensitive to the presence of multi-domain proteins, promiscuous domains or fragmented sequences. Super Paramagnetic Clustering (SPC)[97] (http://www.vcclab.org/lab/spc/) is a different approach that clusters input data based on analogy to the physics of an in-homogeneous ferromagnet; a stepwise implementation of this algorithm, called global SPC (gSPC) was shown to be even more robust than TRIBE-MCL. FlowerPower[98] (http://phylogenomics.berkeley.edu/cgi-bin/flowerpower/input_flowerpower.py) has been designed specifically for the identification of subfamilies with global homology (e.g. from a set of sequences with different domain compositions) using the SCI-PHY algorithm based on HMMs.[99] Finally, unlike other methods that calculate their similarity matrices based on alignments, CLUSS[100] (http://prospectus.usherbrooke.ca/CLUSS/) performs clustering based on a matching amino acid subsequences, which makes it applicable both to alignable and unalignable sequences, e.g. products of circular permutation etc. A number of other clustering approaches have been used to cluster various sequence data sets and construct databases of clusters (see the section on protein family databases); however, the underlying clustering programs have not been made available as standalone applications.

1.7 Multiple Sequence Alignment

As soon as sets of homologous sequences with similar domain composition are identified, or the domain subsequences isolated from non-homologous fragments, they can be aligned together to study sequence conservation across the entire family. Multiple sequence alignment (MSA) is an extension of pairwise alignment, in which multiple related sequences are optimally matched, by bringing the greatest number of similar characters into register in the same column. In this manner, protein sequences are arranged into a rectangular array with the goal that residues in a given column are homologous (derived from a single position in an ancestral sequence), superimposable (in a structural alignment) or play a common functional role. The advantage of the MSA is that it reveals more biological information than a set of pairwise alignments, e.g. conserved patterns and motifs that are common to the whole sequence family and may indicate functionally or structurally important elements. However, finding an optimal alignment of more than two sequences that includes matches, mismatches, and gaps, and that takes into account the degree of variation in all of the sequences at the same time, is very difficult. Usually, an arrangement of amino acid residues that maximizes the sum of similarities for all pairs of sequences (the sum-of-pairs, or SP, score) is sought. Unlike in pairwise alignments, the SP score has no rigorous theoretical foundation for the MSA and, in particular, fails to incorporate an evolutionary model. Moreover, the dynamic programming algorithm used for optimal alignment of pairs of sequences can be extended to multiple sequences, but the computational time and memory required to maximize the SP score has been shown to scale exponentially with the number of sequences and becomes prohibitively expensive for data sets larger than a few proteins.[101] Thus, approximate alternatives are used. The majority of programs (Table 1.4) are based on the 'progressive algorithm' approach, where the MSA

Table 1.4 Methods for calculation of MSAs

Method	URL (http://)	Description
CLUSTALW[110]	www.ebi.ac.uk/clustalw/	Performs pairwise alignments of input sequences, produces a tree based on similarity scores, and realigns sequences sequentially, guided by the tree. Old and inferior to newer methods, but still very popular
DbClustal[111]	bips.u-strasbg.fr/PipeAlign/	Carries out BLAST searches and incorporates local alignment information into a CLUSTAL global alignment in the form of a list of anchor points between pairs of sequences. Allows for incorporation of very long insertions and terminal extensions
SAM[112]	www.soe.ucsc.edu/compbio/sam.html	Employs HMM for MSA. Evolved into a fold-recognition tool SAM-T, current version: SAM-T06 (www.soe.ucsc.edu/research/compbio/SAM.T06/T06-query.html)
HMMer[113]	hmmer.janelia.org/	Employs HMM. Implemented in PFAM database for grouping of sequences into families
MUSCLE[114]	www.drive5.com/muscle/	Rapidly generates a very crude guide tree, generates MSA using a profile function (log-expectation score) and refines it using tree-dependent restricted partitioning. Very fast
PRRN[103]	prrn.hgc.jp align.genome.jp/prrn/	Adopts a doubly nested randomized iterative refinement strategy to make alignment, phylogenetic tree and pair weights mutually consistent. Performs a large number of pairwise group-to-group alignments to gradually improve overall weighted sum-of-pairs score
T-Coffee[104]	www.tcoffee.org/	Employs a consistency measure by considering information from all of the sequences during pairwise sequence alignments, not just those being aligned at that stage. Combines a collection of multiple/pairwise, global/local alignments into a single MSA. Version 2.00 and higher can combine sequences and structures
MAFFT[115]	align.bmr.kyushu-u.ac.jp/mafft/online/server	Uses a fast Fourier transform to generate a guide tree. Refines the alignment by optimizing the weighted sum of pairs (WSP) objective function
PRRN[116]	prrn.hgc.jp/	Optimizes a weighted sum-of-pairs score, in which the weights given to individual sequence pairs are adjusted to compensate for the biased contributions. MSA is refined through partitioning and realignment restricted to the edges of the tree
PRALINE[117]	zeus.cs.vu.nl/programs/pralinewww/	Runs PSI-BLAST for each sequence in the input set to generate a PSSM pre-profile. Pre-profiles are then aligned hierarchically by a profile-profile alignment method
ProbCons[118]	probcons.stanford.edu/	During pairwise alignments employs 3-state HMMs, uses maximum expected accuracy as an objective function, and applies probabilistic consistency transformation to incorporate multiple sequence conservation information
SPEM[108]	sparks.informatics.iupui.edu/Softwares-Services_files/spem.htm	Runs PSI-BLAST and makes secondary structure prediction for each sequence in the input set. Uses SP2 with combined scoring of sequence and structure, then applies probabilistic consistency-based scoring for refinement of pairwise alignments.

Name	URL	Description
MUMMALS[119]	prodata.swmed.edu/mummals/	Employs complex HMMs with multiple match states that capture local structural information. Applies a probabilistic consistency-based scoring function
PROMALS[120]	prodata.swmed.edu/promals/	Runs PSI-BLAST and makes secondary structure prediction for each sequence in the input set. Uses a HMM with combined scoring of sequence and structure, applies probabilistic consistency-based scoring. Slow, but accurate. Works poorly with multidomain proteins
SATCHMO[121]	phylogenomics.berkeley.edu/cgi-bin/satchmo/input_satchmo.py	Simultaneously constructs a tree and a set of MSAs, one for each internal node of the tree (for all sequences within its sub-tree). Generates profile-HMMs at each node; these are used to determine branching order, to align sequences and to predict structurally alignable regions
PRIME[122]	prime.cbrc.jp	Employs doubly nested randomized iterative refinement strategy, based on a group-to-group sequence alignment algorithm with piecewise linear gap cost, instead of traditional affine gap cost
Kalign[123]	msa.cgb.ki.se/	A progressive method, which relies on the Wu-Manber approximate string-matching algorithm in the distance calculation and optionally in the dynamic programming to align the profiles
DIALIGN[124]	bibiserv.techfak.uni-bielefeld.de/dialign/	Constructs pairwise and multiple alignments by comparing segments instead of full-length sequences. Employs a fragment-chaining algorithm
MANGO[125]	www.bioinfo.org.cn/mango/	Identifies motifs shared by two or more sequences, constructs skeletal alignment, extends it to a full MSA, which is iteratively refined
ALIGN-M[126]	bioinformatics.vub.ac.be/software/software.html	Uses a non-progressive local approach to guide a global alignment. Designed to deal with particularly diverged sequences
POA[127]	bioinfo.mbi.ucla.edu/poa2/	Replaces the row-column representation of a MSA with a graph in which each node corresponds to a set of aligned residues. Enables alignment of protein sequences with multiple domains
AliWABA[128]	aba.nbcr.net/	A-Bruijn Alignment represents an alignment as a directed graph, possibly containing cycles. Enables alignment of protein sequences with shuffled and/or repeated domain structure
ProDA[129]	proda.stanford.edu/	Does not assume global alignability; allows repeated, shuffled and absent domains. Clusters alignable regions and returns a collection of local MSAs
ComAlign[130]	www.daimi.au.dk/~ocaprani/ComAlign/programs/	Meta-method. Combines qualitatively good sub-alignments from a set of input MSAs. Software for download. Server: mobyle.pasteur.fr/cgi-bin/MobylePortal/portal.py?form=comalign
M-Coffee[131]	www.tcoffee.org	Meta-method. Runs several methods from the COFFEE family to compute alternative alignments and calculates a consensus MSA

is constructed by a series of pairwise alignments, starting with the most related sequences, followed by progressively adding less related sequences (to construct partial alignments of three or more sequences) or aligning partial alignments with each other.[102]

The knowledge of evolutionary relationships among sequences is a very useful criterion for selecting the order of pairwise alignments. Although the calculation of a phylogenetic tree requires the availability of the MSA (see below), an initial tree for construction of the MSA may be calculated based on preliminary evolutionary distances calculated from pairwise comparisons of sequences. The major problem with progressive alignment programs is the dependence of the ultimate MSA on the initial pairwise sequence alignments. The more distantly related these sequences, the more errors will be made, and these errors will be propagated to the MSA. Two main techniques are utilized to correct or minimize mistakes made in the progressive alignment process. One is iterative refinement of the MSA, e.g. by repeatedly dividing the aligned sequences into subgroups and realigning the subgroups, as implemented in PRRN.[103] The other technique makes a consistency measure among a set of pairwise sequence alignments before the progressive alignment steps.[104] Many methods combine iterative optimization with either progressive algorithm and/or consistency-based scoring (review: [105]). An alternative approach for MSA, which does not require calculation of trees, relies on identification of locally conserved patterns found in the same order in the sequences (e.g. as implemented in the DIALIGN method[106]).

Another possibility is to employ a HMM, a statistical model in which an MSA is represented as a form of directed acyclic graph (also called a partial-order graph), which consists of a series of nodes representing possible entries in the columns of an MSA. In this representation a column that contains the same residue in all sequences is coded as a single node with as many outgoing connections as there are possible characters in the next column of the alignment. Sequences are aligned using the Viterbi algorithm, a variant of a dynamic programming algorithm. Several software programs are available in which variants of HMM-based methods have been implemented, including SAM and HMMER (see Table 1.4). Some of these methods allow for the presence of non-alignable (non-homologous) regions of sequence to be present in the input set. In the approach implemented in AliWABA the graph may contain cycles, which enables alignment of protein sequences with shuffled and/or repeated domain structure.[107]

Currently the best methods for MSA such as SPEM[108] or PROMALS[109] employ PSI-BLAST database searches and secondary structure prediction to construct meta-profiles for all input sequences, then carry out profile-profile alignments (with HMMs or with regular profile methods), often refine these alignments based on consistency scoring, and only then combine the input sequences into an MSA. These methods are therefore much slower than simple (but still very popular) methods like CLUSTAL, but are much more accurate, at least for individual domains. However, they might be more prone to errors in case of data sets comprising proteins of uneven length, e.g. some with single domains, and others with the same domain fused to others. Therefore, comparison of MSAs generated with different methods may provide hints as to reliability of the results. As with most bioinformatics methods, algorithms for MSA rarely generate solutions that ideally reflect the biological reality, especially for large datasets of strongly diverged sequences. However, expert knowledge concerning relationships within a given protein family can be used to improve suboptimal MSAs obtained from automatic software packages. A number of methods exist that allow for graphical visualization and manual editing of MSAs to make them agree with observations that cannot be easily incorporated into the scoring function of most algorithms (e.g. knowledge that particular residues in different sequences that must correspond to each other or agreement of structural patterns obtained from experiment or from predictions). Example tools for displaying and editing protein (often also nucleic acid) sequences and alignments have been listed in Table 1.4.

1.8   Relationship of Multiple Sequence Alignments to Phylogenetic Analysis

A biologically meaningful MSA contains sequences that are all homologous, i.e. derived from a common ancestor sequence. Further, in an ideal MSA, all columns contain amino acid residues that were derived from an ancestral residue in the ancestral sequence (if these conditions are not fulfilled, MSA is 'biologically wrong' and cannot be used for phylogenetic analyses). Within the column are original characters that were present early, as well as other derived characters that appeared later in evolutionary time. In some cases, the position is so important for function that mutational changes are not observed. It is these conserved positions that usually serve as 'anchor points' for producing an alignment. In other cases, the position is less important, and substitutions are observed. Deletions and insertions are also typically more frequent in the variable regions of the alignment. If the sequences in the MSA show evident similarities (e.g. >30% identity and relatively few insertions and deletions), they are likely to be recently derived from a common ancestor sequence. Conversely, sequences with multiple differences are likely to be remotely related. Thus, the number and types of changes in the MSA may be used to infer the mutations that occurred during the evolution of the sequence family. It is also possible to dissect the order of appearance of the sequences during evolution and to relate the relationships between sequences to the relationships between their hosts (organisms). A number of packages for phylogenetic calculations based on user-defined MSAs have been made available, including PHYLIP (http://evolution.genetics.washington.edu/phylip.html),[146] MEGA (http://www.megasoftware.net/),[147] PHYML (http://atgc.lirmm.fr/phyml/),[148] PAML (http://abacus.gene.ucl.ac.uk/software/paml.html),[149] TREE-PUZZLE[150] (http://www.tree-puzzle.de), or MrBayes[151] (http://mrbayes.csit.fsu.edu). Among web resources, MultiPhyl (http://www.cs.nuim.ie/distributed/multiphyl.php)[152] is a particularly useful site, which allows the users to carry out computationally very expensive inference of Maximum Likelihood trees using distributed computing. A review of methods for phylogenetic calculations is outside the scope of this chapter, interested readers should consult reviews: e.g. ref. [153–155]

The result of phylogenetic analysis can be used as a feedback for revising particularly challenging MSAs that are suspect of errors (e.g. sequences may be split it into subgroups and realigned separately or the tree may be used to guide the progressive alignment algorithm). As an example, the SCI-PHY server (http://phylogenomics.berkeley.edu/SCI-PHY/) allows users to upload a MSA for subfamily identification and subfamily HMM construction.[99] Further, analysis of the phylogenetic tree in connection with the known (or assumed) tree of hosts (organisms) can be used to deduce major evolutionary events in the protein family, e.g. gene duplications, gene losses, which provide the basis for discrimination between orthologs and paralogs and may guide functional predictions (review: [156]). Another application of MSA and phylogenetic analysis is the inference of ancestral sequences,

Table 1.4 *Multiple alignment editors*

Name	URL	Description
Jalview[132]	www.jalview.org/	Java tool (OSindependent). Standalone version allows for calculation and manipulation of protein MSAs, (does not work with nucleotide sequences) calculates trees and PCA, displays structures. Web applet version allows visualization of pre-calculated alignments. Coupled with structure prediction server JNet. Used as a default viewer in many web servers and databases
Panta rhei (QAlign2)[133]	gi.cebitec.uni-bielefeld.de/qalign	Standalone tool (Windows and Mac OS X). Allows for manipulations of huge protein and nucleotide datasets in multiple parallel sessions. Calculates MSAs and phylogenetic trees
STRAP[134]	www.charite.de/bioinf/strap/	Standalone Java tool (OS independent) for huge MSAs of protein sequences and structures. Supports annotation of mRNA, intron/exon gene structure. Allows exporting data to Jalview
BioEdit	www.mbio.ncsu.edu/BioEdit/bioedit.html	Standalone tool for MS Windows. Multiple options for manual editing, graphical display and basic analyses of sequence conservation for proteins and nucleic acids, links to external servers. Last update: July 2007. As of February 2008: no longer being reliably maintained, and the documentation is out of date
GeneDoc	www.nrbsc.org/gfx/genedoc/index.html	Standalone tool for MS Windows. Multiple options for manual editing, graphical display and basic analyses of sequence conservation for proteins and nucleic acids. Supports phylogenetic trees and integrates sequence and structure information. Last update: July 2001
JEMBOSS[135]	emboss.sourceforge.net/Jemboss/	Java tool – standalone and web version. A graphical user interface to EMBOSS. Very simple
INTERALIGN[136]	see the right panel for an unusually long link to this program	Java tool (for Linux and Windows) to interactively manipulate and refine multiple sequence alignments using 3D structures. www-dsv.cea.fr/instituts/institut-de-biologie-environnementale-et-biotechnologie-ibeb/unites-de-recherche/service-de-biochimie-et-toxicologie-nucleaire-sbtn/interalign-download-page
CINEMA[137]	utopia.cs.manchester.ac.uk/cinema	Standalone tool for MS Windows, Linux, and Mac OS X. Interactive editor for proteins and nucleic acids. Java-based applets. Serious security issue: data are saved on a remote server and are publicly available. Built into UTOPIA, comprising also protein structure viewer Ambrosia and search and management tool Find-O-Matic allowing for access of remote databases

Tool	URL	Description
Base-By-Base[138]	athena.bioc.uvic.ca/workbench.php?tool=basebybase	Java web tool. Developed for comparative analysis of viral genomes, but handles also proteins. Relatively slow
SQUINT[139]	www.cebl.auckland.ac.nz/index.php?target=software&item=6	Standalone Java tool. Allows for calculation and editing of MSAs both for DNA sequences and the corresponding protein sequences
MACAW[140]	genamics.com/software/downloads/	A standalone (Windows and Mac OS) interactive program for locating, and combining 'blocks' of similar sequence segments. Employs Gibbs sampling and pattern searches
SeaView[141]	pbil.univ-lyon1.fr/software/seaview.html	A standalone application for a variety of systems (including MS Windows, Linux, Mac OS X, and Solaris) as well as a helper application via web browser. Allows for manual editing of the MSAs and basic comparative analyses
ViTO[142]	bioserv.cbs.cnrs.fr/VITO/DOC/	An interactive program coupling a MSA editor with a 3D viewer, especially useful for preparing input files for comparative modeling. Supports macros. Connected to SCWRL and MODELLER for 3D structure modeling (see the chapter by Kosinski et al. in this volume)
POAViz[143]	www.bioinformatics.ucla.edu/poa	A visualization tool for POA alignments (see Table 1.4)
AltAVisT[144]	Bibiserv.techfak.uni-bielefeld.de/altavist/	A web server able to compare two alternative MSAs of a given sequence set to each other. Color-coded regions where MSAs coincide and can be considered to be most reliable
BOXSHADE	www.ch.embnet.org/software/BOX_form.html	Standalone Linux tool and a web server to generate a rendered PostScript, rtf or pict output from an MSA
ESPript[145]	www.lg.ndirect.co.uk/chroma	Standalone Linux tool and a web server to generate a rendered PostScript output from an MSA

with methods such MrBayes or ANCESCON (ftp://iole.swmed.edu/pub/ANCESCON/)[157] (review: ref. [158]).

1.9 Prediction of Domains

It has been reported that around 65% or eukaryotic and around 40% of prokaryotic proteins are composed of two or more globular domains.[159] In addition, 30–60% of eukaryotic proteins are predicted to contain long stretches of disordered residues.[160] Unfortunately, many experimental as well as computational techniques work effectively only on single domains. For instance, experimental structure determination using NMR and in many cases also X-ray crystallography is more successful for isolated globular domains, devoid of disordered regions, rather than for complete multi-domain proteins, unless their constituent parts form a tight complex. Also, many computational methods for protein sequence alignment, phylogenetic analyses (see above), or three-dimensional structure prediction (fold recognition and *de novo* folding – see chapters by Kosinski *et al.* and by Gront *et al.* in this volume) have been designed to work with single domains and may produce erroneous results when presented with multidomain proteins. Thus, identification of domain boundaries from amino acid sequence (hereafter referred to as 1D domain prediction) is an essential step in many protein analyses. However, as mentioned earlier, there is no precise definition of what constitutes a domain even if the structure is known; therefore 1D domain prediction from sequence without structural information presents a great challenge and interpretation of results must consider a certain degree of fuzziness.

Jones and coworkers[161] have classified 1D domain prediction methods into three broad and partially overlapping classes, analogous to 3D structure prediction methods: domain homology prediction, domain recognition (these two classes can be considered 'template-based'), and new domain ('template-free') prediction methods. The most effective way of domain prediction is by detecting its homology to known domain structures (e.g. those classified in SCOP or CATH databases) or to domains from manually curated sequence databases, such as Pfam or CDD (Table 1.1). Main problems in predicting homology occur when the domain is discontinuous (e.g. in the case of insertion of another domain), exhibits circular permutation or forms an evolutionarily conserved module with another associated domain. In this context it must be remembered that some of the entries in domain databases correspond in fact to evolutionarily conserved modules that comprise several structural domains. For sequence regions that cannot be assigned to known domain 'by homology', domain recognition methods can be used. One approach is to apply 3D fold-recognition methods that allow for prediction of structural similarity to known domain structures due to extremely distant homology and sometimes also due to analogy (see Chapter 4 by Kosinski *et al.*) Another approach is to predict secondary structure for the query sequence (see Chapter 2 by Majorek *et al.*) and search for known domains with similar patterns. Finally, new domain prediction rely either on machine learning methods for recognition of sequence features that generally characterize domains or on methods for *de novo* folding (see Chapter 5 by Gront *et al.*) that generate a set of possible tertiary structures, in which compact units are identified. This last class of method is extremely computationally expensive. A list of currently available web servers is shown in Table 1.5; besides, some of domain databases mentioned in Table 1.1 have their own search utilities.

Table 1.5 Domain prediction methods

Method	URL (http://)	Description
		Template-based domain prediction
SSEP-Domain[165]	www.bio.ifi.lmu.de/SSEP	Server input is restricted to 50-600 amino acids. Applies secondary structure element alignment (SSEA) and profile-profile alignment (PPA) in combination with InterPro pattern searches
SBASE[166]	hydra.icgeb.trieste.it/sbase/	Searches a database of known domains and applies SVM to post-process results using a 'similarity network' of inter-sequence similarity scores for known domains
Ginzu[164]	www.robetta.org	Searches for homologous domains in PDB using first PSI-BLAST, then fold-recognition method 3D-Jury, retrieved structures are parsed into domains. In the remaining regions domains are predicted according to the pattern of conservation in PSI-BLAST alignments. Domain boundaries are assigned based on patterns of sequence edges and low-occupied positions in the PSI-BLAST output and secondary structure predicted by PSI-PRED
DOMAINATION[167]	mathbio.nimr.mrc.ac.uk	Infers putative domains and their boundaries in a query sequence from local gapped alignments generated using PSI-BLAST, then submits delineated domains as successive database queries in further iterative steps
Biozon[168]	biozon.org/tools/domains/ (in February 2008 down until further notice)	Analyzes the results of a database search by an ANN, the output is further smoothed and post-processed using a probabilistic model to predict the most likely transition positions between domains
PPRODO[169]	gene.kias.re.kr/~jlee/pprodo (standalone tool available for download)	Analyzes the results of a PSI-BLAST database search by an ANN
DOMpro[170]	www.ics.uci.edu/~baldig/dompro.html	Predicts protein domain boundaries based on bidirectional recurrent ANNs and statistical methods from PSI-BLAST PSSMs, predicted secondary structure and solvent accessibility
		New domain prediction
Globplot[171]	globplot.embl.de/	Identifies putative domains by identifying the globular and non-globular regions within protein sequence based on the amino acid propensities for random coil (disordered) or secondary structure. See also Chapter 2 by Majorek et al. in this volume

(continued overleaf)

Table 1.6 (continued)

Method	URL (http://)	Description
DomCut[172]	www.bork.embl-heidelberg.de/%7Esuyama/domcut/	Identifies domain boundaries by discriminating between regions with amino acid composition characteristic for globular domains and interdomain linkers in multidomain proteins
Scooby-domain[173]	ibivu.cs.vu.nl/programs/scoobywww	Identifies putative globular domains in protein sequence based on the observed lengths and hydrophobicities of domains from proteins with known tertiary structure
CHOPnet[174]	www.rostlab.org/services/CHOP/submit.html	Uses ANN to predicts domain boundaries from sequence conservation, predicted secondary structure, solvent accessibility, amino acid flexibility, and amino acid composition
Meta-servers		
Meta-DP[162]	meta-dp.cse.buffalo.edu/	Meta-server for prediction of globular domains by calculating simple consensus of 10 different primary methods: Adda, Biozon, DomPred-DomSSEA, InterProScan, Mateo, Globplot, ROBETTA-Ginzu, Dopro, Ssep-domain, Dompro
DomPred[161]	bioinf.cs.ucl.ac.uk/dompred/DomPredform.html	Meta-server, consists of: (1) domain homology searches against the Pfam database, (2) DPS method, which predicts domain boundaries from the distribution of termini of sequence matches reported by PSI-BLAST, and (3) DomSSEA, which compares a pattern of secondary structures predicted for the target protein with secondary structure patterns of domains with known 3D structures
DOMAC[163]	www.bioinfotool.org/domac.html	Meta-server, first runs PSI-BLAST to detect similarity to known structures, then builds 3D models using MODELLER, and parses them into domains using PDP[175]. For the remaining regions uses DOMro (see above)

As with most of bioinformatics predictions, the recommended protocol for 1D domain prediction involves application of the consensus rule. A meta-server for domain prediction Meta-DP has been developed[162] that allows for comparison and averaging of results reported by several algorithms. However, the best results are achieved if 1D domain prediction is carried out hierarchically, starting with the template-based methods, followed by the more demanding (and more error-prone) *de novo* methods. This hybrid approach has been already implemented in a few fully automated methods that were shown to outperform individual methods within the framework of the CASP competition. Examples include DOMAC[163] (available as a server, see Table 1.5) and DP_Hybrid (comprising Ginzu and RosettaDOM,[164] components of the Rosetta suite, not available as a standalone server).

1.10 Summary

In this chapter we discussed methods for primary structure analysis of proteins, including identification of short motifs, database searches to detect significantly similar sequences (candidate homologs), sequence clustering to identify protein families regions of homology to sequences, multiple sequence alignment, and identification of globular domains. We have not covered the issue of predicting non-globular or disordered regions and secondary structure prediction, as these analyses are reviewed in depth in another chapter in this volume (Majorek *et al.*). In addition to reviewing theory, we provided tables summarizing different programs dedicated to carry out various types of sequence analyses. These are mostly web servers, and some standalone packages for local installation. We must mention, however, that many databases and methods that have been described in the literature and used to be available as web servers, have now disappeared from the Internet or at least have not been available during preparation of this chapter, therefore were omitted from the tables. It is also expected that with time some of the methods mentioned here will also completely disappear or will move to different websites; on the other hand, new interesting methods will be made available. The readers / potential users are therefore encouraged to consult the periodically updated collections of web servers e.g. the annual special issue of Nucleic Acid Research (http://nar.oxfordjournals.org/) and the Bioinformatics Links Directory, (http://bioinformatics.ca/links_directory/).

There are several considerations in choosing a set of programs to analyze a sequence of interest, including biological accuracy, complexity of the analysis and time required to complete it (without asking a sequence analysis expert for help), and software/hardware usage. In Figure 1.2 we present a flowchart illustrating the recommended protocol of protein sequence analysis, from basic searches to domain prediction, which can be used to generate input data for more subsequent computational or experimental analyses. If the aim is simple, e.g. to obtain an approximate sequence alignment of a few homologs and illustrate the most obvious motifs (both SMs and LMs), then a simple sequence search (e.g. with BLAST) of one of protein family/domain databases is often sufficient to check, whether an annotated data set is already available for download, without the need to carry out new analyses. However, we suggest that web servers for identification of motifs should be queried, as they often provide information that is more up to date than pre-calculated data sets in family databases. In case of novel sequences that are not yet present in major databases, a PSI-BLAST search of one of sequence databases

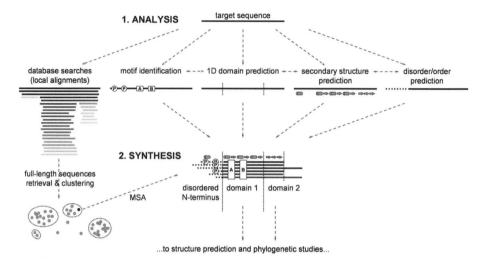

Figure 1.2 *Suggested workflow for protein sequence analysis. Basic sequence analyses involve usually some or all of five tasks: (1) identification of locally similar sequences in databases (here, sequences with decreasing level of similarity are indicated by fading shades of gray) followed by retrieval of full-length sequences and their clustering to identify families, and finally MSA of the family; (2) identification of motifs (LMs and SMs); (3) prediction of putative domains; (4) prediction of secondary structure; (5) prediction of disordered/ordered regions. Tasks (4) and (5) are reviewed in Chapter 2 by Majorek et al. in this volume. Subsequently, results from these analyses (as well as useful data and predictions from other sources, if available) are combined and individual domain families may be subjected to detailed structural and phylogenetic analyses. Alternatively, predicted domain structure may be used to carry out another round of basic analyses, with adjusted parameters (e.g. new database searches, with correction for compositionally biased sequence, and e.g. removed N-terminal region or protein sequence split into individual domains)*

(e.g. nr at the NCBI) is recommended, to be followed by clustering of the extracted homologs and identification of the putative orthologous family, which may be aligned using one of the recent methods for MSA calculation. In parallel, domain databases should be searched by sensitive profile methods to detect potential presence of known domains. If no evident similarity to known protein families or domains is observed, domain prediction methods should be used, preferably in connection with prediction of disordered regions and secondary structure. If the aim of the analysis is an experimental characterization of protein function, such combination of methods is usually sufficient to delineate major domains and conserved regions. However, if an advanced comparative analysis is desired, e.g. calculation of a phylogenetic tree or prediction of protein structure, the MSA must be carefully refined to remove or 'mask' unalignable (e.g. non-homologous) regions. For multidomain proteins domain boundaries must be judiciously localized, and domains should be submitted independently for phylogenetic and modeling calculations, unless there are specific reasons to believe that a set of domains should be analyzed together (e.g. if it forms an evolutionarily conserved module). At all stages of analysis (perhaps with the exception of database searches), we recommend using several alternative methods and comparing their results. As a rule of thumb, consistency between different algorithms indicates higher likelihood that a given result is close to optimal. On the other hand, automatically generated results are seldom ideal and they can be often improved by human experts. Finally, it must be remembered that uncorrected errors tend to accumulate, and 'higher level' methods usually assume that their input is error-free, thus it is very important to carefully check results returned by all automated methods before submitting them to next, usually more time-consuming stages.

Acknowledgements

We thank present and former members of the Bujnicki lab in IIMCB and at the UAM for stimulating discussions and contribution of ideas and information to this article. The authors acknowledge the support from past and current grants for the development of bioinformatics methods from Polish Ministry of Science, NIH, Framework Programme of the EU, EMBO, and HHMI. KHK has worked on this article while being supported by a fellowship from EMBO and a travel grant from Polish Academy of Sciences and JSPS. JMB has worked on this article while being supported by the Institute of Medical Science at the University of Tokyo.

References

1. R.A. Jensen, Enzyme recruitment in evolution of new function, *Annu Rev Microbiol*, **30**, 409–425 (1976).
2. E.V. Koonin, Orthologs, paralogs, and evolutionary genomics, *Annu Rev Genet*, **39**, 309–338 (2005).
3. C. Chothia, and A.M. Lesk, The relation between the divergence of sequence and structure in proteins, *Embo J*, **5**, 823–826 (1986).
4. A. Weichsel, E.M. Maes, J.F. Andersen, *et al.*, Heme-assisted s-nitrosation of a proximal thiolate in a nitric oxide transport protein, *Proc Natl Acad Sci U S A*, **102**, 594–599 (2005).
5. L.N. Kinch, and N.V. Grishin, Evolution of protein structures and functions, *Curr Opin Struct Biol*, **12**, 400–408 (2002).
6. C.A. Orengo, and J.M. Thornton, Protein families and their evolution – a structural perspective, *Annu Rev Biochem*, **74**, 867–900 (2005).
7. D.L. Wheeler, T. Barrett, D.A. Benson, *et al.*, Database resources of the National Center for Biotechnology Information, *Nucleic Acids Res*, **36**, D13-21 (2008).
8. K. Henrick, Z. Feng, W.F. Bluhm, *et al.*, Remediation of the Protein Data Bank Archive, *Nucleic Acids Res*, **36**, D426-433 (2008).
9. M. Perutz, Early days of protein crystallography, *Methods Enzymol*, **114**, 3–18 (1985).
10. A. Elofsson, and G. von Heijne, Membrane protein structure: prediction versus reality, *Annu Rev Biochem*, **76**, 125–140 (2007).
11. C.P. Ponting, and R.R. Russell, The natural history of protein domains, *Annu Rev Biophys Biomol Struct*, **31**, 45–71 (2002).
12. C. Vogel, C. Berzuini, M. Bashton, J. Gough, and S.A. Teichmann, Supra-domains: Evolutionary units larger than single protein domains, *J Mol Biol*, **336**, 809–823 (2004).
13. Y. Lindqvist, and G. Schneider, Circular permutations of natural protein sequences: Structural evidence, *Curr Opin Struct Biol*, **7**, 422–427 (1997).
14. J.C. Wootton, and S. Federhen, Analysis of compositionally biased regions in sequence databases, *Methods Enzymol.*, **266**, 554–571 (1996).
15. P. Romero, Z. Obradovic, X. Li, E.C. Garner, C.J. Brown, and A.K. Dunker, Sequence complexity of disordered protein, *Proteins*, **42**, 38–48 (2001).

16. P. Radivojac, L.M. Iakoucheva, C.J. Oldfield, Z. Obradovic, V.N. Uversky, and A.K. Dunker, Intrinsic disorder and functional proteomics, *Biophys J*, **92**, 1439–1456 (2007).

17. V. Csizmok, Z. Dosztanyi, I. Simon, and P. Tompa, Towards proteomic approaches for the identification of structural disorder, *Curr Protein Pept Sci*, **8**, 173–179 (2007).

18. D.A. Parry, Structural and functional implications of sequence repeats in fibrous proteins, *Adv Protein Chem*, **70**, 11–35 (2005).

19. H. Xie, S. Vucetic, L.M. Iakoucheva, *et al.*, Functional anthology of intrinsic disorder. 3. Ligands, post-translational modifications, and diseases associated with intrinsically disordered proteins, *J Proteome Res*, **6**, 1917–1932 (2007).

20. S. Vucetic, H. Xie, L.M. Iakoucheva, *et al.*, Functional anthology of intrinsic disorder. 2. Cellular components, domains, technical terms, developmental processes, and coding sequence diversities correlated with long disordered regions, *J Proteome Res*, **6**, 1899–1916 (2007).

21. H. Xie, S. Vucetic, L.M. Iakoucheva, *et al.*, Functional anthology of intrinsic disorder. 1. Biological processes and functions of proteins with long disordered regions, *J Proteome Res*, **6**, 1882–1898 (2007).

22. J.E. Walker, M. Saraste, M.J. Runswick, and N.J. Gay, Distantly related sequences in the alpha- and beta-subunits of Atp synthase, myosin, kinases and other Atp-requiring enzymes and a common nucleotide binding fold, *Embo J*, **1**, 945–951 (1982).

23. K. Struhl, Helix-turn-helix, zinc-finger, and leucine-zipper motifs for eukaryotic transcriptional regulatory proteins, *Trends Biochem Sci*, **14**, 137–140 (1989).

24. V. Neduva, and R.B. Russell, Linear motifs: evolutionary interaction switches, *FEBS Lett*, **579**, 3342–3345 (2005).

25. M. Fuxreiter, P. Tompa, and I. Simon, Local structural disorder imparts plasticity on linear motifs, *Bioinformatics*, **23**, 950–956 (2007).

26. T.A. Holland, S. Veretnik, I.N. Shindyalov, and P.E. Bourne, Partitioning protein structures into domains: why is it so difficult?, *J Mol Biol*, **361**, 562–590 (2006).

27. A.G. Murzin, S.E. Brenner, T. Hubbard, and C. Chothia, Scop: A Structural Classification of Proteins Database for the investigation of sequences and structures, *J Mol Biol*, **247**, 536–540 (1995).

28. C.A. Orengo, A.D. Michie, S. Jones, D.T. Jones, M.B. Swindells, and J.M. Thornton, Cath: a hierarchic classification of protein domain structures, *Structure*, **5**, 1093–1108 (1997).

29. R. Apweiler, T.K. Attwood, A. Bairoch, *et al.*, The Interpro Database, an integrated documentation resource for protein families, domains and functional sites, *Nucleic Acids Res*, **29**, 37–40 (2001).

30. N.J. Mulder, R. Apweiler, T.K. Attwood, *et al.*, New developments in the Interpro Database, *Nucleic Acids Res*, **35**, D224–228 (2007).

31. E.L. Sonnhammer, S.R. Eddy, and R. Durbin, Pfam: A comprehensive database of protein domain families based on seed alignments, *Proteins*, **28**, 405–420 (1997).

32. R.D. Finn, J. Tate, J. Mistry, *et al.*, The Pfam Protein Families Database, *Nucleic Acids Res*, **36**, D281–288 (2008).

33. A. Bairoch, Prosite: A dictionary of sites and patterns in proteins, *Nucleic Acids Res*, **19 Suppl**, 2241–2245 (1991).

34. N. Hulo, A. Bairoch, V. Bulliard, *et al.*, The 20 years of prosite, *Nucleic Acids Res*, **36**, D245–249 (2008).

35. A. Marchler-Bauer, A.R. Panchenko, B.A. Shoemaker, P.A. Thiessen, L.Y. Geer, and S.H. Bryant, Cdd: A database of conserved domain alignments with links to domain three-dimensional structure, *Nucleic Acids Res*, **30**, 281–283 (2002).

36. A. Marchler-Bauer, J.B. Anderson, M.K. Derbyshire, *et al.*, Cdd: A conserved domain database for interactive domain family analysis, *Nucleic Acids Res*, **35**, D237–240 (2007).

37. R.L. Tatusov, M.Y. Galperin, D.A. Natale, and E.V. Koonin, The Cog Database: A tool for genome-scale analysis of protein functions and evolution, *Nucleic Acids Res*, **28**, 33–36 (2000).

38. E.V. Kriventseva, W. Fleischmann, E.M. Zdobnov, and R. Apweiler, Clustr: A database of clusters of Swiss-Prot+Trembl proteins, *Nucleic Acids Res*, **29**, 33–36 (2001).

39. N. Kaplan, O. Sasson, U. Inbar, *et al.*, Protonet 4.0: A hierarchical classification of one million protein sequences, *Nucleic Acids Res*, **33**, D216–218 (2005).

40. T. Meinel, A. Krause, H. Luz, M. Vingron, and E. Staub, The Systers Protein Family Database in 2005, *Nucleic Acids Res*, **33**, D226–229 (2005).

41. L.J. Jensen, P. Julien, M. Kuhn, *et al.*, Eggnog: Automated construction and annotation of orthologous groups of genes, *Nucleic Acids Res*, **36**, D250–254 (2008).

42. A.C. Berglund, E. Sjolund, G. Ostlund, and E.L. Sonnhammer, Inparanoid 6: Eukaryotic ortholog clusters with inparalogs, *Nucleic Acids Res*, **36**, D263–266 (2008).

43. E.V. Kriventseva, N. Rahman, O. Espinosa, and E.M. Zdobnov, Orthodb: The hierarchical catalog of eukaryotic orthologs, *Nucleic Acids Res*, **36**, D271–275 (2008).

44. T. Rattei, P. Tischler, R. Arnold, *et al.*, Simap–Structuring the network of protein similarities, *Nucleic Acids Res*, **36**, D289–292 (2008).

45. P. Puntervoll, R. Linding, C. Gemund, *et al.*, Elm server: A new resource for investigating short functional sites in modular eukaryotic proteins, *Nucleic Acids Res*, **31**, 3625–3630 (2003).

46. J.C. Obenauer, L.C. Cantley, and M.B. Yaffe, Scansite 2.0: Proteome-wide prediction of cell signaling interactions using short sequence motifs, *Nucleic Acids Res*, **31**, 3635–3641 (2003).

47. N. Blom, T. Sicheritz-Ponten, R. Gupta, S. Gammeltoft, and S. Brunak, Prediction of post-translational glycosylation and phosphorylation of proteins from the amino acid sequence, *Proteomics*, **4**, 1633–1649 (2004).

48. J. Gough, K. Karplus, R. Hughey, and C. Chothia, Assignment of homology to genome sequences using a library of hidden Markov models that represent all proteins of known structure, *J Mol Biol*, **313**, 903–919 (2001).

49. C. Yeats, M. Maibaum, R. Marsden, *et al.*, Gene3d: Modelling protein structure, function and evolution, *Nucleic Acids Res*, **34**, D281–284 (2006).

50. P.D. Thomas, A. Kejariwal, M.J. Campbell, *et al.*, Panther: A browsable database of gene products organized by biological function, using curated protein family and subfamily classification, *Nucleic Acids Res*, **31**, 334–341 (2003).

51. D.H. Haft, B.J. Loftus, D.L. Richardson, *et al.*, Tigrfams: A protein family resource for the functional identification of proteins, *Nucleic Acids Res*, **29**, 41–43 (2001).

52. C.H. Wu, S. Zhao, and H.L. Chen, A protein class database organized with prosite protein groups and PIR superfamilies, *J Comput Biol*, **3**, 547–561 (1996).

53. E.L. Sonnhammer, and D. Kahn, Modular arrangement of proteins as inferred from analysis of homology, *Protein Sci*, **3**, 482–492 (1994).

54. J. Schultz, F. Milpetz, P. Bork, and C.P. Ponting, Smart, a Simple Modular Architecture Research Tool: Identification of signaling domains, *Proc Natl Acad Sci U S A*, **95**, 5857–5864 (1998).

55. T.K. Attwood, M.E. Beck, A.J. Bleasby, and D.J. Parry-Smith, Prints: a Database of protein motif fingerprints, *Nucleic Acids Res*, **22**, 3590–3596 (1994).

56. S. Balla, V. Thapar, S. Verma, *et al.*, Minimotif miner: A tool for investigating protein function, *Nat Methods*, **3**, 175–177 (2006).

57. O. Emanuelsson, S. Brunak, G. von Heijne, and H. Nielsen, Locating proteins in the cell using Targetp, Signalp and related tools, *Nat Protoc*, **2**, 953–971 (2007).

58. T.L. Bailey, N. Williams, C. Misleh, and W.W. Li, Meme: Discovering and analyzing DNA and protein sequence motifs, *Nucleic Acids Res*, **34**, W369–373 (2006).

59. C.E. Lawrence, S.F. Altschul, M.S. Boguski, J.S. Liu, A.F. Neuwald, and J.C. Wootton, Detecting subtle sequence signals: A Gibbs sampling strategy for multiple alignment, *Science*, **262**, 208–214 (1993).

60. T. Huynh, I. Rigoutsos, L. Parida, D. Platt, and T. Shibuya, The web server of IBM's Bioinformatics and Pattern Discovery Group, *Nucleic Acids Res*, **31**, 3645–3650 (2003).

61. V. Neduva, and R.B. Russell, Dilimot: Discovery of linear motifs in proteins, *Nucleic Acids Res*, **34**, W350–355 (2006).

62. N.E. Davey, D.C. Shields, and R.J. Edwards, Slimdisc: Short, linear motif discovery, correcting for common evolutionary descent, *Nucleic Acids Res*, **34**, 3546–3554 (2006).

63. M. Dogruel, T.A. Down, and T.J. Hubbard, Nestedmica as an ab initio protein motif discovery tool, *BMC Bioinformatics*, **9**, 19 (2008).

64. E. Redhead, and T.L. Bailey, Discriminative motif discovery in DNA and protein sequences using the Deme algorithm, *BMC Bioinformatics*, **8**, 385 (2007).

65. A. Apostolico, M. Comin, and L. Parida, Conservative extraction of over-represented extensible motifs, *Bioinformatics*, **21 Suppl 1**, i9–18 (2005).

66. T.D. Schneider, and R.M. Stephens, Sequence logos: A new way to display consensus sequences, *Nucleic Acids Res*, **18**, 6097–6100 (1990).

67. G.E. Crooks, G. Hon, J.M. Chandonia, and S.E. Brenner, Weblogo: A sequence logo generator, *Genome Res*, **14**, 1188–1190 (2004).

68. L.M. Iakoucheva, P. Radivojac, C.J. Brown, T.R. O'Connor, J.G. Sikes, Z. Obradovic, and A.K. Dunker, The importance of intrinsic disorder for protein phosphorylation, *Nucleic Acids Res*, **32**, 1037–1049 (2004).

69. G. Blackshields, I.M. Wallace, M. Larkin, and D.G. Higgins, Analysis and comparison of benchmarks for multiple sequence alignment, *In Silico Biol*, **6**, 321–339 (2006).

70. S.B. Needleman, and C.D. Wunsch, A general method applicable to the search for similarities in the amino acid sequence of two proteins, *J.Mol.Biol.*, **48**, 443–453 (1970).

71. T.F. Smith, and M.S. Waterman, Identification of common molecular subsequences, *J.Mol.Biol.*, **147**, 195–197 (1981).

72. M.O. Dayhoff, R.M. Schwartz, and B.C. Orcutt, A model of evolutionary change in proteins, in *Atlas of Protein Sequence and Structure*, M.O. Dayhoff (ed.), Natl. Biomed. Res. Found., Washington, DC., 1978.

73. S.A. Benner, M.A. Cohen, and G.H. Gonnet, Amino acid substitution during functionally constrained divergent evolution of protein sequences, *Protein Eng*, **7**, 1323–1332 (1994).

74. D.T. Jones, W.R. Taylor, and J.M. Thornton, The rapid generation of mutation data matrices from protein sequences, *Comput Appl Biosci*, **8**, 275–282 (1992).

75. S. Henikoff, and J.G. Henikoff, Amino acid substitution matrices from protein blocks, *Proc Natl Acad Sci U S A*, **89**, 10915–10919 (1992).

76. W.R. Pearson, and D.J. Lipman, Improved tools for biological sequence comparison, *Proc.Natl.Acad.Sci.U.S.A.*, **85**, 2444–2448 (1988).

77. S.F. Altschul, W. Gish, W. Miller, E.W. Myers, and D.J. Lipman, Basic local alignment search tool, *J Mol Biol*, **215**, 403–410 (1990).

78. W.R. Pearson, Empirical statistical estimates for sequence similarity searches, *J Mol Biol*, **276**, 71–84 (1998).

79. M. Pagni, and C.V. Jongeneel, Making sense of score statistics for sequence alignments, *Brief Bioinform*, **2**, 51–67 (2001).

80. S.F. Altschul, T.L. Madden, A.A. Schaffer, *et al.*, Gapped blast and Psi-blast: A new generation of protein database search programs, *Nucleic Acids Res*, **25**, 3389–3402 (1997).

81. S.R. Eddy, G. Mitchison, and R. Durbin, Maximum discrimination hidden Markov models of sequence consensus, *J Comput Biol*, **2**, 9–23 (1995).

82. J. Park, S.A. Teichmann, T. Hubbard, and C. Chothia, Intermediate sequences increase the detection of homology between sequences, *J Mol Biol*, **273**, 349–354 (1997).

83. J. Park, K. Karplus, C. Barrett, *et al.*, Sequence comparisons using multiple sequences detect three times as many remote homologues as pairwise methods, *J Mol Biol*, **284**, 1201–1210 (1998).

84. W. Li, F. Pio, K. Pawlowski, and A. Godzik, Saturated blast: An automated multiple intermediate sequence search used to detect distant homology, *Bioinformatics*, **16**, 1105–1110 (2000).

85. K.K. Koretke, R.B. Russell, and A.N. Lupas, Fold recognition without folds, *Protein Sci*, **11**, 1575–1579 (2002).

86. S.F. Altschul, and E.V. Koonin, Iterated profile searches with Psi-blast–a tool for discovery in protein databases, *Trends Biochem Sci*, **23**, 444–447 (1998).

87. L. Aravind, and E.V. Koonin, Gleaning non-trivial structural, functional and evolutionary information about proteins by iterative database searches, *J Mol Biol*, **287**, 1023–1040 (1999).

88. A.A. Schaffer, Y.I. Wolf, C.P. Ponting, E.V. Koonin, L. Aravind, and S.F. Altschul, Impala: Matching a protein sequence against a collection of Psi-blast-constructed position-specific score matrices, *Bioinformatics*, **15**, 1000–1011 (1999).

89. G. Yona, and M. Levitt, Within the twilight zone: A sensitive profile-profile comparison tool based on information theory, *J Mol Biol*, **315**, 1257–1275 (2002).

90. R. Sadreyev, and N. Grishin, Compass: A tool for comparison of multiple protein alignments with assessment of statistical significance, *J Mol Biol*, **326**, 317–336 (2003).

91. J. Soding, Protein homology detection by Hmm-Hmm comparison, *Bioinformatics*, **21**, 951–960 (2005).

92. J. Soding, M. Remmert, A. Biegert, and A.N. Lupas, Hhsenser: Exhaustive transitive profile search using Hmm-Hmm comparison, *Nucleic Acids Res*, **34**, W374–378 (2006).

93. T. Frickey, and A. Lupas, Clans: A Java application for visualizing protein families based on pairwise similarity, *Bioinformatics*, **20**, 3702–3704 (2004).

94. A.T. Adai, S.V. Date, S. Wieland, and E.M. Marcotte, Lgl: Creating a map of protein function with an algorithm for visualizing very large biological networks, *J Mol Biol*, **340**, 179–190 (2004).

95. P. Pipenbacher, A. Schliep, S. Schneckener, A. Schonhuth, D. Schomburg, and R. Schrader, Proclust: Improved clustering of protein sequences with an extended graph-based approach, *Bioinformatics*, **18 Suppl 2**, S182–191 (2002).

96. A.J. Enright, S. Van Dongen, and C.A. Ouzounis, An efficient algorithm for large-scale detection of protein families, *Nucleic Acids Res*, **30**, 1575–1584 (2002).

97. I.V. Tetko, A. Facius, A. Ruepp, and H.W. Mewes, Super paramagnetic clustering of protein sequences, *BMC Bioinformatics*, **6**, 82 (2005).

98. N. Krishnamurthy, D. Brown, and K. Sjolander, Flowerpower: Clustering proteins into domain architecture classes for phylogenomic inference of protein function, *BMC Evol Biol*, **7 Suppl 1**, S12 (2007).

99. D.P. Brown, N. Krishnamurthy, and K. Sjolander, Automated protein subfamily identification and classification, *PLoS Comput Biol*, **3**, e160 (2007).

100. A. Kelil, S. Wang, R. Brzezinski, and A. Fleury, Cluss: Clustering of protein sequences based on a new similarity measure, *BMC Bioinformatics*, **8**, 286 (2007).

101. L. Wang, and T. Jiang, On the complexity of multiple sequence alignment, *J Comput Biol*, **1**, 337–348 (1994).

102. P. Hogeweg, and B. Hesper, The alignment of sets of sequences and the construction of phyletic trees: an integrated method, *J Mol Evol*, **20**, 175–186 (1984).

103. O. Gotoh, Significant improvement in accuracy of multiple protein sequence alignments by iterative refinement as assessed by reference to structural alignments, *J Mol Biol*, **264**, 823–838 (1996).

104. C. Notredame, D.G. Higgins, and J. Heringa, T-Coffee: A novel method for fast and accurate multiple sequence alignment, *J Mol Biol*, **302**, 205–217 (2000).

105. I.M. Wallace, O. O'Sullivan, and D.G. Higgins, Evaluation of iterative alignment algorithms for multiple alignment, *Bioinformatics*, **21**, 1408–1414 (2005).

106. B. Morgenstern, K. Frech, A. Dress, and T. Werner, Dialign: Finding local similarities by multiple sequence alignment, *Bioinformatics*, **14**, 290–294 (1998).

107. B. Raphael, D. Zhi, H. Tang, and P. Pevzner, A novel method for multiple alignment of sequences with repeated and shuffled elements, *Genome Res*, **14**, 2336–2346 (2004).

108. H. Zhou, and Y. Zhou, Spem: Improving multiple sequence alignment with sequence profiles and predicted secondary structures, *Bioinformatics*, **21**, 3615–3621 (2005).

109. J. Pei, B.H. Kim, M. Tang, and N.V. Grishin, Promals web server for accurate multiple protein sequence alignments, *Nucleic Acids Res.* (2007).

110. J.D. Thompson, T.J. Gibson, F. Plewniak, F. Jeanmougin, and D.G. Higgins, The Clustal_X Windows Interface: Flexible strategies for multiple sequence alignment aided by quality analysis tools, *Nucleic.Acids Res.*, **25**, 4876–4882 (1997).

111. J.D. Thompson, F. Plewniak, J. Thierry, and O. Poch, Dbclustal: Rapid and reliable global multiple alignments of protein sequences detected by database searches, *Nucleic Acids Res*, **28**, 2919–2926 (2000).

112. R. Hughey, and A. Krogh, Hidden Markov models for sequence analysis: Extension and analysis of the basic method, *Comput Appl Biosci*, **12**, 95–107 (1996).

113. S.R. Eddy, Multiple alignment using hidden Markov models, *Proc Int Conf Intell Syst Mol Biol*, **3**, 114–120 (1995).

114. R.C. Edgar, Muscle: Multiple sequence alignment with high accuracy and high throughput, *Nucleic Acids Res*, **32**, 1792–1797 (2004).

115. K. Katoh, K. Misawa, K. Kuma, and T. Miyata, Mafft: A novel method for rapid multiple sequence alignment based on fast Fourier transform, *Nucleic Acids Res*, **30**, 3059–3066 (2002).

116. O. Gotoh, A weighting system and algorithm for aligning many phylogenetically related sequences, *Comput Appl Biosci*, **11**, 543–551 (1995).

117. V.A. Simossis, J. Kleinjung, and J. Heringa, Homology-extended sequence alignment, *Nucleic Acids Res*, **33**, 816–824 (2005).

118. C.B. Do, M.S. Mahabhashyam, M. Brudno, and S. Batzoglou, Probcons: Probabilistic consistency-based multiple sequence alignment, *Genome Res*, **15**, 330–340 (2005).

119. J. Pei, and N.V. Grishin, Mummals: Multiple sequence alignment improved by using hidden Markov models with local structural information, *Nucleic Acids Res*, **34**, 4364–4374 (2006).

120. J. Pei, and N.V. Grishin, Promals: Towards accurate multiple sequence alignments of distantly related proteins, *Bioinformatics*, **23**, 802–808 (2007).

121. R.C. Edgar, and K. Sjolander, Satchmo: Sequence alignment and tree construction using hidden Markov models, *Bioinformatics*, **19**, 1404–1411 (2003).

122. S. Yamada, O. Gotoh, and H. Yamana, Improvement in accuracy of multiple sequence alignment using novel group-to-group sequence alignment algorithm with piecewise linear gap cost, *BMC Bioinformatics*, **7**, 524 (2006).

123. T. Lassmann, and E.L. Sonnhammer, Kalign: An accurate and fast multiple sequence alignment algorithm, *BMC Bioinformatics*, **6**, 298 (2005).

124. A.R. Subramanian, J. Weyer-Menkhoff, M. Kaufmann, and B. Morgenstern, Dialign-T: An improved algorithm for segment-based multiple sequence alignment, *BMC Bioinformatics*, **6**, 66 (2005).

125. Z. Zhang, H. Lin, and M. Li, Mango: A new approach to multiple sequence alignment, *Comput Syst Bioinformatics Conf*, **6**, 237–247 (2007).

126. I. Van Walle, I. Lasters, and L. Wyns, Align-M: A new algorithm for multiple alignment of highly divergent sequences, *Bioinformatics*, **20**, 1428–1435 (2004).

127. C. Lee, C. Grasso, and M.F. Sharlow, Multiple sequence alignment using partial order graphs, *Bioinformatics*, **18**, 452–464 (2002).

128. N.C. Jones, D. Zhi, and B.J. Raphael, Aliwaba: Alignment on the Web through an a-Bruijn approach, *Nucleic Acids Res*, **34**, W613–616 (2006).

129. T.M. Phuong, C.B. Do, R.C. Edgar, and S. Batzoglou, Multiple alignment of protein sequences with repeats and rearrangements, *Nucleic Acids Res*, **34**, 5932-5942 (2006).

130. K. Bucka-Lassen, O. Caprani, and J. Hein, Combining many multiple alignments in one improved alignment, *Bioinformatics*, **15**, 122–130 (1999).

131. I.M. Wallace, O. O'Sullivan, D.G. Higgins, and C. Notredame, M-Coffee: Combining multiple sequence alignment methods with T-Coffee, *Nucleic Acids Res*, **34**, 1692–1699 (2006).

132. M. Clamp, J. Cuff, S.M. Searle, and G.J. Barton, The Jalview Java Alignment Editor, *Bioinformatics*, **20**, 426–427 (2004).

133. M. Sammeth, T. Griebel, F. Tille, and J. Stoye, Panta Rhei (Qalign2): An open graphical environment for sequence analysis, *Bioinformatics*, **22**, 889–890 (2006).

134. C. Gille, and C. Frommel, Strap: Editor for structural alignments of proteins, *Bioinformatics*, **17**, 377–378 (2001).

135. T.J. Carver, and L.J. Mullan, Jae: Jemboss alignment editor, *Appl Bioinformatics*, **4**, 151–154 (2005).

136. O. Pible, G. Imbert, and J.L. Pellequer, Interalign: Interactive alignment editor for distantly related protein sequences, *Bioinformatics*, **21**, 3166–3167 (2005).

137. D.J. Parry-Smith, A.W. Payne, A.D. Michie, and T.K. Attwood, Cinema: A novel colour interactive editor for multiple alignments, *Gene*, **221**, GC57–63 (1998).

138. R. Brodie, A.J. Smith, R.L. Roper, V. Tcherepanov, and C. Upton, Base-by-Base: Single nucleotide-level analysis of whole viral genome alignments, *BMC Bioinformatics*, **5**, 96 (2004).

139. M.G. Goode, and A.G. Rodrigo, Squint: A multiple alignment program and editor, *Bioinformatics*, **23**, 1553–1555 (2007).

140. G.D. Schuler, S.F. Altschul, and D.J. Lipman, A workbench for multiple alignment construction and analysis, *Proteins*, **9**, 180–190 (1991).

141. N. Galtier, M. Gouy, and C. Gautier, Seaview and Phylo_Win: Two graphic tools for sequence alignment and molecular phylogeny, *Comput Appl Biosci*, **12**, 543–548 (1996).

142. V. Catherinot, and G. Labesse, Vito: Tool for refinement of protein sequence-structure alignments, *Bioinformatics*, **20**, 3694–3696 (2004).

143. C. Grasso, M. Quist, K. Ke, and C. Lee, Poaviz: A partial order multiple sequence alignment visualizer, *Bioinformatics*, **19**, 1446–1448 (2003).

144. B. Morgenstern, S. Goel, A. Sczyrba, and A. Dress, Altavist: Comparing alternative multiple sequence alignments, *Bioinformatics*, **19**, 425–426 (2003).

145. P. Gouet, X. Robert, and E. Courcelle, Espript/Endscript: Extracting and rendering sequence and 3D information from atomic structures of proteins, *Nucleic Acids Res*, **31**, 3320–3323 (2003).

146. J. Felsenstein, Phylip – Phylogeny Inference Package (Version 3.2), *Cladistics*, **5**, 164–166 (1989).

147. K. Tamura, J. Dudley, M. Nei, and S. Kumar, Mega4: Molecular Evolutionary Genetics Analysis (Mega) Software Version 4.0, *Mol Biol Evol*, **24**, 1596–1599 (2007).

148. S. Guindon, and O. Gascuel, A simple, fast, and accurate algorithm to estimate large phylogenies by maximum likelihood, *Syst Biol*, **52**, 696–704 (2003).

149. Z. Yang, Paml 4: Phylogenetic analysis by maximum likelihood, *Mol Biol Evol*, **24**, 1586–1591 (2007).

150. H.A. Schmidt, K. Strimmer, M. Vingron, and A. von Haeseler, Tree-Puzzle: Maximum likelihood phylogenetic analysis using quartets and parallel computing, *Bioinformatics*, **18**, 502–504 (2002).

151. F. Ronquist, and J.P. Huelsenbeck, Mrbayes 3: Bayesian phylogenetic inference under mixed models, *Bioinformatics*, **19**, 1572–1574 (2003).

152. T.M. Keane, T.J. Naughton, and J.O. McInerney, Multiphyl: A high-throughput phylogenomics webserver using distributed computing, *Nucleic Acids Res*, **35**, W33–37 (2007).

153. S. Whelan, P. Lio, and N. Goldman, Molecular phylogenetics: State-of-the-art methods for looking into the past, *Trends Genet*, **17**, 262–272 (2001).

154. J.P. Huelsenbeck, F. Ronquist, R. Nielsen, and J.P. Bollback, Bayesian inference of phylogeny and its impact on evolutionary biology, *Science*, **294**, 2310–2314 (2001).

155. C. Kosiol, L. Bofkin, and S. Whelan, Phylogenetics by likelihood: Evolutionary modeling as a tool for understanding the genome, *J Biomed Inform*, **39**, 51–61 (2006).

156. K. Sjolander, Phylogenomic inference of protein molecular function: Advances and challenges, *Bioinformatics*, **20**, 170–179 (2004).

157. W. Cai, J. Pei, and N.V. Grishin, Reconstruction of ancestral protein sequences and its applications, *BMC Evol Biol*, **4**, 33 (2004).

158. P.D. Williams, D.D. Pollock, B.P. Blackburne, and R.A. Goldstein, Assessing the accuracy of ancestral protein reconstruction methods, *PLoS Comput Biol*, **2**, e69 (2006).

159. S.K. Kummerfeld, and S.A. Teichmann, Relative rates of gene fusion and fission in multidomain proteins, *Trends Genet*, **21**, 25–30 (2005).

160. C.J. Oldfield, Y. Cheng, M.S. Cortese, C.J. Brown, V.N. Uversky, and A.K. Dunker, Comparing and combining predictors of mostly disordered proteins, *Biochemistry*, **44**, 1989–2000 (2005).

161. K. Bryson, D. Cozzetto, and D.T. Jones, Computer-assisted protein domain boundary prediction using the Dompred server, *Curr Protein Pept Sci*, **8**, 181–188 (2007).

162. H.K. Saini, and D. Fischer, Meta-Dp: Domain prediction meta server, *Bioinformatics* (2005).

163. J. Cheng, Domac: An accurate, hybrid protein domain prediction server, *Nucleic Acids Res*, **35**, W354–356 (2007).

164. D.E. Kim, D. Chivian, L. Malmstrom, and D. Baker, Automated prediction of domain boundaries in Casp6 targets using Ginzu and Rosettadom, *Proteins* (2005).

165. J.E. Gewehr, and R. Zimmer, Ssep-Domain: Protein domain prediction by alignment of secondary structure elements and profiles, *Bioinformatics*, **22**, 181–187 (2006).

166. K. Vlahovicek, L. Kajan, V. Agoston, and S. Pongor, The Sbase domain sequence resource, release 12: Prediction of protein domain-architecture using support vector machines, *Nucleic Acids Res*, **33 Database Issue**, D223–225 (2005).

167. R.A. George, and J. Heringa, Protein domain identification and improved sequence similarity searching using Psi-blast, *Proteins*, **48**, 672–681 (2002).

168. N. Nagarajan, and G. Yona, Automatic prediction of protein domains from sequence information using a hybrid learning system, *Bioinformatics*, **20**, 1335–1360 (2004).

169. J. Sim, S.Y. Kim, and J. Lee, Pprodo: Prediction of protein domain boundaries using neural networks, *Proteins*, (2005).

170. J. Cheng, M.J. Sweredoski, and P. Baldi, Dompro: Protein domain prediction using profiles, secondary structure, relative solvent accessibility, and recursive neural networks, *Data Mining and Knowledge Discovery*, **13**, 1–10 (2006).

171. R. Linding, R.B. Russell, V. Neduva, and T.J. Gibson, Globplot: Exploring protein sequences for globularity and disorder, *Nucleic Acids Res*, **31**, 3701–3708 (2003).

172. M. Suyama, and O. Ohara, Domcut: Prediction of inter-domain linker regions in amino acid sequences, *Bioinformatics*, **19**, 673–674 (2003).

173. R.A. George, K. Lin, and J. Heringa, Scooby-Domain: Prediction of globular domains in protein sequence, *Nucleic Acids Res*, **33**, W160–163 (2005).

174. J. Liu, and B. Rost, Sequence-based prediction of protein domains, *Nucleic Acids Res*, **32**, 3522–3530 (2004).

175. N. Alexandrov, and I. Shindyalov, Pdp: Protein domain parser, *Bioinformatics*, **19**, 429–430 (2003).

2

First Steps of Protein Structure Prediction

Karolina Majorek, Łukasz Kozłowski, Marcin Jąkalski and Janusz M. Bujnicki

2.1 Introduction

The famous hypothesis formulated by the Nobel Prize laureate Christian Anfinsen states that

> *The three-dimensional structure of a native protein [...] is determined by the totality of interatomic interactions and hence by the amino acid sequence, in a given environment. In terms of natural selection through the 'design' of macromolecules during evolution, this idea emphasized the fact that a protein molecule only makes stable, structural sense when it exists under conditions similar to those for which it was selected – the so-called physiological state.*[1,2]

On the one hand, essentially all globular domains in proteins studied so far appear to conform to this rule. Most proteins (or at least their major fragments) have been found to fold into unique, well-defined, stable three-dimensional structures under very broadly defined 'physiological conditions' that also include 'laboratory conditions' under which protein samples are prepared for biophysical and biochemical characterization. In agreement with Anfinsen's hypothesis, variations in conditions (e.g. change of pH or addition of a ligand) or changes in a sequence (e.g. due to proteolytic cleavage and removal of a sequence fragment) may result in structural changes that are often functionally relevant, e.g. if a protein's function requires opening and closing of a cavity that is used for binding of another molecule. On the other hand, a growing number of protein sequences (or sequence fragments) have been found to be mostly unstructured (review:[3]). These 'intrinsically disordered proteins' (IDPs) may assume a defined structure only under very specific

Prediction of Protein Structures, Functions, and Interactions   Edited by Janusz Bujnicki
© 2009 John Wiley & Sons, Ltd

conditions, e.g. in the presence of another molecule (e.g. upon binding to another protein or a ligand). In the absence of a stabilizing factor, these sequences exist as an ensemble of rapidly interconverting different conformations. Thus, for many IDPs, the Anfinsen's 'stable physiological state for which the protein was selected' is significantly different from 'standard' conditions assumed for other proteins.

Anfinsen's hypothesis implies that the knowledge of amino acid sequence and of a given environment should be sufficient to infer the native structure of a protein (or to predict the lack of a stable structure). However, despite seemingly solid theoretical foundations, accurate prediction of protein structure from the primary chemical structure of the polypeptide chain remains one of the greatest challenges in biology. Thus far, there have been two major approaches to predict the unknown protein structure from known amino acid sequence: one relies on our knowledge of 'first principles', i.e. the laws of physics, while the other is based on rules inferred from comparative analysis of experimentally solved protein structures. Both approaches had some successes, but neither of them has achieved the final goal.

The 'physical' approach has been successfully applied already in 1951 by Linus Pauling and Robert Corey, who predicted the existence of two periodic structural motifs formally defined by the pattern of hydrogen bonds that may be formed by the protein backbone: the spiral α-helix with 3.6 amino acid residues per turn[4] and the flat β-sheet comprising two or more β-strands having an extended zigzag conformation.[5] These two secondary structure elements (SSEs) are now known to be major features of protein architecture, with >50% of residues of an average protein assuming either helical or extended conformation. Helices and strands provide a natural frame for insightful protein structure visualization (with a helix often represented as a tube or a spiral and strand as an arrow), and are widely used to describe protein three-dimensional folds. They are also used by many programs that use simplified protein structure representation (e.g. SSEs instead of individual amino acids) to speed up calculations, e.g. for superposition of protein structures.

Secondary structure is much more conserved in the evolution than amino acid sequence; therefore accurate prediction of SSEs from sequence would be of great benefit in structural bioinformatics. For instance the knowledge of SSEs can help to guide sequence alignment or improve existing sequence alignment of remotely related sequences with low sequence similarity (see Chapter 1 by Kaminska *et al.* in this volume). Secondary structure prediction is also a good starting point toward elucidating the three-dimensional structure – it serves as an intermediate step in the protein fold-recognition procedure, i.e. identification of templates for comparative modeling (see Chapter 4 by Kosinski *et al.*) and may provide useful restraints both in comparative modeling and in *de novo* modeling (see Chapter 5 by Gront *et al.*). However, it has been found that it is quite difficult to predict accurately, which type of secondary structure is assumed by each amino acid residue of a protein.[6] Computational simulations of peptide and protein folding based on the 'physics-based' approach have been carried out in attempt to predict both local structure and global conformation (review:[7]). The first applications of force field methods to study peptide conformations date back to calculations performed by Nemethy and Scheraga.[8] However, due to an extremely large number of degrees of freedom and very complex calculations of energies, such simulations require extremely large computer resources, such as supercomputers or massively parallel distributed computing.[9] Alas, despite recent advancement in computer hardware and software, physics-based simulation techniques remain incapable of confidently predicting structures of even moderately sized proteins (>100 amino acid residues).

As soon as the first protein structures solved by X-ray crystallography have been determined, it has been observed that secondary structures tend to exhibit regular arrangements of amino acid residues of certain type. The regularities are due to the local periodicity of helical and extended conformations (3.6 and 2 residues per repeated segment, respectively), and the global tendency of a protein to form a well-packed hydrophobic core and to place hydrophilic residues at the surface (at least in water-soluble proteins). For example, SSEs buried in the protein core are composed mainly of hydrophobic residues, while SSEs at the protein surface tend to be amphipathic and show an alternating pattern of hydrophobic and hydrophilic residues (usually 010101 for strands and 0011011 or 0010011 for helices), allowing the respective side chains to be buried in the protein core or exposed to the solvent. Thus, as soon as a sufficient number of protein structures have been determined to make useful statistics, 'knowledge-based' methods have been developed, aiming at predicting SSEs based on calculated tendencies of different residues or peptide fragments to assume particular conformations. The first attempt of predicting secondary structure of polypeptides using the 'knowledge-based' approach dates back to the same time as the afore-mentioned physics-based analyses. It was performed in 1965 by Guzzo,[10] who inferred simple rules for preferences of different amino acid residues to form helical and non-helical regions. With the growing number of available protein structures and sequences the statistics have quickly improved, and new algorithms have been developed, yielding methods that are far from perfect, but achieve useful accuracy of about 80%. Currently the 'knowledge-based' approach is used by essentially all methods for secondary structure prediction, as well as methods for order/disorder prediction and for inference of other simple structural features from the primary sequence. This chapter aims at providing a comprehensive overview of these methods. The 'physical' approach to protein structure prediction is beyond the scope of this chapter and will not be reviewed here – instead, it will be referred to in other chapters, in particular Chapter 5 by Gront *et al*. Because of significant differences between proteins that function as water-soluble and those that are embedded in biological membranes, secondary structure prediction methods for each of two types are different and will be discussed separately. We will also discuss prediction of higher order motifs formed by certain types of SSEs, the so called 'supersecondary' structures (e.g. coiled coils or β hairpins), and prediction of contacts between residues that are remote in primary structure.

2.2 Definition of Secondary Structure and Its Assignment for Known Protein Structures

One caveat of the knowledge-based structure prediction is the requirement of unambiguous definition of secondary structure elements or ordered vs. disordered regions. Statistical methods and machine learning methods require the input data to be appropriately classified in order to make meaningful predictions. Unfortunately, defining the boundaries between disordered and ordered regions or between helix, sheet, and coil structures is arbitrary, and commonly accepted standard assignments do not exist. Therefore, various researchers employed different criteria that in some cases have lead to considerably different assignments.

Secondary structure is formally defined by the hydrogen bonds, but the hydrogen bonding is correlated with other features, such as dihedral angels generally adopted by particular types of secondary structure, which has given rise to less formal definitions of SSEs. To standardize secondary structure assignment, the Dictionary of Secondary Structure in Proteins (DSSP) was designed.[11] It was the first method for protein secondary structure assignment available as a computer program, and has remained most popular until now. DSSP classifies each amino acid residue in a protein with known 3D structure into one of 8 types of secondary structure, based on recognition of hydrogen bonding patterns and geometrical features defined in terms of the concepts torsion and curvature of differential geometry. The 7 types of SSEs include the α-helix (H), the β-strand (E, for 'extended'), and less frequent types of secondary structure namely non-α helices: 3/10 (G) and π (I), isolated β-bridge (B), highly curved bend (S) and hydrogen-bonded turn (T), while the remaining residues are classified as outside of SSEs.[11]

Another quite popular method for secondary structure assignment named STRIDE was proposed later by Frishman and Argos.[12] STRIDE is based on the combined use of hydrogen bond energy and statistically derived backbone torsional angle information, with parameters of the pattern recognition procedure optimized to improve (compared to DSSP) agreement with manual designations provided by the crystallographers as a standard-of-truth. More recently developed methods for secondary structure assignment include P-SEA,[13] SECSTR,[14] KAKSI,[15] SEGNO,[16] and PALSSE.[17] These methods are based on either the hydrogen-bond pattern, geometric features, expert knowledge or their combinations. However, they often disagree on their assignments, up to 25%. The discrepancy among different methods is caused by nonideal configurations of helices and sheets in experimentally solved structures and by different definitions of helices and strands. Of particular interest is PALSSE, which identifies only two types of SSEs that can be approximated by vectors: helix and strand. In contrast to other algorithms, which identify a secondary structure state for every residue in a protein chain, PALSSE attributes residues to SSEs in such a way that consecutive elements may overlap, thus allowing residues located at the overlapping region to have more than one secondary structure type. This method is robust to coordinate errors and can be used to define SSEs even in poorly refined and low-resolution structures (e.g. if only C-α atoms are available, thus if no hydrogen-bonds are present). PALSSE usually assigns a larger fraction of residues to SSEs as compared to other methods, e.g. 80% vs. 53% in the case of DSSP.[17]

Discrepancies with structural assignment concern not only algorithms. Protein structures are dynamic objects with some regions more mobile than others. Local conformations near the ends of secondary structures vary under native conditions, but may be forced to assume a single conformation in crystals due to packing constraints, hence secondary structure assignments differ by about 5–15 percentage points between different X-ray versions or different NMR models for the same protein[6] (Figure 2.1). This inherent protein flexibility is the main reason why the theoretical upper limit of secondary structure prediction accuracy is about 90%, for a particular SSE assignment method. Recently it has been proposed that instead of relying on single structures, structure assignment methods should be assessed based on the similarity of the secondary structures assigned to established pairwise sequence-alignment benchmarks, where these benchmarks are determined by prior structural alignments of the protein pairs. The use of this criterion has led to identification of STRIDE and KAKSI as the most robust methods (PALSSE was not included in that

Figure 2.1   *Illustration of secondary structure, conformational variability, and order vs disorder. Three-dimensional structure of small protein SUMO-1 solved by NMR (1a5r in Protein Data Bank), 10 alternative models shown in the 'cartoon' representation. Black spirals indicate α-helices, dark grey arrows indicate β-strands, white coils indicate loops. While the central part shows relatively ordered structure (with only some fluctuations at one end of the helix), the N- and C-terminal regions (left and right, respectively) show 'intrinsic disorder'. Interestingly, a short helical region persists in the disordered N-terminal tail, demonstrating the presence of secondary structure despite the absence of stable tertiary structure*

comparison), and to development of a consensus of STRIDE, KAKSI, SECSTR, and P-SEA, called SKSP, which is 2–3% higher in agreement with structurally aligned residues than DSSP for three established alignment benchmarks.[18]

Summarizing, assignments of secondary structure for one particular protein may vary, depending on the method used. Thus, in theoretical protein structure prediction it is important to select, which type of assignment is going to be predicted. Thus far, DSSP has been used as the 'golden standard' because of its popularity among crystallographers, but it is likely that methods for secondary structure assignment that are more consistent with the 3D structure alignments (e.g. SKSP) may lead to improved secondary structure prediction. Another caveat of secondary structure prediction methods is that they are aimed at predicting only the three basic classes of local structure: (H)elix, (E)xtended, and (C)oil; thus, e.g. an 8-letter 'structural alphabet' used in the DSSP notation is reduced to a 3-letter alphabet. In different algorithms it is done according to different conversion rules, which may yield an apparent increase of accuracy, but cause errors when the predicted secondary structure is used to predict 3D structure.[6] Karplus and coworkers found that the replacement of simple alphabets of secondary structure with highly informative, detailed alphabets can improve detection and alignment of structurally similar, but remotely related proteins.[19] Examples of such alphabets include STR, an enhanced version of DSSP that subdivides DSSP letter E (strand) into six letters, according to properties of a residue's relationship to its strand partners (number of partners and their parallel/antiparallel character) or Protein Blocks, a set of overlapping protein backbone fragments of length 5 amino acid residues.[20] HMMSTR[21] implements yet another solution to this problem: in this method protein structure is represented by short structural fragments taken from the database of known structures; for secondary structure it uses an alphabet of 11 conformation states, 10 corresponding to Φ-Ψ angle regions and one for cis-peptide bonds. A recently developed method Real-SPINE[22,23] predicts real values of torsion angles from a given sequence.

2.3 Prediction of Secondary Structure and Solvent Accessibility for Water-soluble Proteins

The first empirical prediction system aiming at predicting SSEs from protein sequence, based on statistics calculated from structures solved by X-ray crystallography, was developed by Fasman and Chou.[24,25] This very simple method is based on analysis of the relative frequency of each amino acid in helices, strands and coils and on assumption that single residue individually influences secondary structure. Subsequently, a more sophisticated GOR method has been developed, which is based on information theory and Bayesian statistics, and takes into account not only one residue but also adjacent positions in the sequence. These methods have been built-in into commercial software packages for protein sequence analysis and structure modeling and have become very popular among biologists despite their accuracy was only slightly better than random. The main limitation of these early methods was a small amount of known 3D structures, from which parameters could be derived. Besides, these methods did not utilize any evolutionary information and were applicable to single sequences rather than to multiple sequence alignments (MSAs) of homologous sequences.[26] Although over the years, both the Chou-Fasman[27] and the GOR methods[28] have been improved, the level of their accuracy is inferior to the best modern methods.

A significant improvement in prediction accuracy (>70%) has been achieved by 'second generation' methods such as PHD,[29] SAM-T98,[30] and PSIPRED,[31] which utilized MSA-derived information concerning sequence conservation, often combined with machine learning techniques such as artificial neural networks (ANNs), nearest-neighbor search (NNS) methods, and support vector machines (SVMs), or advanced statistical methods such as Hidden Markov Models (HMM) (review:[32]). These methods were also made available as web servers (Table 2.1). MSA, provided by the user or generated by an internal routine of an algorithm, is usually based on identification of homologs by searches of protein sequence databases (see Chapter 1 by Kaminska *et al.* in this volume). It is important to note that PSIPRED was the first method, in which iterative PSI-BLAST sequence searches have been introduced, compared to single-pass searches in earlier methods. Currently, iterative database searches to obtain the input MSA for prediction methods are considered a standard. Typically, patterns in sequence variability observed in MSA provide information on conservation of core elements (hydrophobic core and regions important for protein function), while the location of insertions and deletions (indels) hints at a position of surface-exposed loops. Incorporated machine learning techniques allow training the methods on known structures to learn characteristic sequence-structure patterns and then use those patterns to predict the secondary structure of the query protein. Most of the SSE prediction methods of the above-mentioned generation, or their derivatives developed later, have been associated with predictors of solvent accessibility used to identify residues that are buried to different extents in the hydrophobic core (Table 2.2).

In addition to methods for predicting the three main types of SEEs, there are several methods based on sequence profile analysis for predicting certain types of local structure, including various hairpin structures,[52] or specialized in α-turns,[71,72] β-turns,[51,73] γ-turns,[50] and π-turns.[74] There are also methods to predict conformation of individual residues, e.g. the trans/cis state of Pro.[75,76] To our knowledge, these types of methods have not yet been

Table 2.1   *Software for secondary structure prediction*

Program	URL (http://)
Three-state (Helix/Extended/Coil) prediction	
IPSSP[33] (for single sequences)	exon.gatech.edu/genemark/ipssp/webIPSSP.cgi
PSIPRED[31]	bioinf.cs.ucl.ac.uk/psipred/
SSPRO[34]	scratch.proteomics.ics.uci.edu/
PHD[29]	www.predictprotein.org/
PROFsec[35]	www.predictprotein.org/
PRED2ARY[36]	alexander.ucsf.edu/~jmc/pred2ary/
APSSP2[37]	www.imtech.res.in/raghava/apssp2/
PREDATOR[38]	ftp://ftp.ebi.ac.uk/pub/software/unix/predator/
HMMSTR[21]	www.bioinfo.rpi.edu/~bystrc/hmmstr/
NNPREDICT[39]	www.cmpharm.ucsf.edu/~nomi/nnpredict.html
PORTER[40]	distill.ucd.ie/porter/
HYPROSPII[41]	bioinformatics.iis.sinica.edu.tw/HYPROSPII/
SAM-T06[42]	www.soe.ucsc.edu/compbio/SAM_T06/T06-query.html
JNET[43]	www.compbio.dundee.ac.uk/Software/JNet/jnet.html
SABLE[44]	sable.cchmc.org/
YASSPP[45]	glaros.dtc.umn.edu/yasspp/
YASPIN[46]	ibivu.cs.vu.nl/programs/yaspinwww/
CRNPred[47]	ftp.bioinformatics.org/pub/crnpred/
JUFO3D[48]	www.meilerlab.org/index.php
SPINE[22]	sparks.informatics.iupui.edu/SPINE/spine.html
Other types of secondary and supersecondary structure, and other types of local conformation	
TURNS (α, β, γ)[49,50]	imtech.res.in/raghava/
β-Turn[51]	serine.umdnj.edu/~zhangq3/betaturn/prediction.htm
TURNPRED[52]	www.meilerlab.org/index.php
COILS[53]	www.ch.embnet.org/software/COILS_form.html
MARCOIL[54]	www.isrec.isb-sib.ch/webmarcoil/webmarcoilC1.html
PCOILS[55]	toolkit.tuebingen.mpg.de/pcoils
PairCoil2[56]	groups.csail.mit.edu/cb/paircoil2/paircoil2.html
MultiCoil[57]	groups.csail.mit.edu/cb/multicoil/cgi-bin/multicoil.cgi
LearnCoil[58]	groups.csail.mit.edu/cb/learncoil-vmf/cgi-bin/vmf.cgi
'Meta-servers' for secondary structure prediction	
JPRED[59]	www.compbio.dundee.ac.uk/~www-jpred/
NPS@[60]	npsa-pbil.ibcp.fr
META-PP[61]	www.predictprotein.org/meta.php
PROTEUS[62]	wishart.biology.ualberta.ca/proteus
DISTILL[63]	distill.ucd.ie
GeneSilico[64]	genesilico.pl/meta2/

integrated into metaservers for secondary or tertiary structure prediction and their practical utility for protein modeling and function prediction remains to be established.

Currently the recommended approach to secondary structure prediction involves combining the results of different methods; it may involve advanced machine learning approaches, such as voting, linear discrimination, neural networks or decision trees[77] or even simple

Table 2.2 *Software for solvent accessibility prediction*

Program	URL (http://)
Jnet[43]	www.compbio.dundee.ac.uk/Software/JNet/jnet.html
PHDacc[35]	www.predictprotein.org/
PROFacc[35]	www.predictprotein.org/
SABLE[65]	sable.cchmc.org
NETASA[66]	www.netasa.org
MLRprdsec[67]	spg.biosci.tsinghua.edu.cn/
WESA[68]	pipe.scs.fsu.edu/wesa.html
ACCpro[69]	scratch.proteomics.ics.uci.edu/
SARpred[70]	www.imtech.res.in/raghava/sarpred/
SPINE[22]	sparks.informatics.iupui.edu/SPINE/spine.html
PaleAle[63]	distill.ucd.ie/paleale/

consensus.[78] The idea of combining different prediction methods was first implemented in JPRED,[59] a consensus meta-server that standardizes input and output requirements of a range of secondary structure prediction algorithms, each representing a different prediction strategy, and computes a consensus of PHD, NNSSP, DSC, and PREDATOR secondary structure predictions. In addition, the output of the JPRED server includes predictions of solvent accessibility by the JNET method,[43] as well as predictions of coiled-coil regions and transmembrane helices (see below), which however are not directly incorporated in the calculation of the secondary structure.

The most recent class of meta-approaches, exemplified by PROTEUS[62] exploits the observation that if an experimentally determined three-dimensional structure of a closely related protein is known, then copying the secondary structure assignment from the known structure provides a better result than by predicting it *de novo*. PROTEUS initially carries out a sequence similarity search against the PDB database in order to determine if the whole or a part of the query sequence is significantly similar to a known structure, and if such a template structure is found, secondary structure mapping is carried out from the template to the query based on a sequence alignment. For the sequence segments that are not covered by template structures, *de novo* secondary structure prediction is carried out with three different, high quality neural network approaches (PSIPRED, JNET and TRANSSEC), whose results are combined into a consensus prediction by the fourth neural network. Merging template-based predictions and *de novo* predictions allows PROTEUS to yield a full sequence prediction, regardless of the extent of sequence overlap to a PDB hit (when complete 3D-to-2D mapping is achieved, when only partial coverage is provided and when no homolog with known structure can be found), and to achieve high average accuracy of >80% per residue. A similar approach of merging template-based and *de novo* predictions of secondary structure and solvent accessibility has been implemented in the DISTILL suite.[63]

While the early methods of secondary structure prediction were about 60–65% accurate, with accuracy for β-strands only slightly better than random,[6] the best modern methods reach about 80% accuracy per residue,[22] with ~10% lower accuracy for β-strands. The difference between theoretical upper limit of prediction accuracy and actual secondary structure prediction accuracy, and between level of prediction accuracy of α-helices and β-strands, is mainly due to difficult to detect long-range interactions that may influence secondary structure formation. It has been shown that the same amino acid sequence of substantial length may fold as α-helix when in one position in primary protein sequence but as β-sheet when in another sequence context.[79] Besides, during the folding process, a certain fragment of a protein might first adopt a secondary structure preferred by the local sequence and later, because of non-local interactions, be transformed to another secondary structure. The latter concern has been addressed in the method 3D-JUFO,[48] which combines iterative *de novo* secondary structure prediction using an approach similar to PSIPRED with tertiary structure prediction with the ROSETTA method (see Chapter 5 by Gront *et al.*), followed by re-prediction of SSEs based on local environment of particular residues observed in models of tertiary structure. 3D-JUFO achieves remarkable accuracy of 80%, with notable improvement of accuracy for β-strand prediction (76%) over sequence-only methods. Another interesting recently developed method that brings the accuracy of secondary structure prediction close to the theoretical limit combines bioinformatics methodology with experimental techniques of circular dichroism (CD) and Fourier transform infrared (FTIR) spectroscopy for assessing the overall secondary structure content.[80]

2.4 Prediction of Secondary Structure for Transmembrane Proteins

Membrane proteins are different from water-soluble proteins in that a large fraction of their surface is hydrophobic to enable stability in the environment of a lipid bilayer. They constitute about 20–30% of all proteins in the fully sequenced genomes, and are typically involved in cell signaling, molecular pumping and energy transduction. Integral membrane proteins consist of one or more transmembrane (TM) segments and can be divided into two structural classes: the α-helical TM proteins and the β-barrel TM proteins, varying in structure, localization and physicochemical features. Typical TM proteins of the more abundant α-helical class are present in all types of biological membranes including outer membranes. They comprise one or more hydrophobic α-helical membrane spanning regions separated by hydrophilic loops that are exposed into the solvent (review:[81]). TM β-barrel proteins are found only in outer membranes of Gram-negative bacteria, cell wall of Gram-positive bacteria, and outer membranes of mitochondria and chloroplasts. They consist of different number of antiparallel, membrane spanning β-strands with a simple up-and-down topology.[82]

TM proteins aggregate and precipitate in water and require detergents or nonpolar solvents for extraction, therefore they are much more difficult to analyze experimentally than their soluble counterparts, in relation to all steps from overexpression to high-resolution structure determination. Although TM proteins represent the most important drug targets, their structure determination has lagged behind that for soluble proteins; currently they represent less than 1% of available crystal structures.[83] On the one hand, this situation generates a great deal of pressure to develop effective methods for predicting the structure of TM proteins. On the other hand, the paucity of structural data hampers the development of knowledge-based approaches. Nonetheless, for both types of TM proteins specialized structure predictors have been designed, but due to the relatively easily detectable patterns of hydrophobic residues forming α-helical TM segments and much smaller amount of known β-barrel TM proteins structures, the majority of them was focused on the α-helical TM proteins until quite recently (review:[84]).

Prediction of TM helices should be intuitively easy due to their hydrophobic nature. However, predictions based solely on hydrophobicity profiles have high error rates. Besides, hydrophobic signal peptides may be confused with TM helices. It is also a consecutive challenge to predict TM proteins topology. Prediction of the way in which TM segments cross the membrane (inside-out or outside-in) is done mainly by considering the different charge distribution between the inside (cytoplasmic) and outside (extracellular) regions connecting TM segments, and by application of the so-called 'positive-inside rule'[85] based on the observation that there is an overrepresentation of positively charged residues in the intracellular loops of TM proteins. Contemporary approaches usually predict both localization of TM segments and their orientation (topology). The best methods such as PHOBIUS[86] or MEMSAT3[87] utilize evolutionary information as well as discriminate against signal peptides. Prediction of TM segments for β-barrel proteins is more difficult, because the strands are amphipathic. They contain 10–22 residues with alternating hydrophobic side chains facing the lipid bilayer and hydrophilic side chains facing the internal pore. To predict the β-barrel type of TM proteins a small number of specialized algorithms have been developed based on standard statistical and machine learning techniques including HMMs, ANNs, or SVMs (Table 2.3).

As in the case of secondary structure prediction for globular soluble proteins, consensus methods perform much better compared to each individual prediction method separately and the recommended strategy for identification membrane spanning segments and their orientation in membranes is to use many different methods and combine results into a consensus prediction. Examples of 'metaservers' that combine the results of several individual methods, providing a more accurate consensus prediction, include BPROMPT,[100] ConPredII,[110] and PONGO[111] for α-helical TM proteins, and ConBBPRED[114] for β-barrel TM proteins. The newest trends in TM structure prediction include meta-predictions that utilize predictions of solvent accessibility and secondary structure propensity typical for globular proteins in the form of 'structural profiles'.[102] There have also been attempts to make concurrent prediction of secondary and tertiary structure by simulating folding in lipid membranes, e.g. with modified versions of *de novo* structure prediction methods FRAGFOLD[115] and ROSETTA.[116,117]

In addition to predictors specific for TM proteins, a new method MeTaDoR has been recently proposed that predicts membrane-binding peripheral proteins that do not form an integral part of the membrane, but bind to it mostly in a reversible manner and thereby function in various important processes, including cell signaling and membrane trafficking.[113]

2.5   Prediction of Supersecondary Structure

Individual SSEs may be arranged in simple geometrical shapes forming recurring supersecondary structures. There is a number of well-defined α–α, β–β, α–β and β–α structural motifs that serve as 'building blocks' of tertiary structure. Prediction of supersecondary structures can be an important step toward building a tertiary structure from the specified secondary structure elements.[118] The β–hairpin, comprising two adjacent antiparallel

Table 2.3 *Software for prediction of TM regions in proteins*

Program	URL (http://)
	α-TM proteins
HMMTOP[88]	www.enzim.hu/hmmtop/
DAS[89]	www.sbc.su.se/~miklos/DAS/
PHDhtm[90]	www.predictprotein.org/
TMAP[91]	bioinfo4.limbo.ifm.liu.se/tmap/index.html
TMHMM[92]	www.cbs.dtu.dk/services/TMHMM/
Tmpred[93]	www.ch.embnet.org/software/TMPRED_form.html
MEMSAT3[87]	bioinf.cs.ucl.ac.uk/memsat
TopPred2[94]	bioweb.pasteur.fr/seqanal/interfaces/toppred.html
WHAT[95]	saier-144-37.ucsd.edu/what.html
THUMBUP[96]	sparks.informatics.iupui.edu/Softwares-Services_files/thumbup.htm
UMDHMM[96]	sparks.informatics.iupui.edu/Softwares-Services_files/umdhmm.htm
PRED-TMR[97]	athina.biol.uoa.gr/PRED-TMR/
HMM-TM[98]	biophysics.biol.uoa.gr/HMM-TM/
ORIENTM[99]	athina.biol.uoa.gr/orienTM/
BROMPT[100]	www.jenner.ac.uk/BPROMPT
LOCALIZOME[101]	localizome.org
PHOBIUS[86]	phobius.sbc.su.se/
MINNOU[102]	minnou.cchmc.org
	β-Transmembrane proteins
BBF*[103]	www-biology.ucsd.edu/~msaier/transport/software/bbfsource.tar.gz
HMM-B2TMR**[104]	gpcr.biocomp.unibo.it/biodec/
MINNOU[102]	minnou.cchmc.org
B2TMPRED[105]	gpcr.biocomp.unibo.it/cgi/predictors/outer/pred_outercgi.cgi
PRED-TMBB[106]	bioinformatics.biol.uoa.gr/PRED-TMBB/
PROFtmb[107]	cubic.bioc.columbia.edu/services/proftmb/
TMBETA-NET[108]	psfs.cbrc.jp/tmbeta-net/
BOMP[109]	www.bioinfo.no/tools/bomp
	Metaservers
BPROMPT (α)[100]	www.jenner.ac.uk/BPROMPT
ConPredII (α)[110]	bioinfo.si.hirosaki-u.ac.jp/~ConPred2/
PONGO (α)[111]	pongo.biocomp.unibo.it/pongo/
TUPS (α)[112]	sparks.informatics.iupui.edu/Softwares-Services_files/tups.htm
ConBBPRED (β)[106]	bioinformatics.biol.uoa.gr/ConBBPRED/
	Membrane-binding peripheral proteins
MeTaDoR[113]	proteomics.bioengr.uic.edu/metador

*The BBF program is freely available to academic users upon request to the corresponding author.
**HMM-B2TMR is a commercial program, demo version is available.

hydrogen bonded β-strands, is an example of the frequently occurring motif for which predictors have been developed. BhairPred[119] is an example of a method for discriminating hairpins from non-hairpins; obviously it achieves high accuracy only if the prediction of secondary structure is correct. Coiled coils are another type of super-secondary structure characterized by a bundle of two or more α-helices wrapping around each other. Coiled coil structures have been implicated in inter- and intraprotein interactions, and may be formed by helices formed by segments of sequence distant in the primary structure or even contributed by different proteins. Thus, coiled coils allow monomeric building blocks to form complex assemblages that can serve as molecular motors and springs (review: ref.[120]). The helices forming coiled coils have a unique pattern of hydrophobicity, which repeats every seven residues (five hydrophobic and two hydrophilic). This sequence periodicity has prompted the development of special algorithms to predict the location of α-helices that form coiled coils. According to the recent benchmark, the two best computational methods are a HMM-based MARCOIL[54] and PCOILS,[55] followed by PairCoil2[56] (Table 2.2).

2.6   Disorder Prediction

During the past decade, the literature has exploded with reports on intrinsically unstructured proteins (IDPs). Currently it is estimated that 30–60% of proteins are predicted to contain long stretches of disordered residues. Many of the disordered regions have been confirmed experimentally; they have been often found to be essential for protein function. Interestingly, intrinsic disorder appears to be significantly correlated with certain terms from functional ontologies and with specific functional motifs.[121–124] In particular, linear motifs[125] that harbor sites of posttranslational modification, such as phosphorylation, or sites of protein–protein interactions, often fall into regions that are locally disordered or undergo order-disorder transition in different, biologically relevant situations.[126,127] (see Chapter 1 by Kaminska *et al.* in this volume). With respect to molecular/biochemical function, IDPs have been frequently implicated in protein-nucleic acid interactions as transcription factors or in protein–protein interactions as e.g. regulators of enzyme activity. With respect to cellular roles, they have been implicated in regulatory processes, in particular in regulation of gene expression on the level of transcription and RNA processing, and in cellular signaling. On a more general level, IDPs are crucial for cell survival, proliferation, differentiation and apoptosis. Dysfunctions of IDPs may therefore lead to cancer, which makes them particularly important from a biomedical point of view. On the other hand, disordered regions often prevent crystallization of proteins, or the generation of interpretable NMR data, and in protein bioinformatics – they introduce compositional biases that hamper comparison of sequences of ordered regions. Recognition of disordered regions in a protein is therefore important for delineating boundaries of stably folded protein domains for structural and functional studies and for reducing bias in sequence similarity analyses by avoiding alignment of disordered regions against ordered ones (reviews:[128,129]). Detection of disordered regions may also facilitate identification of domains (see Chapter 1 by Kaminska *et al.*). A very important resource for disorder is the DISPROT database (http://www.disprot.org).[130] It links structure and function information for proteins that contains at least one experimentally determined disordered region.

The relatively frequent occurrence of IDPs and their importance in understanding protein structure-function relationships and cellular processes make it worthwhile to develop predictors of protein disordered regions. Since the SEG algorithm for identification of low-complexity regions that are typically associated with molecular disorder was developed in 1994,[131] an increasing number of groups have been developing such methods. However, as with secondary structure, it is not immediately clear how to unambiguously define 'disorder'. The lack of stable structure and conformational heterogeneity can manifest itself either at the secondary or tertiary level, and may include sites with varying extent of residual secondary structure and conformational mobility: molten globules, pre-molten globules, liquid-like collapsed-disordered state, or gas-like extended-disordered state.[132] Various researchers employed different criteria for defining disorder, resulting in numerous predictors that attempt to identify different features. Thus, depending on a research question being asked, using a single disorder predictor may be insufficient to achieve a meaningful prediction. Ferron *et al.*[128] presented an informative review of a number of methods published until 2006, highlighting their advantages and drawbacks. Table 2.4 presents succinct descriptions of disorder predictors, taking into account also the most recently published methods.

According to our own benchmark focused at accuracy of predicting regions of short disorder (using the criterion employed in CASP-7, i.e. the absence of resolved coordinates in crystal structures[153]), the best methods include POODLE,[147] DisPSSMP,[138] and iPDA.[142] We have developed a meta-predictor that reports the results of two primary coiled-coil predictors (COILS and Marcoil; see above and Table 2.2), and 10 primary disorder predictors (DISOPRED2, GlobPlot, Spritz, DISPROT (VSL2), IUPred, POODLE-L, POODLE-S, iPDA, PrDOS, and DisPSSMP, see Table 2.4). It also calculates a consensus prediction. The disorder meta-predictor is available via the gateway of the GeneSilico metaserver[64] at http://genesilico.pl/meta2/.

2.7   Prediction of Long-range Contacts between Amino Acid Residues

In addition to predicting local structure, a number of methods have been developed to predict contacts between residues that are remote in primary structure. This type of information is of particular interest, because it has been shown that it is possible to directly infer three-dimensional protein structures, if a sufficiently large number of contacts are known with sufficient accuracy. It has been estimated that as few as one contact on average per seven residues may be sufficient.[154] Various measures of distance and various thresholds may be used to define a contact between two residues (see e.g. ref.[155]), however the most common definition of contact used in prediction methods is as a $C\beta$-$C\beta$ pair ($C\alpha$ in the case of Gly residue) less than or equal to 8 Å apart.[156] According to the recent benchmark within the framework of the CASP7 experiment, the best contact predictor is an ANN associated with the SAM-T06 structure prediction server.[157] Other well-performing programs (according to the CASP7 benchmark or to other tests published by their authors) that are available as web-servers have been summarized in Table 2.5.

Special kinds of methods for long-range contact prediction are those for identification of Cys residues involved in disulfide bond formation (review: ref.[158]). Disulfide bonds are primary covalent cross-links between two Cys residues in proteins that play critical roles in stabilizing the protein structures. They can impose a substantial distance and angular constraint on the backbone of protein, thus making a large contribution to the stabilization of protein tertiary structures. A number of proposed algorithms for prediction of disulfide bonding states of Cys (involved in disulfide formation or not), as well as prediction of disulfide connectivity patterns (with the prior knowledge of disulfide bonding states) have been implemented as freely available web servers (Table 2.5). Most of these methods

Table 2.4 *Software for disorder prediction*

Program	URL (http://)	Short description
DisEMBL™[133]	Dis.embl.de/	ANN trained to predict classic loops (DSSP), flexible loops with high B-factors, missing coordinates in X-ray structures, regions of low-complexity and prone to aggregation.
DISOPRED2[134]	bioinf.cs.ucl.ac.uk/disopred/ disopred.html	SVM trained to predict residues with missing coordinates. Standalone version available.
DISpro[135]	www.ics.uci.edu/~baldig/ dispro.html	Recursive neural networks (RNNs) trained to predict missing coordinates.
DISPROT[136,137]	www.ist.temple.edu/disprot/ predictor.php	VL2 (least-squares regression) and VL3 (ANN) predict long disorder, VSL2 predicts both short and long disorder. Standalone version available.
DisPSSMP[138]	biominer.bime.ntu.edu.tw/ dispssmp/	Radial Basis Function Network (RBFN) trained to predict missing coordinates.
DRIP-PRED[139]	sbcweb.pdc.kth.se/cgi-bin/ maccallr/disorder/submit.pl	Self-organizing maps (SOMs) trained to predict missing coordinates.
FoldIndex©[140]	Bip.weizmann.ac.il/fldbin/findex	Simple method to predict whether a given protein will fold or not, based on average hydrophobicity and net charge.
FoldUnfold[141]	skuld.protres.ru/~mlobanov/ ogu/ogu.cgi	A statistical method to predict regions of weak packing density (less than 8 Å between heavy atoms of non-adjacent residues).
GlobPlot2[133]	globplot.embl.de/	A simple method based on several hydrophobicity scales to predict regions of missing coordinates and loops with high B-factors.
iPDA[142]	biominer.cse.yzu.edu.tw/ipda	A successor of DisPSSMP. Incorporates information about sequence conservation, predicted secondary structure, sequence complexity and hydrophobic clusters.
IUPred[143]	iupred.enzim.hu/	Estimates pairwise interaction energies using a statistical potential. Disordered regions tend to exhibit poor inter-residue contact capacity.
NORSp[144]	www.rostlab.org/services/NORSp/	Predicts long regions exposed to the solvent, with no regular secondary structure.
PONDR®[145]	www.pondr.com/	A commercial package containing several predictors based on FFNNs.

Table 2.4 *(continued)*

Program	URL (http://)	Short description
POODLE (S,L,W)[146,147]	mbs.cbrc.jp/poodle/poodle.html	L predicts long disorder using an SVM. S adds analysis of PSSMs generated by PSI-BLAST to detect short disorder. W uses Joachims' spectral graph transducer (SGT) to classify entire proteins as either disordered or ordered.
PrDOS[148]	prdos.hgc.jp	Predicts missing coordinates using SVM and PSSMs from PSI-BLAST.
PreLink[149]	genomics.eu.org/prelink/	Identifies regions with biased composition and poor in hydrophobic clusters to predict regions with missing coordinates.
RONN[150]	www.strubi.ox.ac.uk/RONN	Predicts missing coordinates using an ANN.
SEG[131]	mendel.imp.ac.at/METHODS/ seg.server.html	Ancient precursor of modern disorder predictors, identifies regions of low sequence complexity.
SPRITZ[151]	distill.ucd.ie/spritz/	Predicts long and short disorder (missing coordinates) using two separate SVMs. Utilizes secondary structure predicted by PORTER.
Grishin Lab Disorder Predictor[152]	prodata.swmed.edu/disorder/ disorder_prediction/predict.cgi	Predicts missing coordinates based on a PSSM and optimized propensities of amino acid residues toward disorder.
GeneSilico[64]	genesilico.pl/meta2/	A metaserver that predicts different types of disorder using weighted consensus of several methods.

employ combinations of various machine-learning techniques and utilize information from multiple sequence alignments (e.g. to identify correlated Cys pairs) and predicted secondary structure. Their reported prediction accuracy reaches 80%, but different methods have not been compared directly with each other on the same test set.

2.8 Summary

To obtain better quality of secondary structure prediction, when no related structures are known, it is advisable to follow some general rules:

First, it is important to use multiple sequence information, but if target sequence shows high similarity to none or to only a few other proteins it is worth trying to search different databases (e.g. not only the non-redundant database at the NCBI, but also protein sequences deduced from unfinished genomes and environmental sequencing projects) to find moderately divergent sequences that can be used to build MSA (see also Chapter 1

Table 2.5 *Software for prediction of long-range contacts and disulfide bonds*

Program	URL (http://)
SAM-T06[157]	www.soe.ucsc.edu/research/compbio/SAM_T06/T06-query.html
GPCPred[159]	sbcweb.pdc.kth.se/cgi-bin/maccallr/gpcpred/submit.pl
PROFcon[160]	www.predictprotein.org/submit_profcon.html
CONpro[161]	www.ics.uci.edu/~baldig/scratch/
SVMcon[162]	www.bioinfotool.org/svmcon.html
CRNPRED[47]	ftp.bioinformatics.org/pub/crnpred/
CMAPpro[69]	scratch.proteomics.ics.uci.edu/
Distill[151]	distill.ucd.ie/distill/
PoCM[163]	foo.maths.uq.edu.au/~nick/Protein/contact.html
CMA[164]	ligin.weizmann.ac.il/cma/
HMMSTR-CM[165]	www.bioinfo.rpi.edu/~bystrc/hmmstr/server.php
BETApro (β)[166]	www.ics.uci.edu/~baldig/betasheet.html
DiANNA (C-C)[167]	clavius.bc.edu/~clotelab/DiANNA/
DIpro 2.0 (C-C)[168]	contact.ics.uci.edu/bridge.html
DISULFIND (C-C)[169]	cassandra.dsi.unifi.it/disulfind/index.php
GDAP (C-C)[170]	www.doe-mbi.ucla.edu/~boconnor/GDAP/
DCON (C-C)[171]	gpcr.biocomp.unibo.it/cgi/predictors/cys-cys/pred_dconcgi.cgi
DISULFIDE (C-C)[172]	foo.maths.uq.edu.au/~huber/disulfide/

by Kaminska *et al.*). If the sequence is a true 'ORFan' with no homologs, a specialized method IPSSP[33] (Table 2.1) may be used. Alignments that include remotely related sequences should be inspected in the most divergent regions, and sequences that cannot be aligned with confidence should be removed. In case of secondary structure prediction algorithms that do not accept an MSA as an input (but e.g. construct one from scratch by themselves), secondary structure may be predicted independently for a few homologous sequences and checked for mutual consistency. Correctly aligned positions should display similar structure; therefore regions of low sequence similarity with different predictions should be checked for possible errors in MSA.

Second, we recommend using meta-servers for disorder and secondary structure prediction, because combining results of several best prediction methods into a consensus prediction is more reliable than relying on any individual method alone. Agreement between methods usually indicates confident prediction, while disagreement may indicate various things: different peculiarities of methods used, poor alignment in the input data, and/or non-standard type of secondary structure, such as surface-exposed β-strands with bulges that are often mispredicted as helices due to their irregular pattern of hydrophobic and hydrophilic residues. It is also important to remember that specialized methods are usually better for predicting particular types of structure than general-purpose methods for secondary structure prediction. Therefore, it may be useful to use methods for prediction of TM regions to pre-screen sequence for non-globular elements and 'mask' them before considering regular secondary structure prediction.

Third, when selecting 'best' methods for consensus prediction it is important to remember that many authors use different benchmarks to assess their methods and that many published accuracies have been shown to be overestimated, when these methods were assessed in rigorous blind tests on standard benchmarks, such as within CASP[173] or EVA.[174] Although secondary structure predictions are no longer assessed in CASP, the EVA website (http://cubic.bioc.columbia.edu/eva/) is updated automatically each week, to cope with the large number of existing prediction servers and the constant changes in the prediction methods. EVA currently assesses servers for secondary structure prediction, contact prediction, comparative protein structure modeling and threading/fold recognition. The identity of the test set assures that the competition is fair, while a large sample of targets assures that methods are compared reliably.

Fourth, we recommend making simultaneous predictions of secondary structure, solvent accessibility, and disorder, as they usually reinforce each other (e.g. regions of disorder usually exhibit little tendency to form secondary structure and their residues are predicted to be largely solvent-exposed). However, discrepancies in this regard (i.e. presence of confidently predicted secondary structure and/or buried residues within regions of disorder) may indicate interesting structural and functional elements, such as partially folded molten globule-like structures or candidates for linear motifs (see Chapter 1 by Kaminska *et al.* in this volume). Thus, again we recommend using meta-servers for making predictions on the level of primary and secondary structure, in particular if they are going to be used as restraints for modeling of protein tertiary structure (see articles by Kosinski *et al.* and Gront *et al.* in this volume).

Acknowledgements

We thank present and former members of the Bujnicki lab in IIMCB and at the UAM for stimulating discussions and collaboration in development of some of the methods mentioned in this article. The authors acknowledge the support from past and current grants for the development of structure prediction methods from Polish Ministry of Science, NIH, Framework Programme of the EU, EMBO, and HHMI. JMB has worked on this article while being supported by the Institute of Medical Science at the University of Tokyo.

References

1. C.J. Epstein, R.F. Goldberger, and C.B. Anfinsen, The genetic control of tertiary protein structure. Model systems. *Cold Spring Harb Symp Quant Biol*, **28**, 439–449 (1963).
2. C.B. Anfinsen, Studies on the principles that govern the folding of protein chains. Nobel lecture, December 11, 1972, in *Nobel Lectures, Chemistry 1971–1980*, T. Frängsmyr (Ed), World Scientific Publishing Co., Singapore, 1993.
3. H.J. Dyson, and P.E. Wright, Intrinsically unstructured proteins and their functions, *Nat Rev Mol Cell Biol*, **6**, 197–208 (2005).
4. L. Pauling, R.B. Corey, and H.R. Branson, The structure of proteins; Two hydrogen-bonded helical configurations of the polypeptide chain, *Proc Natl Acad Sci U S A*, **37**, 205–211 (1951).
5. L. Pauling, and R.B. Corey, The pleated sheet, a new layer configuration of polypeptide chains, *Proc Natl Acad Sci U S A*, **37**, 251–256 (1951).
6. B. Rost, Review: Protein secondary structure prediction continues to rise, *J Struct Biol*, **134**, 204–218 (2001).
7. G.A. Chasse, A.M. Rodriguez, M.L. Mak, *et al.*, Peptide and protein folding, *J Mol Struct THEOCHEM*, **537**, 319–361 (2001).

8. G. Nemethy, and H.A. Scheraga, Theoretical determination of sterically allowed conformations of a polypeptide chain by a computer method, *Biopolymers*, **3**, 155–181 (1965).

9. H.A. Scheraga, M. Khalili, and A. Liwo, Protein-folding dynamics: Overview of molecular simulation techniques, *Annu Rev Phys Chem*, **58**, 57–83 (2007).

10. A.V. Guzzo, The influence of amino-acid sequence on protein structure, *Biophys J*, **5**, 809–822 (1965).

11. W. Kabsch, and C. Sander, Dictionary of protein secondary structure: Pattern recognition of hydrogen-bonded and geometrical features, *Biopolymers*, **22**, 2577–2637 (1983).

12. D. Frishman, and P. Argos, Knowledge-based protein secondary structure assignment, *Proteins*, **23**, 566–579 (1995).

13. G. Labesse, N. Colloc'h, J. Pothier, and J.P. Mornon, P-Sea: A new efficient assignment of secondary structure from C alpha trace of proteins, *Comput Appl Biosci*, **13**, 291–295 (1997).

14. M.N. Fodje, and S. Al-Karadaghi, Occurrence, conformational features and amino acid propensities for the Pi-helix, *Protein Eng*, **15**, 353–358 (2002).

15. J. Martin, G. Letellier, A. Marin, J.F. Taly, A.G. de Brevern, and J.F. Gibrat, Protein secondary structure assignment revisited: A detailed analysis of different assignment methods, *BMC Struct Biol*, **5**, 17 (2005).

16. M.V. Cubellis, F. Cailliez, and S.C. Lovell, Secondary structure assignment that accurately reflects physical and evolutionary characteristics, *BMC Bioinformatics*, **6 Suppl 4**, S8 (2005).

17. I. Majumdar, S.S. Krishna, and N.V. Grishin, Palsse: A program to delineate linear secondary structural elements from protein structures, *BMC Bioinformatics*, **6**, 202 (2005).

18. W. Zhang, A.K. Dunker, and Y. Zhou, Assessing secondary structure assignment of protein structures by using pairwise sequence-alignment benchmarks, *Proteins*, (2007).

19. R. Karchin, M. Cline, Y. Mandel-Gutfreund, and K. Karplus, Hidden Markov models that use predicted local structure for fold recognition: Alphabets of backbone geometry, *Proteins*, **51**, 504–514 (2003).

20. C. Etchebest, C. Benros, S. Hazout, and A.G. de Brevern, A structural alphabet for local protein structures: Improved prediction methods, *Proteins*, (2005).

21. C. Bystroff, V. Thorsson, and D. Baker, Hmmstr: A hidden Markov model for local sequence-structure correlations in proteins, *J Mol Biol*, **301**, 173–190 (2000).

22. O. Dor, and Y. Zhou, Achieving 80% ten-fold cross-validated accuracy for secondary structure prediction by large-scale training, *Proteins*, **66**, 838–845 (2007).

23. B. Xue, O. Dor, E. Faraggi, and Y. Zhou, Real-value prediction of backbone torsion angles, *Proteins*, (2008).

24. P.Y. Chou, and G.D. Fasman, Prediction of protein conformation, *Biochemistry*, **13**, 222–245 (1974).

25. P.Y. Chou, and G.D. Fasman, Conformational parameters for amino acids in helical, beta-sheet, and random coil regions calculated from proteins, *Biochemistry*, **13**, 211–222 (1974).

26. J.M. Levin, S. Pascarella, P. Argos, and J. Garnier, Quantification of secondary structure prediction improvement using multiple alignments, *Protein Eng*, **6**, 849–854 (1993).

27. H. Chen, F. Gu, and Z. Huang, Improved Chou-Fasman method for protein secondary structure prediction, *BMC Bioinformatics*, **7 Suppl 4**, S14 (2006).

28. A. Kloczkowski, K.L. Ting, R.L. Jernigan, and J. Garnier, Combining the Gor V algorithm with evolutionary information for protein secondary structure prediction from amino acid sequence, *Proteins*, **49**, 154–166 (2002).

29. B. Rost, C. Sander, and R. Schneider, Phd: An automatic mail server for protein secondary structure prediction, *Comput Appl Biosci*, **10**, 53–60 (1994).

30. J. Park, K. Karplus, C. Barrett, *et al.*, Sequence comparisons using multiple sequences detect three times as many remote homologues as pairwise methods, *J Mol Biol*, **284**, 1201–1210 (1998).

31. D.T. Jones, Protein secondary structure prediction based on position-specific scoring matrices, *J Mol Biol*, **292**, 195–202 (1999).

32. J. Heringa, Computational methods for protein secondary structure prediction using multiple sequence alignments, *Curr Protein Pept Sci*, **1**, 273–301 (2000).

33. Z. Aydin, Y. Altunbasak, and M. Borodovsky, Protein secondary structure prediction for a single-sequence using hidden semi-Markov models, *BMC Bioinformatics*, **7**, 178 (2006).
34. G. Pollastri, D. Przybylski, B. Rost, and P. Baldi, Improving the prediction of protein secondary structure in three and eight classes using recurrent neural networks and profiles, *Proteins*, **47**, 228–235 (2002).
35. B. Rost, G. Yachdav, and J. Liu, The Predictprotein Server, *Nucleic Acids Res*, **32**, W321–326 (2004).
36. J.M. Chandonia, and M. Karplus, Neural networks for secondary structure and structural class predictions, *Protein Sci*, **4**, 275–285 (1995).
37. G.P.S. Raghava. Apssp2: A combination method for protein secondary structure prediction based on neural network and example based learning. Casp5 Online Abstract. A-132, 2002.
38. D. Frishman, and P. Argos, Seventy-five percent accuracy in protein secondary structure prediction, *Proteins*, **27**, 329–335 (1997).
39. D.G. Kneller, F.E. Cohen, and R. Langridge, Improvements in protein secondary structure prediction by an enhanced neural network, *J Mol Biol*, **214**, 171–182 (1990).
40. G. Pollastri, and A. McLysaght, Porter: A new, accurate server for protein secondary structure prediction, *Bioinformatics*, **21**, 1719–1720 (2005).
41. H.N. Lin, J.M. Chang, K.P. Wu, T.Y. Sung, and W.L. Hsu, Hyprosp Ii-a knowledge-based hybrid method for protein secondary structure prediction based on local prediction confidence, *Bioinformatics*, **21**, 3227–3233 (2005).
42. K. Karplus, S. Katzman, G. Shackleford, *et al.*, Sam-T04: What is new in protein-structure prediction for Casp6, *Proteins*, **61 Suppl 7**, 135–142 (2005).
43. J.A. Cuff, and G.J. Barton, Application of multiple sequence alignment profiles to improve protein secondary structure prediction, *Proteins*, **40**, 502–511 (2000).
44. R. Adamczak, A. Porollo, and J. Meller, Combining prediction of secondary structure and solvent accessibility in proteins, *Proteins*, **59**, 467–475 (2005).
45. G. Karypis, Yasspp: Better kernels and coding schemes lead to improvements in protein secondary structure prediction, *Proteins*, **64**, 575–586 (2006).
46. K. Lin, V.A. Simossis, W.R. Taylor, and J. Heringa, A simple and fast secondary structure prediction method using hidden neural networks, *Bioinformatics*, **21**, 152–159 (2005).
47. A.R. Kinjo, and K. Nishikawa, Crnpred: Highly accurate prediction of one-dimensional protein structures by large-scale critical random networks, *BMC Bioinformatics*, **7**, 401 (2006).
48. J. Meiler, and D. Baker, Coupled prediction of protein secondary and tertiary structure, *Proc Natl Acad Sci U S A*, **100**, 12105–12110 (2003).
49. H. Kaur, and G.P. Raghava, Prediction of beta-turns in proteins from multiple alignment using neural network, *Protein Sci*, **12**, 627–634 (2003).
50. H. Kaur, and G.P. Raghava, A neural-network based method for prediction of gamma-turns in proteins from multiple sequence alignment, *Protein Sci*, **12**, 923–929 (2003).
51. Q. Zhang, S. Yoon, and W.J. Welsh, Improved method for predicting {beta}-turn using support vector machine, *Bioinformatics*, (2005).
52. M. Kuhn, J. Meiler, and D. Baker, Strand-loop-strand motifs: prediction of hairpins and diverging turns in proteins, *Proteins*, **54**, 282–288 (2004).
53. A. Lupas, M. Van Dyke, and J. Stock, Predicting coiled coils from protein sequences, *Science*, **252**, 1162–1164 (1991).
54. M. Delorenzi, and T. Speed, An Hmm model for coiled-coil domains and a comparison with Pssm-based predictions, *Bioinformatics*, **18**, 617–625 (2002).
55. M. Gruber, J. Soding, and A.N. Lupas, Comparative analysis of coiled-coil prediction methods, *J Struct Biol*, **155**, 140–145 (2006).
56. A.V. McDonnell, T. Jiang, A.E. Keating, and B. Berger, Paircoil2: Improved prediction of coiled coils from sequence, *Bioinformatics*, **22**, 356–358 (2006).
57. E. Wolf, P.S. Kim, and B. Berger, Multicoil: A program for predicting two- and three-stranded coiled coils, *Protein Sci*, **6**, 1179–1189 (1997).
58. M. Singh, B. Berger, and P.S. Kim, Learncoil-Vmf: Computational evidence for coiled-coil-like motifs in many viral membrane-fusion proteins, *J Mol Biol*, **290**, 1031–1041 (1999).

59. J.A. Cuff, M.E. Clamp, A.S. Siddiqui, M. Finlay, and G.J. Barton, Jpred: A consensus secondary structure prediction server, *Bioinformatics*, **14**, 892–893 (1998).

60. Y. Guermeur, C. Geourjon, P. Gallinari, and G. Deleage, Improved performance in protein secondary structure prediction by inhomogeneous score combination, *Bioinformatics*, **15**, 413–421 (1999).

61. V.A. Eyrich, and B. Rost, Meta-Pp: Single interface to crucial prediction servers, *Nucleic Acids Res*, **31**, 3308–3310 (2003).

62. S. Montgomerie, S. Sundararaj, W.J. Gallin, and D.S. Wishart, Improving the accuracy of protein secondary structure prediction using structural alignment, *BMC Bioinformatics*, **7**, 301 (2006).

63. G. Pollastri, A.J. Martin, C. Mooney, and A. Vullo, Accurate prediction of protein secondary structure and solvent accessibility by consensus combiners of sequence and structure information, *BMC Bioinformatics*, **8**, 201 (2007).

64. M.A. Kurowski, and J.M. Bujnicki, Genesilico protein structure prediction meta-server, *Nucleic Acids Res*, **31**, 3305–3307 (2003).

65. R. Adamczak, A. Porollo, and J. Meller, Accurate prediction of solvent accessibility using neural networks-based regression, *Proteins*, **56**, 753–767 (2004).

66. S. Ahmad, and M.M. Gromiha, Netasa: Neural network based prediction of solvent accessibility, *Bioinformatics*, **18**, 819–824 (2002).

67. S. Qin, Y. He, and X.M. Pan, Predicting protein secondary structure and solvent accessibility with an improved multiple linear regression method, *Proteins*, **61**, 473–480 (2005).

68. H. Chen, and H.X. Zhou, Prediction of solvent accessibility and sites of deleterious mutations from protein sequence, *Nucleic Acids Res*, **33**, 3193–3199 (2005).

69. J. Cheng, A.Z. Randall, M.J. Sweredoski, and P. Baldi, Scratch: A protein structure and structural feature prediction server, *Nucleic Acids Res*, **33**, W72–76 (2005).

70. A. Garg, H. Kaur, and G.P. Raghava, Real value prediction of solvent accessibility in proteins using multiple sequence alignment and secondary structure, *Proteins*, **61**, 318–324 (2005).

71. H. Kaur, and G.P. Raghava, Prediction of alpha-turns in proteins using Psi-blast profiles and secondary structure information, *Proteins*, **55**, 83–90 (2004).

72. Y. Wang, Z. Xue, and J. Xu, Better prediction of the location of alpha-turns in proteins with support vector machine, *Proteins*, **65**, 49–54 (2006).

73. H. Kaur, and G.P. Raghava, A neural network method for prediction of beta-turn types in proteins using evolutionary information, *Bioinformatics*, **20**, 2751–2758 (2004).

74. Y. Wang, Z.D. Xue, X.H. Shi, and J. Xu, Prediction of Pi-turns in proteins using Psi-blast profiles and secondary structure information, *Biochem Biophys Res Commun*, **347**, 574–580 (2006).

75. M.L. Wang, W.J. Li, M.L. Wang, and W.B. Xu, Support vector machines for prediction of peptidyl prolyl Cis/trans isomerization, *J Pept Res*, **63**, 23–28 (2004).

76. J. Song, K. Burrage, Z. Yuan, and T. Huber, Prediction of Cis/trans isomerization in proteins using Psi-blast profiles and secondary structure information, *BMC Bioinformatics*, **7**, 124 (2006).

77. R.D. King, M. Ouali, A.T. Strong, *et al.*, Is it better to combine predictions?, *Protein Eng*, **13**, 15–19 (2000).

78. M. Albrecht, S.C. Tosatto, T. Lengauer, and G. Valle, Simple consensus procedures are effective and sufficient in secondary structure prediction, *Protein Eng*, **16**, 459-462 (2003).

79. D.L. Minor, Jr., and P.S. Kim, Context-dependent secondary structure formation of a designed protein sequence, *Nature*, **380**, 730–734 (1996).

80. J.G. Lees, and R.W. Janes, Combining sequence-based prediction methods and circular dichroism and infrared spectroscopic data to improve protein secondary structure determinations, *BMC Bioinformatics*, **9**, 24 (2008).

81. J.L. Popot, and D.M. Engelman, Helical membrane protein folding, stability, and evolution, *Annu Rev Biochem*, **69**, 881–922 (2000).

82. G.E. Schulz, Transmembrane beta-barrel proteins, *Adv Protein Chem*, **63**, 47–70 (2003).

83. K. Lundstrom, Structural genomics for membrane proteins, *Cell Mol Life Sci*, **63**, 2597–2607 (2006).

84. A. Elofsson, and G. von Heijne, Membrane protein structure: Prediction versus reality, *Annu Rev Biochem*, **76**, 125–140 (2007).
85. L. Sipos, and G. von Heijne, Predicting the topology of eukaryotic membrane proteins, *Eur J Biochem*, **213**, 1333–1340 (1993).
86. L. Kall, A. Krogh, and E.L. Sonnhammer, A combined transmembrane topology and signal peptide prediction method, *J Mol Biol*, **338**, 1027–1036 (2004).
87. D.T. Jones, Improving the accuracy of transmembrane protein topology prediction using evolutionary information, *Bioinformatics*, **23**, 538–544 (2007).
88. G.E. Tusnady, and I. Simon, The Hmmtop transmembrane topology prediction server, *Bioinformatics*, **17**, 849–850 (2001).
89. M. Cserzo, E. Wallin, I. Simon, G. von Heijne, and A. Elofsson, Prediction of transmembrane alpha-helices in prokaryotic membrane proteins: The dense alignment surface method, *Protein Eng*, **10**, 673–676 (1997).
90. B. Rost, P. Fariselli, and R. Casadio, Topology prediction for helical transmembrane proteins at 86% accuracy, *Protein Sci.*, **5**, 1704–1718 (1996).
91. F. Milpetz, P. Argos, and B. Persson, Tmap: A new email and WWW service for membrane-protein structural predictions, *Trends Biochem Sci*, **20**, 204–205 (1995).
92. E.L. Sonnhammer, G. von Heijne, and A. Krogh, A hidden Markov model for predicting transmembrane helices in protein sequences, *Proc Int Conf Intell Syst Mol Biol*, **6**, 175–182 (1998).
93. K. Hofmann, and W. Stoffel, Tmbase – a database of membrane spanning proteins segments, *Biol Chem Hoppe Seyler*, **347**, 166 (1993).
94. G. von Heijne, Membrane protein structure prediction. Hydrophobicity analysis and the positive-inside rule, *J Mol Biol*, **225**, 487–494 (1992).
95. Y. Zhai, and M.H. Saier, Jr., A web-based program (What) for the simultaneous prediction of hydropathy, amphipathicity, secondary structure and transmembrane topology for a single protein sequence, *J Mol Microbiol Biotechnol*, **3**, 501–502 (2001).
96. H. Zhou, and Y. Zhou, Predicting the topology of transmembrane helical proteins using mean burial propensity and a hidden-Markov-model-based method, *Protein Sci*, **12**, 1547–1555 (2003).
97. C. Pasquier, and S.J. Hamodrakas, An hierarchical artificial neural network system for the classification of transmembrane proteins, *Protein Eng*, **12**, 631–634 (1999).
98. P.G. Bagos, T.D. Liakopoulos, and S.J. Hamodrakas, Algorithms for incorporating prior topological information in Hmms: Application to transmembrane proteins, *BMC Bioinformatics*, **7**, 189 (2006).
99. T.D. Liakopoulos, C. Pasquier, and S.J. Hamodrakas, A novel tool for the prediction of transmembrane protein topology based on a statistical analysis of the Swissprot database: The Orientm algorithm, *Protein Eng*, **14**, 387–390 (2001).
100. P.D. Taylor, T.K. Attwood, and D.R. Flower, Bprompt: A consensus server for membrane protein prediction, *Nucleic Acids Res*, **31**, 3698–3700 (2003).
101. S. Lee, B. Lee, I. Jang, S. Kim, and J. Bhak, Localizome: A server for identifying transmembrane topologies and Tm helices of eukaryotic proteins utilizing domain information, *Nucleic Acids Res*, **34**, W99–103 (2006).
102. B. Cao, A. Porollo, R. Adamczak, M. Jarrell, and J. Meller, Enhanced recognition of protein transmembrane domains with prediction-based structural profiles, *Bioinformatics*, **22**, 303–309 (2006).
103. Y. Zhai, and M.H. Saier, Jr., The Beta-Barrel Finder (Bbf) Program, Allowing identification of outer membrane beta-barrel proteins encoded within prokaryotic genomes, *Protein Sci*, **11**, 2196–2207 (2002).
104. P.L. Martelli, P. Fariselli, A. Krogh, and R. Casadio, A sequence-profile-based Hmm for predicting and discriminating beta barrel membrane proteins, *Bioinformatics*, **18 Suppl 1**, S46–53 (2002).
105. I. Jacoboni, P.L. Martelli, P. Fariselli, V. De Pinto, and R. Casadio, Prediction of the transmembrane regions of beta-barrel membrane proteins with a neural network-based predictor, *Protein Sci*, **10**, 779–787 (2001).

106. P.G. Bagos, T.D. Liakopoulos, I.C. Spyropoulos, and S.J. Hamodrakas, Pred-Tmbb: A web server for predicting the topology of beta-barrel outer membrane proteins, *Nucleic Acids Res*, **32**, W400–404 (2004).

107. H.R. Bigelow, D.S. Petrey, J. Liu, D. Przybylski, and B. Rost, Predicting transmembrane beta-barrels in proteomes, *Nucleic Acids Res*, **32**, 2566–2577 (2004).

108. M.M. Gromiha, S. Ahmad, and M. Suwa, Neural network-based prediction of transmembrane beta-strand segments in outer membrane proteins, *J Comput Chem*, **25**, 762–767 (2004).

109. F.S. Berven, K. Flikka, H.B. Jensen, and I. Eidhammer, Bomp: A program to predict integral beta-barrel outer membrane proteins encoded within genomes of gram-negative bacteria, *Nucleic Acids Res*, **32**, W394–399 (2004).

110. M. Arai, H. Mitsuke, M. Ikeda, *et al.*, Conpred Ii: A consensus prediction method for obtaining transmembrane topology models with high reliability, *Nucleic Acids Res*, **32**, W390–393 (2004).

111. M. Amico, M. Finelli, I. Rossi, *et al.*, Pongo: A web server for multiple predictions of all-alpha transmembrane proteins, *Nucleic Acids Res*, **34**, W169–172 (2006).

112. H. Zhou, C. Zhang, S. Liu, and Y. Zhou, Web-based toolkits for topology prediction of transmembrane helical proteins, fold recognition, structure and binding scoring, folding-kinetics analysis and comparative analysis of domain combinations, *Nucleic Acids Res*, **33**, W193–197 (2005).

113. N. Bhardwaj, R.V. Stahelin, R.E. Langlois, W. Cho, and H. Lu, Structural bioinformatics prediction of membrane-binding proteins, *J Mol Biol*, **359**, 486–495 (2006).

114. P.G. Bagos, T.D. Liakopoulos, and S.J. Hamodrakas, Evaluation of methods for predicting the topology of beta-barrel outer membrane proteins and a consensus prediction method, *BMC Bioinformatics*, **6**, 7 (2005).

115. M. Pellegrini-Calace, A. Carotti, and D.T. Jones, Folding in lipid membranes (film): A novel method for the prediction of small membrane protein 3D structures, *Proteins*, **50**, 537–545 (2003).

116. V. Yarov-Yarovoy, J. Schonbrun, and D. Baker, Multipass membrane protein structure prediction using Rosetta, *Proteins*, **62**, 1010–1025 (2006).

117. P. Barth, J. Schonbrun, and D. Baker, Toward high-resolution prediction and design of transmembrane helical protein structures, *Proc Natl Acad Sci U S A*, **104**, 15682–15687 (2007).

118. X. de la Cruz, E.G. Hutchinson, A. Shepherd, and J.M. Thornton, Toward predicting protein topology: An approach to identifying beta hairpins, *Proc Natl Acad Sci U S A*, **99**, 11157–11162 (2002).

119. M. Kumar, M. Bhasin, N.K. Natt, and G.P. Raghava, Bhairpred: Prediction of beta-hairpins in a protein from multiple alignment information using Ann and Svm techniques, *Nucleic Acids Res*, **33**, W154–159 (2005).

120. A.N. Lupas, and M. Gruber, The structure of alpha-helical coiled coils, *Adv Protein Chem*, **70**, 37–78 (2005).

121. A. Lobley, M.B. Swindells, C.A. Orengo, and D.T. Jones, Inferring function using patterns of native disorder in proteins, *PLoS Comput Biol*, **3**, e162 (2007).

122. H. Xie, S. Vucetic, L.M. Iakoucheva, *et al.*, Functional anthology of intrinsic disorder. 3. Ligands, post-translational modifications, and diseases associated with intrinsically disordered proteins, *J Proteome Res*, **6**, 1917–1932 (2007).

123. S. Vucetic, H. Xie, L.M. Iakoucheva, *et al.*, Functional anthology of intrinsic disorder. 2. Cellular components, domains, technical terms, developmental processes, and coding sequence diversities correlated with long disordered regions, *J Proteome Res*, **6**, 1899–1916 (2007).

124. H. Xie, S. Vucetic, L.M. Iakoucheva, *et al.*, Functional anthology of intrinsic disorder. 1. Biological processes and functions of proteins with long disordered regions, *J Proteome Res*, **6**, 1882–1898 (2007).

125. V. Neduva, and R.B. Russell, Linear motifs: Evolutionary interaction switches, *FEBS Lett*, **579**, 3342–3345 (2005).

126. M. Fuxreiter, P. Tompa, and I. Simon, Local structural disorder imparts plasticity on linear motifs, *Bioinformatics*, **23**, 950–956 (2007).

127. Y. Zhang, B. Stec, and A. Godzik, Between order and disorder in protein structures: Analysis of 'dual personality' fragments in proteins, *Structure*, **15**, 1141–1147 (2007).

128. F. Ferron, S. Longhi, B. Canard, and D. Karlin, A practical overview of protein disorder prediction methods, *Proteins*, **65**, 1–14 (2006).

129. Z. Dosztanyi, M. Sandor, P. Tompa, and I. Simon, Prediction of protein disorder at the domain level, *Curr Protein Pept Sci*, **8**, 161–171 (2007).

130. M. Sickmeier, J.A. Hamilton, T. LeGall, *et al*., Disprot: The database of disordered proteins, *Nucleic Acids Res*, **35**, D786–793 (2007).

131. J.C. Wootton, Non-globular domains in protein sequences: Automated segmentation using complexity measures, *Comput.Chem.*, **18**, 269–285 (1994).

132. V.N. Uversky, Natively unfolded proteins: A point where biology waits for physics, *Protein Sci*, **11**, 739–756 (2002).

133. R. Linding, L.J. Jensen, F. Diella, P. Bork, T.J. Gibson, and R.B. Russell, Protein disorder prediction: Implications for structural proteomics, *Structure*, **11**, 1453–1459 (2003).

134. J.J. Ward, J.S. Sodhi, L.J. McGuffin, B.F. Buxton, and D.T. Jones, Prediction and functional analysis of native disorder in proteins from the three kingdoms of life, *J Mol Biol*, **337**, 635–645 (2004).

135. J. Cheng, M. Sweredoski, and P. Baldi, Accurate prediction of protein disordered regions by mining protein structure data, *Data Mining and Knowledge Discovery*, **11**, 213–222 (2005).

136. S. Vucetic, C.J. Brown, A.K. Dunker, and Z. Obradovic, Flavors of protein disorder, *Proteins*, **52**, 573–584 (2003).

137. Z. Obradovic, K. Peng, S. Vucetic, P. Radivojac, and A.K. Dunker, Exploiting heterogeneous sequence properties improves prediction of protein disorder, *Proteins*, **61 Suppl 7**, 176–182 (2005).

138. C.T. Su, C.Y. Chen, and Y.Y. Ou, Protein disorder prediction by condensed Pssm considering propensity for order or disorder, *BMC Bioinformatics*, **7**, 319 (2006).

139. R.M. MacCallum. Order/disorder prediction with self organising maps. CASP 6 meeting, Online paper. http://www.forcasp.org/paper2127.html

140. J. Prilusky, C.E. Felder, T. Zeev-Ben-Mordehai, *et al*., Foldindex: A simple tool to predict whether a given protein sequence is intrinsically unfolded, *Bioinformatics*, **21**, 3435–3438 (2005).

141. O.V. Galzitskaya, S.O. Garbuzynskiy, and M.Y. Lobanov, Foldunfold: Web server for the prediction of disordered regions in protein chain, *Bioinformatics*, **22**, 2948–2949 (2006).

142. C.T. Su, C.Y. Chen, and C.M. Hsu, Ipda: Integrated protein disorder analyzer, *Nucleic Acids Res*, **35**, W465–472 (2007).

143. Z. Dosztanyi, V. Csizmok, P. Tompa, and I. Simon, Iupred: Web server for the prediction of intrinsically unstructured regions of proteins based on estimated energy content, *Bioinformatics*, **21**, 3433–3434 (2005).

144. J. Liu, and B. Rost, Norsp: Predictions of long regions without regular secondary structure, *Nucleic Acids Res*, **31**, 3833–3835 (2003).

145. P. Romero, Z. Obradovic, X. Li, E.C. Garner, C.J. Brown, and A.K. Dunker, Sequence complexity of disordered protein, *Proteins*, **42**, 38–48 (2001).

146. K. Shimizu, S. Hirose, and T. Noguchi, Poodle-S: Web application for predicting protein disorder by using physicochemical features and reduced amino acid set of a position-specific scoring matrix, *Bioinformatics*, **23**, 2337–2338 (2007).

147. S. Hirose, K. Shimizu, S. Kanai, Y. Kuroda, and T. Noguchi, Poodle-L: A two-level Svm prediction system for reliably predicting long disordered regions, *Bioinformatics*, **23**, 2046–2053 (2007).

148. T. Ishida, and K. Kinoshita, Prdos: Prediction of disordered protein regions from amino acid sequence, *Nucleic Acids Res*, **35**, W460–464 (2007).

149. K. Coeytaux, and A. Poupon, Prediction of unfolded segments in a protein sequence based on amino acid composition, *Bioinformatics*, **21**, 1891–1900 (2005).

150. Z.R. Yang, R. Thomson, P. McNeil, and R.M. Esnouf, Ronn: The bio-basis function neural network technique applied to the detection of natively disordered regions in proteins, *Bioinformatics*, **21**, 3369–3376 (2005).

151. A. Vullo, O. Bortolami, G. Pollastri, and S.C. Tosatto, Spritz: A server for the prediction of intrinsically disordered regions in protein sequences using kernel machines, *Nucleic Acids Res*, **34**, W164–168 (2006).

152. N.B. Holladay, L.N. Kinch, and N.V. Grishin, Optimization of linear disorder predictors yields tight association between crystallographic disorder and hydrophobicity, *Protein Sci*, **16**, 2140–2152 (2007).

153. L. Bordoli, F. Kiefer, and T. Schwede, Assessment of disorder predictions in Casp7, *Proteins*, **69 Suppl 8**, 129–136 (2007).

154. J. Skolnick, A. Kolinski, and A.R. Ortiz, Monsster: A method for folding globular proteins with a small number of distance restraints, *J Mol Biol*, **265**, 217–241 (1997).

155. M.J. Pietal, I. Tuszynska, and J.M. Bujnicki, Protmap2d: Visualization, comparison, and analysis of 2D maps of protein structure, *Bioinformatics*, (2007).

156. J.M. Izarzugaza, O. Grana, M.L. Tress, A. Valencia, and N.D. Clarke, Assessment of intramolecular contact predictions for Casp7, *Proteins*, **69 Suppl 8**, 152–158 (2007).

157. G. Shackelford, and K. Karplus, Contact prediction using mutual information and neural nets, *Proteins*, **69 Suppl 8**, 159–164 (2007).

158. C.H. Tsai, C.H. Chan, B.J. Chen, C.Y. Kao, H.L. Liu, and J.P. Hsu, Bioinformatics approaches for disulfide connectivity prediction, *Curr Protein Pept Sci*, **8**, 243–260 (2007).

159. R.M. MacCallum, Striped Sheets and Protein Contact Prediction, *Bioinformatics*, **20 Suppl 1**, i224–231 (2004).

160. M. Punta, and B. Rost, Profcon: Novel prediction of long-range contacts, *Bioinformatics*, **21**, 2960–2968 (2005).

161. G. Pollastri, P. Baldi, P. Fariselli, and R. Casadio, Prediction of coordination number and relative solvent accessibility in proteins, *Proteins*, **47**, 142–153 (2002).

162. J. Cheng, and P. Baldi, Improved residue contact prediction using support vector machines and a large feature set, *BMC Bioinformatics*, **8**, 113 (2007).

163. N. Hamilton, K. Burrage, M.A. Ragan, and T. Huber, Protein contact prediction using patterns of correlation, *Proteins*, **56**, 679–684 (2004).

164. V. Sobolev, E. Eyal, S. Gerzon, *et al.* Space: A suite of tools for protein structure prediction and analysis based on complementarity and environment, *Nucleic Acids Res*, **33**, W39–43 (2005).

165. Y. Shao, and C. Bystroff, Predicting interresidue contacts using templates and pathways, *Proteins*, **53 Suppl 6**, 497–502 (2003).

166. J. Cheng, and P. Baldi, Three-stage prediction of protein beta-sheets by neural networks, alignments and graph algorithms, *Bioinformatics*, **21 Suppl 1**, i75–84 (2005).

167. F. Ferre, and P. Clote, Dianna 1.1: An extension of the Dianna web server for ternary cysteine classification, *Nucleic Acids Res*, **34**, W182–185 (2006).

168. J. Cheng, H. Saigo, and P. Baldi, Large-scale prediction of disulphide bridges using kernel methods, two-dimensional recursive neural networks, and weighted graph matching, *Proteins*, **62**, 617–629 (2006).

169. A. Ceroni, A. Passerini, A. Vullo, and P. Frasconi, Disulfind: A disulfide bonding state and cysteine connectivity prediction server, *Nucleic Acids Res*, **34**, W177–181 (2006).

170. B.D. O'Connor, and T.O. Yeates, Gdap: A web tool for genome-wide protein disulfide bond prediction, *Nucleic Acids Res*, **32**, W360–364 (2004).

171. P. Fariselli, P. Martelli, and R. Casadio, A neural network based method for predicting the disulfide connectivity in proteins, in *Knowledge Based Intelligent Information Engineering Systems and Allied Technologies*, E. Damiani (ed.), International Conference on Knowledge-Based Intelligent Engineering (Kes 2002), IOS Press, Amsterdam, 2002.

172. J. Song, Z. Yuan, H. Tan, T. Huber, and K. Burrage, Predicting disulfide connectivity from protein sequence using multiple sequence feature vectors and secondary structure, *Bioinformatics*, **23**, 3147–3154 (2007).

173. P. Aloy, A. Stark, C. Hadley, and R.B. Russell, Predictions without templates: New folds, secondary structure, and contacts in Casp5, *Proteins*, **53 Suppl 6**, 436–456 (2003).

174. I.Y. Koh, V.A. Eyrich, M.A. Marti-Renom, *et al.* Eva: Evaluation of protein structure prediction servers, *Nucleic Acids Res*, **31**, 3311–3315 (2003).

3

Automated Prediction of Protein Function from Sequence

Meghana Chitale, Troy Hawkins and Daisuke Kihara

3.1 Introduction

Investigation of protein gene function is a central question in molecular biology, biochemistry, and genetics. Because genes evolved from the same ancestral gene retain similarity in their function in most cases, finding known genes which have sufficient sequence similarity is a powerful way for predicting function. In this chapter we review computational techniques and resources for gene function prediction from sequence. We start with an overview of widely used homology search tools, such as BLAST, and extend discussion to more recently developed methods.

3.2 Principle of Inferring Function from Sequence Similarity

The driving forces of the evolution of life include complete or partial genome duplication and rearrangement,[1] and also duplications which occur on a gene basis,[2,3] that lead speciation of organisms. While active exchange of a portion of genomes between organisms such as lateral gene transfer makes ancestral relationship of organisms far more complicated that previously thought,[4,5] on the individual gene level it is generally true that duplicated or transferred genes within or between organisms retain significant sequence similarity. Genes that have evolved from a single ancestral gene are referred as *homologous* with each other.[6] Two types of homology are distinguished. *Orthologous* genes are those are diverged from speciation events of a common gene of an ancestral organism and thus reside

Prediction of Protein Structures, Functions, and Interactions   Edited by Janusz Bujnicki
© 2009 John Wiley & Sons, Ltd

in different organisms. In contrast, *paralogous* genes refer to those which are duplicated in a same organism thus locate at different positions in a same genome. Thus sequence similarity is an effective way to detect homology between genes (reviewed in detail in Chapter 1 by Kaminska *et al.* in this volume).

A pair of genes which share significant sequence similarity may have diverged quite recently in the history of the evolution, or there may have been an evolutionary pressure which kept the sequence unchanged over the course of a long evolution time. Another possibility is that the two sequences converged to be similar because of structural or functional constraints. In either case, functions of such two genes usually share significant similarity considering the evolutionary scenario behind it. Thus sequence similarity between two genes strongly indicates homology, which implies functional similarity in most of the cases. However, caution is needed because there are exceptions that homologous proteins have very different functions. Recent works discuss such interesting examples.[7,8]

The relationship between the sequence similarity and function similarity is also well understood in the light of the tertiary structure of proteins (reviewed in detail in chapters by Majorek *et al.* (Chapter 2) and Kosinski *et al.* (Chapter 4) in this volume). The widely accepted Anfinsen's dogma claims that the protein sequence determines the tertiary structure of the protein.[9] Moreover, from the observation of a growing number of solved protein structures, it is well established that proteins with a similar sequence generally have a similar overall fold.[10,11] Considering that the structure of a protein has crucial roles in realizing function, e.g. to catalyze chemical reaction at an active site binding a substrate or to interact with other proteins, having the same fold can be strong evidence that the proteins share functional similarity. (But there are notable counter examples, *e.g. superfolds,* which are protein folds adopted by different protein families.[12])

3.3 Homology Search Methods

The strategy of a sequence-based protein function prediction for a target protein is to find known protein genes which share a significant sequence similarity from a database (reviewed in detail in Chapter 1 by Kaminska *et al.* in this volume) and make prediction with function terms associated with the protein genes found. The sequence similarity of two proteins is effectively and rigorously computed by using a dynamic programming algorithm.[13,14] The SSEARCH program[15] performs rigorous local sequence alignment by the Smith-Waterman algorithm[14] between a target sequence and each sequence in a database and lists retrieved sequences sorted by their statistical significance score, E-value. As computing rigorous local sequence alignments against a current large database by SSEARCH take a considerable amount of time on a regular desktop computer, FASTA[15] and BLAST,[16] both of which employ faster algorithms than dynamic programming algorithm for computing alignments, are more widely used. FASTA reduces computational time by restricting computation of a pairwise alignment only within highly similar regions using a lookup table, while BLAST starts with finding precomputed similar 'words' of a fixed length taken from a target sequence in the framework of the finite automaton. Benchmark studies show FASTA and BLAST deteriorate the sensitivity of database search in the tradeoff for the speed compared to SSEARCH,[11,17] but all three methods will not miss obvious homologous sequences with significant sequence similarity. A search result will also depends on parameters used, such as the amino acid similarity matrix and gap penalties.[18]

The conventional way of using these homology search tools is to extract function annotation from top hit sequences which have a significant score either in terms of the E-value or the Smith-Waterman (SW) alignment score. The commonly used threshold for the E-value is 0.01 (or 0.001), and 200 for the SW score, which were originally established on benchmark datasets of a limited size.[19,20] This strategy is commonly used in gene function annotation in genome sequencing projects.[21,22] The advantage of using a unique threshold value is that it is easy to process automatically for a large number of genes. On the other hand, problems of this strategy include that it does not take into account that each protein family has a different degree of sequence conservation[7] and also a large portion of genes in a genome are usually left as unknown because of the rather conservative function assignment.[23]

Several interesting ideas have been proposed to identify further distantly related homologs using the homology search tools. For example, an intermediate sequence found in an initial search is used to reach further distant homologs in the second run of the search[24,25] and consensus of different methods is shown to improve search performance.[26,27]

The three homology search methods introduced above perform sequence-to-sequence comparisons. In contrast, PSI-BLAST performs profile-to-sequence comparisons, making a very sensitive database search possible.[28] PSI-BLAST iterates searches, at each time constructing a profile (multiple sequence alignment, MSA) with a target and retrieved sequences, which is used for a search in the next iteration. The iteration is halted to make the final function prediction when retrieved sequences are saturated or the predefined maximum time of iterations is reached. A profile can enhance family specific conserved sequence information in a query sequence. The flip-side of PSI-BLAST's extreme sensitivity is that it occasionally produces false positives.[29] Thus, PSI-BLAST is often used with a conservative (strict) parameter setting.[30]

Profiles can also be precomputed for sequences in a database, and a target sequence is matched against them (sequence-to-profile comparison).[31] BLOCKS[32] and ProDom[33] are databases of profiles of protein domains, where a user can search known functional domains in a sequence. A protein fingerprint is a group of conserved regions used to characterize a protein family. PRINTS[34] is a collection of such protein fingerprints. Pfam[35] and SUPERFAMILY[36] are databases which store profiles of protein domains in the form of hidden Markov models (HMMs), which are statistical representations of sequence profiles.[37] Finally, both a target sequence and database sequences are precomputed into profiles and the target profile is aligned with profiles in the database. Profile-to-profile comparison methods have been shown to be very sensitive and used not only for protein function prediction[38] but also for protein structure prediction (i.e. predicting protein fold).[39,40] Numerous methods for constructing and comparing profiles have been proposed, including ways to select sequences to be included in a profile, ways to score an alignment of two profiles, and how to handle gaps.[39–41]

3.4   Predicting Function from the Other Types of Information

Besides using sequence, various other features of genes can be used for function prediction. The global tertiary structure of proteins can indicate very distant evolutionary relationships between proteins,[42] and detecting local structure similarity is aimed to predict function by identifying functionally importance sites, such as active sites of enzymes.[43,44] Known pathway information is used as a template for finding missing genes which fit to holes in known pathways.[45] Use of Microarray gene expression data[46] and protein–protein interaction data[47] is actively investigated in function prediction. Now that many different types of databases are established and more new experimental data are made available, combination of heterogeneous data has become an interesting and promising direction for function prediction. However, as the focus of this chapter is sequence-based approaches, refer to recent review articles[23,48 49] and also the other chapters of this book for more information.

3.5   Limitations and Problems of Function Prediction from Sequence

A practical convenience of predicting function from sequence is that most of the function information of genes resides in sequence databases, such as UniProt,[50] Pfam,[35] and also protein domain and motif databases (reviewed in Chapter 1 by Kaminska *et al.* in this volume), e.g. PROSITE,[51] BLOCKS,[32] and PRINTS.[34] A consequent intrinsic limitation is that any method can essentially only extract function information which exists in a database and it is very difficult to make a prediction which goes beyond available function description of retrieved sequences. By the same reason, if function information of a gene in a database is wrong, that wrong information will be transferred to a target gene. Thus, erroneous annotation may be propagated by being reused in subsequent function assignments.[52,53] Incorrect function prediction can happen even with having genes with correct function description because of various reasons, such as ignoring multi-domain organization of genes and non-orthologous gene displacement.[54] Indeed erroneous function annotations are frequently reported.[55] To amend wrong annotations, the research community of *Escherichia coli* has held a meeting to manually curate gene annotations.[56] A recent interesting approach is a community based annotation using wiki, allowing any researcher to participate in annotating genes.[57]

3.6   Controlled Vocabularies for Gene Function Annotation

Automation of protein function prediction requires a well-established controlled vocabulary describing the annotations, which is unified across different species and research communities. If arbitrary terms are used for describing a biological function, for example, if a gene involved in 'bacterial protein synthesis' is described as involved in 'translation' in one database and as 'protein synthesis' in another, an automatic procedure would easily miss the similarity between the two annotations. Even for manual annotation, non-critical use of annotations from existing database entries is a major cause of erroneous function assignment.[54] Thus we need a universal way to describe gene function in structured manner which avoids ambiguity. To allow uniform referencing for functional annotations across databases several ontologies (vocabularies) have been developed. Those ontologies include Gene Ontology (GO),[58] Enzyme Commission (EC) number[59] and MIPS functional catalogue (FunCat).[60] These ontologies provide the basis for computational prediction of protein functions as they constitute the exhaustive organized space that will be searched in order to assign the most probably function to an un-annotated protein.

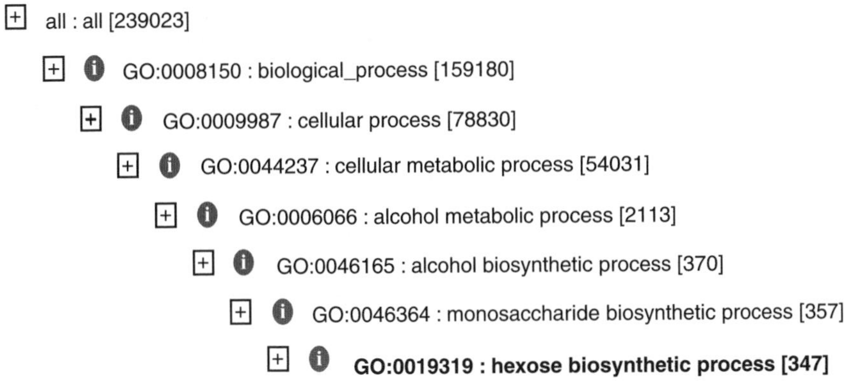

⊞ all : all [239023]

   ⊞ ⓘ GO:0008150 : biological_process [159180]

      ⊞ ⓘ GO:0009987 : cellular process [78830]

         ⊞ ⓘ GO:0044237 : cellular metabolic process [54031]

            ⊞ ⓘ GO:0006066 : alcohol metabolic process [2113]

               ⊞ ⓘ GO:0046165 : alcohol biosynthetic process [370]

                  ⊞ ⓘ GO:0046364 : monosaccharide biosynthetic process [357]

                    ⊞ ⓘ **GO:0019319 : hexose biosynthetic process [347]**

Figure 3.1 *Hierarchical organization for term GO:0019319 in Gene Ontology as displayed by Amigo(http://www.geneontology.org/) tool for searching and browsing Gene Ontology.*

3.7 Gene Ontology

GO consists of hierarchically structured vocabulary divided into three basic subcategories: molecular function, biological process and cellular component. Each term in GO is referred by an identifier of the form GO:xxxxxxx, a subcategory, and an associated textual description for that term. For example, the identifier GO:0019319 is of subcategory biological process and has short description as 'hexose biosynthetic process' (Figure 3.1). GO organizes the terms in a directed acyclic graph (DAG) structure where terms are associated by is_a or part_of relationships. The is_a classifier represents a subclass relationship where 'A is_a B' means A is description of B but at higher depth or more narrower description. 'A part_of B' indicates that whenever A is present it is part of B.

A gene can be described as performing one or more molecular functions, being part of one or more biological process and located in one or more cellular components. Another important feature of GO is that it supports association of an evidence code with each annotation indicating the nature of evidence sources that are used to support that annotation. Examples of the evidence codes are IDA (Inferred from Direct Assay), which indicates that a direct assay was carried out to determine the function, and ISS (Inferred from Sequence or Structural Similarity), which clarifies that any analysis based on sequence alignment, structure comparison, or evaluation of sequence features such as composition is performed.

3.8 Other Functional Ontologies

EC numbers are used for classifying enzymes based on the reactions they catalyze. The nomenclature of enzyme number has the form of EC x.x.x.x, consisting of four level hierarchies describing the activity of the enzyme. Partial EC numbers with only initial parts out of the four subparts will be used to refer to a class of enzymes describing a biochemical activity at a broader level. The FunCat scheme for functional description of proteins divides the annotations into 28 main categories that cover general fields. The FunCat version 2.1

includes 1362 functional categories where main categories are further subdivided up to six levels with increase in the specificity. A difference between FunCat and GO is that FunCat is organized in a hierarchical tree, while GO is structured into a DAG. A difference of enzymatic function description between FunCat and EC number is that EC number classifies catalytic activities based on the chemical reaction, while FunCat classification is based on the pathway where an enzyme acts. TCDB (Transport classification database)[61] is a database of Transporter Classification (TC) system that gives detailed comprehensive IUBMB (International Union of Biochemistry and Molecular Biology) approved classification system for membrane transport proteins. The TC system is analogous to the Enzyme Commission system for classification of enzymes, but additionally incorporates phylogenetic information. It consists of a set of representative protein sequences, most of which have been functionally characterized. These transporters are classified with a five-character designation, as follows: $D_1.L.D_2.D_3.D_4$. The letters in sequence correspond to transporter class, subclass, family, subfamily and transporter itself. The TCDB website also offers several tools specifically designed for analyzing the unique characteristics of transport proteins. The KEGG orthology (KO)[62] is both an ontology arranged around binary relations and an ontology giving annotations of class of gene products. KO decomposes the universe of all genes in all organisms into groups of functionally identical genes (orthologs). They define relationships between KEGG database objects such as reactions, substrates and products; relationships between enzyme and its location in the pathway; relationship between enzyme and protein super family to which it belongs.

3.9   Quantifying Functional Similarity

To compute the prediction accuracy of a function prediction we need to compare the similarity of predicted and actual ontology terms. The hierarchical nature of GO provides natural mechanism for comparing the terms. The basic idea is to consider the closest common parental node between predicted and correct GO terms. The scoring scheme used in the function prediction category in Critical Assessment of Techniques for Protein Structure Prediction 7 (CASP7) computes fraction of the path depth of the common parent compared with the path depth of the correct annotated GO term.[63] Resnik uses the maximum information content computed as maximum negative logarithm of any common ancestor term probability for pair of GO terms being compared.[64] Probability of occurrence of each term is defined as frequency of its occurrence in the annotation database as compared to the frequency of root term in the GO. Lord *et al.*[65] were first ones to compute the semantic similarity between a pair of proteins using Resnik's measure. Semantic similarity between two proteins was computed as the average similarity of the GO terms that annotate both the proteins. Schlicker *et al.*[66] further extend the Resnik's measure to include probabilities of both terms being compared for normalizing the semantic similarity score and also use the relevance (that decreases with probability) of the common ancestor term. Poze *et al.*[67] take a completely different approach to compute a functional distance between a pair of GO terms based on co-occurrence of terms in a same set of Interpro entries. A profile is constructed for GO terms representing its association with a set of Interpro domains taking into account the is_a relationships for GO terms and its ancestors. The profiles are used to generate a matrix of co-occurrences between GO terms.

3.10   Automated GO Term Prediction Methods

Recent years have observed development of new generation of function prediction algorithms. It is triggered by the growing need of function annotation of genes in an increasing number of newly sequenced genomes and newly solved protein tertiary structures. Moreover, large scale experimental data, such as protein–protein interaction and gene expression data, further add the urgency of developing different techniques to predict reliable annotations even at broad levels of detail for new genes. Many of the new generation of function prediction algorithms have some common features. First, they take advantage of controlled vocabulary of Gene Ontology, which facilitates computational handling of function terms. Second, most of them use BLAST or PSI-BLAST search results as the primary source of function information, realizing (or expecting) that homology search results contain more information than conventionally extracted by applying a unique E-value threshold to select significant hits. Third, some of the methods employ machine learning techniques, such as Support Vector Machines (SVM), that have recently become popular in bioinformatics area. Below we will discuss some of such methods.

Goblet[68,69] provides a web platform which assists users to analyze a BLAST search result of an input protein sequence in terms of GO terms. GO terms of retrieved sequences are displayed on the GO tree, which facilitates comparison of the GO terms. GOFigure[70] uses an idea of a minimum covering graph (MCG), which is a graph on the GO tree rooted at the GO terms that subsumes all extracted GO annotations from BLAST hits for a query sequence. The score assigned to each GO term is a weighted score of all the hits that map to it as well as the scores of all its children term. As a consequence of using MCG, not only the GO terms which directly associate to the retrieved BLAST hits but also their children terms have possibility of being final GO prediction to the query sequence. Verspoor *et al.*[71] use an ontology categorizer named POset Ontology Categorizer, which summarizes weighted collection of GO terms taken from PSI-BLAST hits. The weight of a GO term reflects the E-value of the sequence hit. For an evaluation metric of prediction, they introduce hierarchical precision and recall, which considers accuracy at each ancestral node of predicted and actual GO term.

GOtcha[72] runs BLAST for a query sequence, and GO terms are extracted from each BLAST hit. The set of GO terms and all ancestral terms are assigned a score of negative logarithm of the E-value of the BLAST hit (R-score). The sum of the R-score for all matches is normalized to the total R-score of the root node of each category in the GO tree.

GOPET[73] employs SVMs to analyze a BLAST search for a query sequence. GO terms are extracted from each retrieved sequence with attached features, including the E-value, the bit-score, the sequence identity, the coverage score, the alignment length, GO term frequency, and the evidence code of GO annotation, all of which are used as input parameters to SVMs. 99 SVM classifiers, each of which predicts a particular GO term, are constructed. An advantage of using SVM is that many different properties of retrieved sequences can be considered. On the other hand, a drawback is that a limited number of GO terms can be predicted by this implementation because a SVM needs to be constructed for individual GO term, and a sufficient number of instances (sequences) are needed for training a SVM.

ProtFun[74] is an interesting method of protein function prediction that is not based on sequence similarity but on sequence based protein features such as predicted post translational modifications, protein sorting signals, and physical/chemical properties calculated from amino acid composition. They use the InterPro database which maps protein families to GO terms. For each GO class a standard feed-forward neural network with a single layer of hidden neurons was trained with different combinations of sequence derived features.

JAFA[75] is protein function meta-server that provides joint assembly of function predictions from five different prediction servers, namely, GOFigure,[70] Gotcha,[72] Goblet,[68] InterProScan,[76] and PhydBac2.[77] The score provided with each GO terms is the product of the GO level multiplied by the fraction of agreeing servers. Hence the scoring function rewards the predictions that are more specific and predicted by multiple servers.

SIFTER[78] models a phylogenomics procedure of annotating molecular function of genes in a probabilistic method. For a given query protein, a rooted phylogenetic tree is constructed using homologs taken from the Pfam database. Annotated GO terms to the proteins in the tree are represented as a vector, and the probabilities with which known GO terms are propagated to descendants are computed.

Another approach by Cai *et al.*[79] for predicting enzyme subclasses is based on the amino acid composition of a protein sequence. This is particularly useful when it is not possible to identify a subfamily class for protein using the sequence similarity approach. They have developed FunD-PseAA Intimate Sorting (ISort) predictor using domain information obtained from InterPro database and amino acid frequencies in the sequence.

Pattern analysis of the distributions of disordered regions has shown that functions of intrinsically disordered proteins are both length and position dependent. Lobley *et al.*[80] used location descriptors to encode the position of disordered regions in proteins and showed their correlations with GO categories by calculating the average frequency of disordered residues within different location windows for proteins sequences annotated by GO term. Their results suggest that disorder regions are more indicative of biological process than the molecular function and the information content of disorder feature set is comparably lower than that for secondary structure or amino acid composition.

3.11 Protein Function Prediction (PFP) Algorithm

Our group has developed PFP algorithm for function prediction which extends a conventional PSI-BLAST search[81] (Figure 3.2). Along with strong PSI-BLAST hits which have significant E-value, PFP also uses weak hits that are not generally considered for transferring annotations. Weakly similar hits that are not recognized as homologous to the query sequence are also used in PFP because they often share common functional domains or some functional similarity at a broader level. GO terms extracted from retrieved sequences are ranked according to the following equation considering the E-value assigned to the retrieved sequences. Currently sequences of an E-value of up to 100 are used:

$$s(f_a) = \sum_{i=1}^{N} \sum_{j=1}^{Nfunc(i)} \left((-\log(E\_value(i)) + b)P(f_a|f_j) \right), \qquad (3.1)$$

where $s(f_a)$ is the final score assigned to the GO term, f_a, N is the number of the similar sequences retrieved by PSI-BLAST, $Nfunc(i)$ is the number of GO terms assigned to sequence j, $E\_value(i)$ is the E-value given to the sequence i, f_j is a GO term assigned to

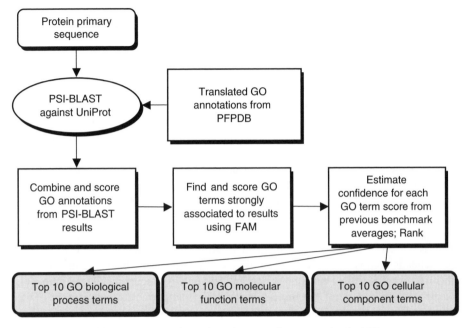

Figure 3.2   *Flowchart describing prediction method of PFP.*

the sequence i, and b is the constant value, 2 (=$\log_{10} 100$), which keeps the score positive. $P(f_a|f_j)$ is the conditional probability that f_a is associated with f_j. This conditional probability is computed from co-occurrence of GO terms in single sequences in the UniProt database and stored in a two dimensional matrix named Function Association Matrix (FAM):

$$P(f_a|f_j) = \frac{c(f_a, f_j) + \varepsilon}{c(f_j) + \mu \cdot \varepsilon},$$

(3.2)

$c(f_a, f_j)$ is number of times f_a and f_j are assigned simultaneously to each sequence in UniProt, and $c(f_j)$ is the total number of times f_j appeared in UniProt, μ is the size of one dimension of the FAM (i.e. the total number of unique GO terms), and ε is the pseudo-count.

The pre-computed FAM allows PFP to extract information about strongly associated terms in the database across the categories of GO which may be intuitive for biologists but not directly retrieved from the sequence database searched. For example, the (GO: 0008234) 'cysteine-type peptidase activity' in the molecular function category shows high association score with biological process term (GO:0006508) 'proteolysis' in the biological process. And molecular function (GO:0015662) 'ATPase activity, coupled to trans-membrane movement of ions, phosphorylative mechanism' is highly associated with the cellular component term (GO:0016020) membrane.

Moreover, scores given to each GO term are propagated to parent terms in the GO tree according to the number of genes associated to the predicted term relative to the parent term:

$$s(f_p) \sum_{i=1}^{N_c} \left(s(f_{ci}) \left(\frac{c(f_{ci})}{c(f_p)} \right) \right). \tag{3.3}$$

where $s(f_p)$ is the score of the parent term f_p, N_c is the number of child GO term which belong to the parent term f_p, $s(f_{ci})$ is the score of a child term ci, and $c(f_{ci})$ and $c(f_p)$ is the number of known genes which are annotated with function term f_{ci} and f_p in the Gene Ontology Annotation (GOA) database released at the European Bioinformatics Institute (EBI).

Since prediction crucially depends on available GO term annotations assigned to sequences in the database to be searched, we enriched annotated GO terms in the GOA database by adding GO terms from other databases including HAMAP, InterPro,[82] Pfam,[35] PRINTS,[34] ProDom,[33] PROSITE.[51] SMART,[83] and TIGRFam[84] as well as SwissProt Key Words.

Once a raw score of a GO term is obtained according to the equations above, its statistical significance is computed in terms of the P-value by considering the score distribution of that GO term taken from a benchmark dataset. And finally, predicted GO terms are ranked by their P-value in each of the three categories. It is important to consider the P-value rather than a raw score because some GO terms occur more frequently in a database, and thus tend to have a high raw score. For example, GO terms at a higher level in the GO tree (thus have more general function) have a high score also because scores given to its child terms are propagated to it.[85]

3.12   PFP Benchmark Results

In the paper published in 2006, we have benchmarked PFP on a set of randomly selected 2000 proteins from UniProt[81] (Figure 3.3). Three methods are compared: PFP using FAM to incorporate the GO term associations, PFP without using FAM, and transferring GO annotations from the top PSI-BLAST hit (top PSI-BLAST method). For the PFP predictions, five GO terms with the highest raw scores are predicted, and the top PSI-BLAST method predicts all the GO terms assigned to the top hit sequence. The performance was compared in terms of the sequence coverage, which reports the percentage of sequences for which correct biological process (sharing a common parent with a target annotation at GO depth ≥ 4) were predicted. To mimic a realistic situation that no significant homologs are found for a query protein sequence, the most significant sequence hits up to several E-value cutoffs in a PSI-BLAST search are ignored and only sequences with an E-value of the cutoff or larger (E-value > 0, 0.01, 0.1, 1, 2, 3, 5,10, 15, 20, 25, 50, 100) were used.

When all retrieved sequences are used, PFP with FAM correctly predicted biological process over 80 % of the tested query sequences, while PFP without FAM and top PSI-BLAST method made correct prediction to approximately 72 % of the query sequences. The strength of PFP is more evident when top hit sequences up to a certain E-value are not used. When only retrieved sequences with an E-value of 10 or higher are used, PFP with FAM made correct predictions to around 50 % of the query sequences, which is about five times larger than the top PSI-BLAST method. Interestingly, the sequence coverage by PFP

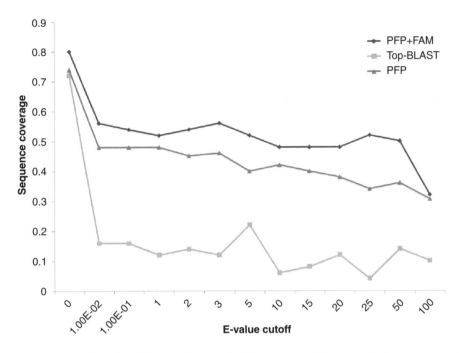

Figure 3.3   *Benchmark of PFP on a data set of 2000 sequences. Three methods are compared, PFP with FAM, PFP without FAM, and the top PSI-BLAST method. The data used in Figure 1 of our paper in 2006[81] is replotted.*

with FAM stays almost the same when the sequence hits of even larger E-value > 10 are used.

A characteristic advantage of PFP is that it can often predict a broader function or a 'low-resolution' function by identifying consensus GO terms which occur in retrieved sequences with a wide range of E-value by PSI-BLAST. Note that it is not trivial for conventional methods to make this kind of low-resolution function prediction, because there are no apparent sequence patterns for low-resolution functions. Conventional ways of using (PSI-)BLAST or motif searches are rather yes/no type prediction methods, meaning that a prediction is made when a clear functional sequence pattern is found, but no prediction is made otherwise. In contrast, PFP is able to make low-resolution function prediction when detailed function prediction cannot be made by taking consensus between function annotations of weakly similar sequences. In other words, PFP tries to give some functional clue to a query sequence by lowering resolution of function when necessary without sacrificing accuracy. An important point revealed by the benchmark study (Figure 3.3) is that the top hit by PSI-BLAST is not necessarily accurate and PFP outperforms the top PSI-BLAST method even when all retrieved sequences (with an E-value ≥ 0) are used. The pitfall of relying on only the top hit sequence has been pointed out by Galperin and Koonin.[54] PFP can often avoid transferring irrelevant annotations of the top hit sequence in a search by summarizing consensus GO terms which occur in a large number of hits in a PSI-BLAST search.

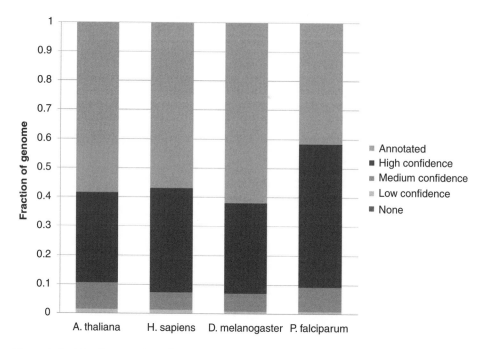

Figure 3.4 *Distribution of predictions done by PFP for four genomes classified based on the confidence score for the predicted annotations. A. thaliana, H. sapiens, D. melanogaster, and P. falciparum. Annotations of theses genomes are taken from the GOA database.*

A practical strength of PFP is that it can give function annotation to a larger number of genes in a genome by predicting low resolution functions, while typically BLAST searches can cover up to half of genes in a genome.[23] Very general function, e.g. transporter or enzyme, is not very helpful for designing biochemical experiments, but may be helpful for interpreting a large-scale data, such as gene expression data or protein–protein interaction data.[86] In Figure 3.4, fractions of genes with PFP annotations along with annotated genes in the GOA database for four organisms are shown. Predictions made by PFP are classified into three groups according to confidence level of the predictions, which are estimated by the correlation with the P-value and the accuracy in a benchmark dataset used.[85] For these genomes, PFP can provide function predictions to an additional 30–50 % of the total genes in a genome with a high confidence.

3.13   Comparative Genomics Based Methods

Completely different approaches for sequence-based function prediction use the genomic context of genes, taking advantage of the increasing number of available complete genomes. There are three major methods for this category. The first approach is to examine conservation of gene clusters in multiple genomes. Because gene locations tend to be dynamically shuffled during evolution,[87] if proximity between genes is evolutionarily conserved across species (conserved gene clusters), there is a high likelihood of functional association between the genes.[88,89] Bacterial genomes have operon structures, which is a transcription unit with multiple genes,[90] but more conserved gene clusters are found which are not known operons. Another evidence of functional association of genes is domain fusion events.[91] If two separate genes in one organism are seen as fused domains occurring in a single protein in another organism, apparently the fusion does not interfere with function of the two genes, and most likely the two genes are involved in the same functional context. Similarity in the pattern of existence and absence of orthologous genes in genomes, which is called phylogenetic profiles,[92] also indicates functional association of genes. Bork's group has implemented these three comparative genomics based approaches in the STRING server.[93]

These comparative genomics-based methods will become more useful as the number of sequenced genomes will further increase. However, what can be predicted by these methods is functional association of genes but not functional terms of each gene. Thus, homology-based function prediction is still needed for the starting point of a genome scale annotation.

3.14   Subcellular Localization Prediction

Subcellular localization can be considered as a type of gene function. Indeed the Gene Ontology organizes terms for describing localization in a DAG named cellular component. Some proteins have a signal peptide typically at its N-terminal region, which are recognized by a transporting protein and later often cleaved off. Therefore a direct way to predict subcellular localization is to recognize these signals.[94] Since molecular protein sorting mechanism differs in prokaryotes and eukaryotes, prediction methods is usually specifically designed for either one of them or for a sub-category, such as plants. PSORT is one of the earliest prediction methods, which uses multiple sequence features including signal peptides, amino acid composition, sequence motifs, and predicted trans-membrane domains in the form of a decision rule or a classifier.[95,96] They have an extensive collection of links to prediction methods and related resources at their web site, http://www.psort.org,[97] Nair et al.[98] demonstrate that cellular localization is an evolutionarily conserved property and homologs tend to occur at the same cellular sites. Proteome Analyst[99] obtains annotations corresponding to homologous sequences detected using BLAST and then uses them with an organism specific Bayesian classifier to classify the query protein to localization sites. Some methods[100–102] use SVM to classify proteins across different cellular components based on the frequency of twenty amino acids. The phylogenetic profile can be also used to predict localization.[103]

3.15   Identification of Functionally Important Residues

Usually molecular function of proteins, such as catalytic activity of enzymes, is carried out by a small number of residues in a protein sequence. These functionally indispensable residues are identified experimentally by constructing point mutation/deletion or domain deletion mutants, or from the tertiary structure in a ligand bound form solved by X-ray crystallography or NMR. Databases such as PROSITE[51] and ELM[104] (for eukaryotes) store such short sequence motifs. Since a local alignment of these short motifs does not result in an alignment score which yields a significant E-value in a BLAST search, searching against a motif database is a complementary method to homology search for function prediction. If the tertiary structure of the target protein is known, conservation of residues which are not close on the sequence but locate in spatial proximity can be further detected and compared against a database of three-dimensional motifs.[44,105-107] See Chapter 7 by Kinoshita in this volume for more details on structure-based function prediction.

Functionally important residues are generally well conserved among orthologous proteins, thus, selecting conserved residues from a carefully constructed MSA of a protein family is a fundamental procedure of identifying functionally important residues.[108-112] Besides sequence conservation, combining local structure information helps accurately identifying functionally important residues.[113] Some methods are developed that identify residue positions in a MSA which discriminate predefined subfamilies thus considered to be functional residues specific to subfamilies.[114-116] In contrast, Pei's method starts with constructing a phylogenetic tree for a given set of sequences, and identifies residue positions in the MSA which have a high likelihood that follows evolution along the tree.[117] Casari *et al.*[118] apply principal component analysis to a matrix representing sequences of a family to identify groups of residues that are conserved in the whole family and also those which are specific to subfamilies. MINER is based on the finding by La and Livesay that sequence regions which show a mutation pattern that conserves the overall familial phylogeny correspond to functional sites.[119,120]

3.16   Function Prediction Competitions

Responding to the increasing need of automatic function prediction, the bioinformatics community has held function prediction contests in the last few years. Friedberg, Godzik, and their co-workers have held the Automated Function Prediction Special Interest Group meeting at ISMB 2005,[121] where they summarized the results of a blind prediction contest of protein gene function. The participants were required to set up an automatic web server which accepts protein sequences, to which the organizers submitted target sequences and evaluated returned predictions. The Critical Assessment of Techniques for Protein Structure Prediction (CASP) competitions included a function prediction category in CASP6 (2004)[122] and CASP7 (2006).[63] Target protein sequences were given to predict EC numbers, GO terms or active site/ligand binding site residues. In both AFP-SIG and CASP7, PFP had the highest overall score[63] (no ranking was given in CASP6). Objective evaluation of existing methods is essential for enhancing continuous improvements of the methods and for keeping the field active. A larger number of participants are expected to participate in these competitions in the future.

3.17   Summary

We have reviewed recent advances of sequence-based function prediction methods. Figure 3.5 summarizes different techniques for predicting function from sequence. The

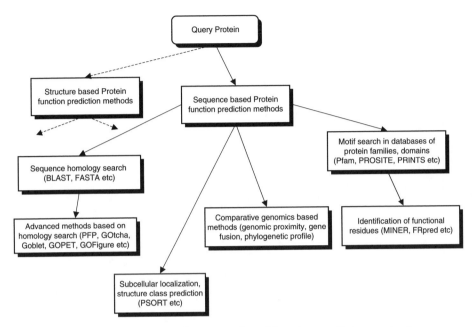

Figure 3.5   *Summary of sequence-based function prediction methods.*

first step is to perform homology search using BLAST, PSI-BLAST or FASTA. Also it is recommended to perform motif and domain searches, such as Pfam and PROSITE. If significant hits are not found, some of recent methods which expand homology search, such as PFP, could be performed. If reasonable results are still not obtained, we recommend the STRING server, which performs comparative genomics based approaches. However, note that comparative genomics methods don't predict specific functional terms of a query protein, rather shows a set of proteins which are predicted to be functionally related to the query protein. If knowing a broad class of protein is useful, subcellular localization prediction and some local structure class predictions, such as prediction of transmembrane proteins[123] will be worthwhile to try. Finally, functional residue prediction methods, e.g. MINER, will be informative for some purposes, but note that these methods are aimed to select residues for function, not to predict functional terms. Refer to Table 3.1 for available online tools.

The need of function prediction is increasing, especially for interpreting large-scale omics data. This situation is very different from more than ten years ago when BLAST, FASTA, and PSI-BLAST were developed. Automatic function prediction methods will evolve in harmony with new developments of experimental methods by incorporating those experimental data in prediction algorithms and by helping biological reasoning of experimental data. More advances in this field are expected in the near future keeping pace with the other bioinformatics areas described in the other chapters in this book.

Table 3.1 Protein function prediction methods

Name	WWW Address	Description
BLAST[16], PSI-BLAST[28]	http://www.ncbi.nlm.nih.gov/blast/	Sequence homology search
FASTA[15]	http://www.ebi.ac.uk/fasta33/	Sequence homology search
PFP[81]	http://dragon.bio.purdue.edu/pfp/	BLAST-based GO term prediction + association mining
GOtcha[72]	http://www.compbio.dundee.ac.uk/gotcha/gotcha.php	BLAST-based GO term prediction
GOblet[68, 69]	http://goblet.molgen.mpg.de/	BLAST-based GO term prediction
GOPET[73]	http://genius.embnet.dkfz-heidelberg.de/menu/biounit/open-husar	BLAST-based GO term prediction by SVM
ProtFun[74]	http://www.cbs.dtu.dk/services/ProtFun/	Sequence feature based function classification
OntoBlast[124]	http://functionalgenomics.de/ontogate/	BLAST-based GO term prediction
FIGENIX[125]	http://sites.univ-provence.fr/evol/figenix/	Genomic annotation using phylogenomic approaches
JAFA[75]	http://jafa.burnham.org/	GO term prediction metaserver
Pfam[35]	http://pfam.sanger.ac.uk/	Protein family HMM database
SMART[126]	http://smart.embl-heidelberg.de/	Sequence fingerprint scanning
ProDom[33]	http://prodom.prabi.fr	Protein domain sequence database
BLOCKS[32]	http://blocks.fhcrc.org/	Protein domain sequence database
PRINTS[34]	http://www.bioinf.manchester.ac.uk/dbbrowser/PRINTS/	Protein fingerprint database
ELM[104]	http://elm.eu.org/	Functional motif
PROSITE[51]	http://ca.expasy.org/prosite/	Database of protein domains, families and functional sites
InterProScan[82]	http://www.ebi.ac.uk/InterProScan/	Functional motif search
ScanProsite[127]	http://www.expasy.ch/prosite/	Functional motif scanning
STRING[93]	http://string.embl.de/	Comparative genomics approaches
FRpred[128]	http://toolkit.tuebingen.mpg.de/frpred	Prediction of protein functional residues
MINER[120]	http://coit-apple01.uncc.edu/MINER/	Functional residue prediction
PSORT[97]	http://www.psort.org/	PSORT family of programs for subcellular localization prediction
SignalP[94]	http://www.cbs.dtu.dk/services/SignalP/	Prediction of the presence and location of signal peptide cleavage sites

Tool	URL	Description
CELLO[100]	http://cello.life.nctu.edu.tw/	subCELlular LOcalization predictor
SubLoc[102]	http://www.bioinfo.tsinghua.edu.cn/SubLoc/	Prediction of Protein Subcellular Localization by Support Vector Machine
LOCtree[129]	http://cubic.bioc.columbia.edu/cgi/var/nair/loctree/query	Prediction of Protein Subcellular Localization by Support Vector Machine
TMHMM[123]	http://www.cbs.dtu.dk/services/TMHMM-2.0/	Prediction of transmembrane helices in proteins
BOMP[130]	http://www.bioinfo.no/tools/bomp	Tool for prediction of beta-barrel integral outer membrane proteins
PROFtmb[131]	http://rostlab.org/cgi-bin/var/bigelow/proftmb/query	Per-residue and whole-proteome prediction of bacterial transmembrane beta barrels
LipoP[132]	http://www.cbs.dtu.dk/services/LipoP/	Prediction of lipoproteins and signal peptides in Gram negative bacteria

Acknowledgement

This work is partially supported by National Institute of General Medical Sciences of the National Institutes of Health (U24 GM077905 and R01GM075004), and the National Science Foundation (DMS 0604776).

References

1. J.P. Gogarten, and L. Olendzenski, Orthologs, paralogs and genome comparisons, *Curr Opin Genet Dev*, **9**, 630–636 (1999).
2. Z. Gu, L.M. Steinmetz, X. Gu, C. Scharfe, R.W. Davis, and W.H. Li, Role of duplicate genes in genetic robustness against null mutations, *Nature*, **421**, 63–66 (2003).
3. S. Ohno, *Evolution by Gene Duplication*, George Allen & Unwin, London, 1970.
4. Y. Boucher, C.J. Douady, R.T. Papke, et al. Lateral gene transfer and the origins of prokaryotic groups, *Annu Rev Genet*, **37**, 283–328 (2003).
5. W.F. Doolittle, Phylogenetic classification and the universal tree, *Science*, **284**, 2124–2129 (1999).
6. W.M. Fitch, Distinguishing homologous from analogous proteins, *Syst Zool*, **19**, 99–113 (1970).
7. W. Tian, and J. Skolnick, How well is enzyme function conserved as a function of pairwise sequence identity? *J. Mol. Biol.*, **333**, 863–882 (2003).
8. Y. Van de Peer, Evolutionary genetics: When duplicated genes don't stick to the rules, *Heredity*, **96**, 204–205 (2006).
9. C.B. Anfinsen, Principles that govern the folding of protein chains, *Science*, **181**, 223–230 (1973).
10. C. Chothia, and A.M. Lesk, The relation between the divergence of sequence and structure in proteins, *EMBO J.*, **5**, 823–826 (1986).
11. S.E. Brenner, C. Chothia, and T.J. Hubbard, Assessing sequence comparison methods with reliable structurally identified distant evolutionary relationships, *Proc. Natl. Acad. Sci. U.S.A*, **95**, 6073–6078 (1998).
12. C.A. Orengo, D.T. Jones, and J.M. Thornton, Protein superfamilies and domain superfolds, *Nature*, **372**, 631–634 (1994).
13. S.B. Needleman, and C.D. Wunsch, A general method applicable to the search for similarities in the amino acid sequence of two proteins, *J. Mol. Biol.*, **48**, 443–453 (1970).
14. T.F. Smith, and M.S. Waterman, Identification of common molecular subsequences, *J. Mol. Biol.*, **147**, 195–197 (1981).
15. W.R. Pearson, and D.J. Lipman, Improved tools for biological sequence comparison, *Proc Natl Acad Sci U S A*, **85**, 2444–2448 (1988).
16. S.F. Altschul, W. Gish, W. Miller, E.W. Myers, and D.J. Lipman, Basic local alignment search tool, *J Mol Biol*, **215**, 403–410 (1990).
17. T. Hulsen, J. de Vlieg, and P.M. Groenen, Phylopat: Phylogenetic pattern analysis of eukaryotic genes, *BMC Bioinformatics*, **7**, 398 (2006).
18. W.R. Pearson, Comparison of methods for searching protein sequence databases, *Protein Sci*, **4**, 1145–1160 (1995).
19. W.R. Pearson, Effective protein sequence comparison, *Methods Enzymol.*, **266**, 227–258 (1996).
20. W.R. Pearson, Flexible sequence similarity searching with the Fasta3 program package, *Methods Mol Biol*, **132**, 185–219 (2000).
21. E.S. Lander, L.M. Linton, B. Birren, *et al.*, Initial sequencing and analysis of the human genome, *Nature*, **409**, 860–921 (2001).
22. S.G. Oliver, Q.J. Van Der Aart, M.L. Agostoni-Carbone, et al. The complete DNA sequence of yeast chromosome Iii, *Nature*, **357**, 38–46 (1992).
23. T. Hawkins, and D. Kihara, Function prediction of uncharacterized proteins, *J. Bioinform. Comput. Biol.*, **5**, 1–30 (2007).

24. B. John, and A. Sali, Detection of homologous proteins by an intermediate sequence search, *Protein Sci*, **13**, 54–62 (2004).
25. J. Park, S.A. Teichmann, T. Hubbard, and C. Chothia, Intermediate sequences increase the detection of homology between sequences, *J Mol Biol*, **273**, 349–354 (1997).
26. I. Alam, A. Dress, M. Rehmsmeier, and G. Fuellen, Comparative homology agreement search: An effective combination of homology-search methods, *Proc Natl Acad Sci U S A*, **101**, 13814–13819 (2004).
27. C. Webber, and G.J. Barton, Increased coverage obtained by combination of methods for protein sequence database searching, *Bioinformatics*, **19**, 1397–1403 (2003).
28. S.F. Altschul, T.L. Madden, A.A. Schaffer, *et al.*, Gapped Blast and Psi-Blast: A new generation of protein database search programs, *Nucleic Acids Res*, **25**, 3389–3402 (1997).
29. W.R. Pearson, and M.L. Sierk, The limits of protein sequence comparison? *Curr Opin Struct Biol*, **15**, 254–260 (2005).
30. A.A. Schaffer, L. Aravind, T.L. Madden, *et al.*, Improving the accuracy of Psi-Blast protein database searches with composition-based statistics and other refinements, *Nucleic Acids Res*, **29**, 2994–3005 (2001).
31. A.A. Schaffer, Y.I. Wolf, C.P. Ponting, E.V. Koonin, L. Aravind, and S.F. Altschul, Impala: Matching a protein sequence against a collection of Psi-Blast-constructed position-specific score matrices, *Bioinformatics*, **15**, 1000–1011 (1999).
32. J.G. Henikoff, E.A. Greene, S. Pietrokovski, and S. Henikoff, Increased coverage of protein families with the Blocks database servers, *Nucleic Acids Res.*, **28**, 228–230 (2000).
33. C. Bru, E. Courcelle, S. Carrere, Y. Beausse, S. Dalmar, and D. Kahn, The Prodom database of protein domain families: More emphasis on 3D, *Nucleic Acids Res.*, **33**, D212–D215 (2005).
34. T.K. Attwood, P. Bradley, D.R. Flower, *et al.*, Prints and its automatic supplement, Preprints, *Nucleic Acids Res.*, **31**, 400–402 (2003).
35. R.D. Finn, J. Mistry, B. Schuster-Bockler, *et al.*, Pfam: Clans, web tools and services, *Nucleic Acids Res.*, **34**, D247–D251 (2006).
36. D. Wilson, M. Madera, C. Vogel, C. Chothia, and J. Gough, The Superfamily database in 2007: Families and functions, *Nucleic Acids Res*, **35**, D308–313 (2007).
37. S.R. Eddy, Hidden Markov models, *Curr Opin Struct Biol*, **6**, 361–366 (1996).
38. R.I. Sadreyev, D. Baker, and N.V. Grishin, Profile-profile comparisons by Compass predict intricate homologies between protein families, *Protein Sci*, **12**, 2262–2272 (2003).
39. K. Ginalski, N.V. Grishin, A. Godzik, and L. Rychlewski, Practical lessons from protein structure prediction, *Nucleic Acids Res.*, **33**, 1874–1891 (2005).
40. L. Rychlewski, L. Jaroszewski, W. Li, and A. Godzik, Comparison of sequence profiles. Strategies for structural predictions using sequence information, *Protein Sci.*, **9**, 232–241 (2000).
41. R.L. Dunbrack, Jr., Sequence comparison and protein structure prediction, *Curr. Opin. Struct. Biol.*, **16**, 374–384 (2006).
42. D. Kihara, and J. Skolnick, Microbial genomes have over 72 % structure assignment by the threading algorithm prospector_Q, *Proteins*, **55**, 464–473 (2004).
43. K. Kinoshita, and H. Nakamura, Protein informatics towards function identification, *Curr. Opin. Struct. Biol.*, **13**, 396–400 (2003).
44. J.S. Fetrow, A. Godzik, and J. Skolnick, Functional analysis of the escherichia coli genome using the sequence- to-structure-to-function paradigm: Identification of proteins exhibiting the glutaredoxin/thioredoxin disulfide oxidoreductase activity, *J Mol Biol*, **282**, 703–711 (1998).
45. M.L. Green, and P.D. Karp, A Bayesian method for identifying missing enzymes in predicted metabolic pathway databases, *BMC. Bioinformatics*, **5**, 76 (2004).
46. R.K. Curtis, M. Oresic, and A. Vidal-Puig, pathways to the analysis of microarray data, *Trends Biotechnol*, **23**, 429–435 (2005).
47. R. Sharan, I. Ulitsky, and R. Shamir, Network-based prediction of protein function, *Mol Syst Biol*, **3**, 88 (2007).
48. J.D. Watson, R.A. Laskowski, and J.M. Thornton, Predicting protein function from sequence and structural data, *Curr. Opin. Struct. Biol.*, **15**, 275–284 (2005).
49. D. Kihara, D.Y. Yang, and T. Hawkins, Bioinformatics resources for cancer research with an emphasis on gene function and structure prediction tools, *Cancer Informatics*, **2**, 25–35 (2006).

50. C.H. Wu, R. Apweiler, A. Bairoch, *et al.*, The Universal Protein Resource (Uniprot): An expanding universe of protein information, *Nucleic Acids Res*, **34**, D187–191 (2006).

51. N. Hulo, A. Bairoch, V. Bulliard, *et al.*, The 20 years of prosite, *Nucleic Acids Res*, **36**, D245–249 (2008).

52. S.E. Brenner, Errors in genome annotation, *Trends Genet*, **15**, 132–133 (1999).

53. W.R. Gilks, B. Audit, D. de Angelis, S. Tsoka, and C.A. Ouzounis, Percolation of annotation errors through hierarchically structured protein sequence databases, *Math Biosci*, **193**, 223–234 (2005).

54. M.Y. Galperin, and E.V. Koonin, Sources of systematic error in functional annotation of genomes: domain rearrangement, non-orthologous gene displacement and operon disruption, *In Silico Biol*, **1**, 55–67 (1998).

55. D. Devos, and A. Valencia, Intrinsic errors in genome annotation, *Trends Genet*, **17**, 429–431 (2001).

56. M. Riley, T. Abe, M.B. Arnaud, *et al.*, Escherichia Coli K-12: A cooperatively developed annotation snapshot – 2005, *Nucleic Acids Res*, **34**, 1–9 (2006).

57. S.L. Salzberg, Genome re-annotation: A Wiki solution? *Genome Biol*, **8**, 102 (2007).

58. M.A. Harris, J. Clark, A. Ireland, *et al*, The Gene Ontology (Go) database and informatics resource, *Nucleic Acids Res*, **32**, D258–261 (2004).

59. Nomenclature Committee of the International Union of Biochemistry and Molecular Biology (Nc-Iubmb), Enzyme Supplement 5 (1999), *Eur J Biochem*, **264**, 610–650 (1999).

60. A. Ruepp, A. Zollner, D. Maier, *et al.*, The Funcat, a functional annotation scheme for systematic classification of proteins from whole genomes, *Nucleic Acids Res*, **32**, 5539–5545 (2004).

61. M.H. Saier, Jr., C.V. Tran, and R.D. Barabote, Tcdb: The Transporter Classification Database for membrane transport protein analyses and information, *Nucleic Acids Res.*, **34**, D181–186 (2006).

62. M. Kanehisa, M. Araki, S. Goto, *et al.*, Kegg for linking genomes to life and the environment, *Nucleic Acids Res*, **36**, D480–484 (2008).

63. G. Lopez, A. Rojas, M. Tress, and A. Valencia, Assessment of predictions submitted for the Casp7 function prediction category, *Proteins*, **69 Suppl 8**, 165–174 (2007).

64. P. Resnik, Using information content to evaluate semantic similarity in a taxonomy, *Proceedings of the 14th International Joint Conference on Artificial Intelligence* (1995).

65. P.W. Lord, R.D. Stevens, A. Brass, and C.A. Goble, Investigating semantic similarity measures across the gene ontology: The relationship between sequence and annotation, *Bioinformatics*, **19**, 1275–1283 (2003).

66. A. Schlicker, F.S. Domingues, J. Rahnenfuhrer, and T. Lengauer, A new measure for functional similarity of gene products based on gene ontology, *BMC Bioinformatics*, **7**, 302 (2006).

67. A. Del Pozo, F. Pazos, and A. Valencia, Defining functional distances over gene ontology, *BMC Bioinformatics*, **9**, 50 (2008).

68. D. Groth, H. Lehrach, and S. Hennig, Goblet: A platform for gene ontology annotation of anonymous sequence data, *Nucleic Acids Res*, **32**, W313–317 (2004).

69. S. Hennig, D. Groth, and H. Lehrach, Automated gene ontology annotation for anonymous sequence data, *Nucleic Acids Res*, **31**, 3712–3715 (2003).

70. S. Khan, G. Situ, K. Decker, and C.J. Schmidt, Gofigure: Automated gene ontology annotation, *Bioinformatics*, **19**, 2484–2485 (2003).

71. K. Verspoor, J. Cohn, S. Mniszewski, and C. Joslyn, A categorization approach to automated ontological function annotation, *Protein Sci*, **15**, 1544–1549 (2006).

72. D.M. Martin, M. Berriman, and G.J. Barton, Gotcha: A new method for prediction of protein function assessed by the annotation of seven genomes, *BMC Bioinformatics*, **5**, 178 (2004).

73. A. Vinayagam, C. del Val, F. Schubert, *et al.*, Gopet: A tool for automated predictions of gene ontology terms, *BMC Bioinformatics*, **7**, 161 (2006).

74. L.J. Jensen, R. Gupta, H.H. Staerfeldt, and S. Brunak, Prediction of human protein function according to gene ontology categories, *Bioinformatics*, **19**, 635–642 (2003).

75. I. Friedberg, T. Harder, and A. Godzik, Jafa: A protein function annotation meta-server, *Nucleic Acids Res*, **34**, W379–381 (2006).

76. E.M. Zdobnov, and R. Apweiler, Interproscan: An integration platform for the signature-recognition methods in Interpro, *Bioinformatics*, **17**, 847–848 (2001).
77. F. Enault, K. Suhre, and J.M. Claverie, Phydbac 'Gene Function Predictor': A gene annotation tool based on genomic context analysis, *BMC Bioinformatics*, **6**, 247 (2005).
78. B.E. Engelhardt, M.I. Jordan, K.E. Muratore, and S.E. Brenner, Protein molecular function prediction by Bayesian phylogenomics, *PLoS Comput Biol*, **1**, e45 (2005).
79. Y.D. Cai, and K.C. Chou, Predicting enzyme subclass by functional domain composition and pseudo amino acid composition, *J Proteome Res*, **4**, 967–971 (2005).
80. A. Lobley, M.B. Swindells, C.A. Orengo, and D.T. Jones, Inferring function using patterns of native disorder in proteins, *PLoS Comput Biol*, **3**, e162 (2007).
81. T. Hawkins, S. Luban, and D. Kihara, Enhanced automated function prediction using distantly related sequences and contextual association by PFP, *Protein Sci*, **15**, 1550–1556 (2006).
82. N.J. Mulder, R. Apweiler, T.K. Attwood, *et al.*, New developments in the Interpro Database, *Nucleic Acids Res*, **35**, D224–228 (2007).
83. I. Letunic, R.R. Copley, S. Schmidt, *et al.*, Smart 4.0: Towards genomic data integration, *Nucleic Acids Res*, **32**, D142–144 (2004).
84. D.H. Haft, J.D. Selengut, and O. White, The Tigrfams database of protein families, *Nucleic Acids Res.*, **31**, 371–373 (2003).
85. T. Hawkins, M. Chitale, S. Luban, and D. Kihara, PFP: Automated prediction of gene ontology functional annotations with confidence scores, *Proteins*, Epub. (2008). DOI 10.1002/prot.22172
86. T. Hawkins, M. Chitale, and D. Kihara, New paradigm in protein function prediciton for large scale omics analysis, *Molecular BioSystems*, **4**, 223–231 (2008).
87. H. Watanabe, H. Mori, T. Itoh, and T. Gojobori, Genome plasticity as a paradigm of eubacteria evolution, *J. Mol. Evol.*, **44 Suppl 1**, S57–64 (1997).
88. R. Overbeek, M. Fonstein, M. D'Souza, G.D. Pusch, and N. Maltsev, The use of gene clusters to infer functional coupling, *Proc. Natl. Acad. Sci. U.S.A*, **96**, 2896–2901 (1999).
89. T. Dandekar, B. Snel, M. Huynen, and P. Bork, Conservation of gene order: A fingerprint of proteins that physically interact, *Trends Biochem. Sci.*, **23**, 324–328 (1998).
90. H. Salgado, G. Moreno-Hagelsieb, T.F. Smith, and J. Collado-Vides, Operons in escherichia coli: genomic analyses and predictions, *Proc Natl Acad Sci U S A*, **97**, 6652–6657 (2000).
91. A.J. Enright, I. Iliopoulos, N.C. Kyrpides, and C.A. Ouzounis, Protein interaction maps for complete genomes based on gene fusion events, *Nature*, **402**, 86–90 (1999).
92. M. Pellegrini, E.M. Marcotte, M.J. Thompson, D. Eisenberg, and T.O. Yeates, Assigning protein functions by comparative genome analysis: Protein phylogenetic profiles, *Proc. Natl. Acad. Sci. U.S.A*, **96**, 4285–4288 (1999).
93. B. Snel, G. Lehmann, P. Bork, and M.A. Huynen, String: A web-server to retrieve and display the repeatedly occurring neighbourhood of a gene, *Nucleic Acids Res.*, **28**, 3442–3444 (2000).
94. O. Emanuelsson, S. Brunak, G. von Heijne, and H. Nielsen, Locating proteins in the cell using Targetp, Signalp and related tools, *Nat Protoc*, **2**, 953–971 (2007).
95. K. Nakai, and P. Horton, Psort: A program for detecting sorting signals in proteins and predicting their subcellular localization, *Trends Biochem Sci*, **24**, 34–36 (1999).
96. J.L. Gardy, C. Spencer, K. Wang, *et al.*, Psort-B: Improving protein subcellular localization prediction for gram-negative bacteria, *Nucleic Acids Res.*, **31**, 3613–3617 (2003).
97. J.L. Gardy, and F.S. Brinkman, Methods for predicting bacterial protein subcellular localization, *Nat Rev Microbiol*, **4**, 741–751 (2006).
98. R. Nair, and B. Rost, Sequence conserved for subcellular localization, *Protein Sci*, **11**, 2836–2847 (2002).
99. Z. Lu, D. Szafron, R. Greiner, *et al.*, Predicting subcellular localization of proteins using machine-learned classifiers, *Bioinformatics*, **20**, 547–556 (2004).
100. C.S. Yu, C.J. Lin, and J.K. Hwang, Predicting subcellular localization of proteins for gram-negative bacteria by support vector machines based on N-peptide compositions, *Protein Sci*, **13**, 1402–1406 (2004).
101. J. Wang, W.K. Sung, A. Krishnan, and K.B. Li, protein subcellular localization prediction for gram-negative bacteria using amino acid subalphabets and a combination of multiple support vector machines, *BMC. Bioinformatics*, **6**, 174 (2005).

102. S. Hua, and Z. Sun, Support vector machine approach for protein subcellular localization prediction, *Bioinformatics*, **17**, 721–728 (2001).
103. E.M. Marcotte, I. Xenarios, A.M. Van Der Bliek, and D. Eisenberg, Localizing proteins in the cell from their phylogenetic profiles, *Proc Natl Acad Sci U S A*, **97**, 12115–12120 (2000).
104. P. Puntervoll, R. Linding, C. Gemund, *et al.* Elm server: A new resource for investigating short functional sites in modular eukaryotic proteins, *Nucleic Acids Res*, **31**, 3625–3630 (2003).
105. J.W. Torrance, G.J. Bartlett, C.T. Porter, and J.M. Thornton, Using a library of structural templates to recognise catalytic sites and explore their evolution in homologous families, *J. Mol. Biol.*, **347**, 565–581 (2005).
106. O. Lichtarge, and M.E. Sowa, Evolutionary predictions of binding surfaces and interactions, *Curr. Opin. Struct. Biol.*, **12**, 21–27 (2002).
107. S. Jones, and J.M. Thornton, Searching for functional sites in protein structures, *Curr Opin Chem Biol*, **8**, 3-7 (2004).
108. W. Tian, A.K. Arakaki, and J. Skolnick, Eficaz: A comprehensive approach for accurate genome-scale enzyme function inference, *Nucleic Acids Res.*, **32**, 6226–6239 (2004).
109. M.N. Wass, and M.J. Sternberg, Confunc – functional annotation in the twilight zone, *Bioinformatics*, (2008).
110. J.A. Capra, and M. Singh, Predicting functionally important residues from sequence conservation, *Bioinformatics*, **23**, 1875–1882 (2007).
111. S. Chakrabarti, and C.J. Lanczycki, Analysis and prediction of functionally important sites in proteins, *Protein Sci*, **16**, 4–13 (2007).
112. B. Sterner, R. Singh, and B. Berger, Predicting and annotating catalytic residues: An information theoretic approach, *J Comput Biol*, **14**, 1058–1073 (2007).
113. J.D. Fischer, C.E. Mayer, and J. Soding, Prediction of protein functional residues from sequence by probability density estimation, *Bioinformatics*, **24**, 613–620 (2008).
114. O.V. Kalinina, A.A. Mironov, M.S. Gelfand, and A.B. Rakhmaninova, Automated selection of positions determining functional specificity of proteins by comparative analysis of orthologous groups in protein families, *Protein Sci*, **13**, 443–456 (2004).
115. S.S. Hannenhalli, and R.B. Russell, Analysis and prediction of functional sub-types from protein sequence alignments, *J Mol Biol*, **303**, 61–76 (2000).
116. F. Pazos, A. Rausell, and A. Valencia, Phylogeny-independent detection of functional residues, *Bioinformatics*, **22**, 1440–1448 (2006).
117. J. Pei, W. Cai, L.N. Kinch, and N.V. Grishin, Prediction of functional specificity determinants from protein sequences using log-likelihood ratios, *Bioinformatics*, **22**, 164–171 (2006).
118. G. Casari, C. Sander, and A. Valencia, A method to predict functional residues in proteins, *Nat Struct Biol*, **2**, 171–178 (1995).
119. D. La, and D.R. Livesay, Predicting functional sites with an automated algorithm suitable for heterogeneous datasets, *BMC. Bioinformatics.*, **6**, 116 (2005).
120. D. La, B. Sutch, and D.R. Livesay, Predicting protein functional sites with phylogenetic motifs, *Proteins*, **58**, 309–320 (2005).
121. I. Friedberg, M. Jambon, and A. Godzik, New avenues in protein function prediction, *Protein Sci*, **15**, 1527–1529 (2006).
122. S. Soro, and A. Tramontano, The prediction of protein function at Casp6, *Proteins*, **61 Suppl 7**, 201–213 (2005).
123. A. Krogh, B. Larsson, G. von Heijne, and E.L. Sonnhammer, Predicting transmembrane protein topology with a hidden Markov model: Application to complete genomes, *J Mol Biol*, **305**, 567–580 (2001).
124. G. Zehetner, Ontoblast function: From sequence similarities directly to potential functional annotations by ontology terms, *Nucleic Acids Res.*, **31**, 3799–3803 (2003).
125. P. Gouret, V. Vitiello, N. Balandraud, A. Gilles, P. Pontarotti, and E.G. Danchin, Figenix: Intelligent automation of genomic annotation: Expertise integration in a new software platform, *BMC Bioinformatics*, **6**, 198 (2005).
126. I. Letunic, R.R. Copley, B. Pils, S. Pinkert, J. Schultz, and P. Bork, Smart 5: Domains in the context of genomes and networks, *Nucleic Acids Res.*, **34**, D257–260 (2006).

127. N. Hulo, A. Bairoch, V. Bulliard, L. Cerutti, C.E. De, P.S. Langendijk-Genevaux, M. Pagni, and C.J. Sigrist, The Prosite Database, *Nucleic Acids Res.*, **34**, D227–230 (2006).

128. J.D. Fischer, C.E. Mayer, and J. Soding, Prediction of protein functional residues from sequence by probability density estimation, *Bioinformatics*, (2008).

129. R. Nair, and B. Rost, Mimicking cellular sorting improves prediction of subcellular localization, *J Mol Biol*, **348**, 85–100 (2005).

130. F.S. Berven, K. Flikka, H.B. Jensen, and I. Eidhammer, Bomp: A program to predict integral beta-barrel outer membrane proteins encoded within genomes of gram-negative bacteria, *Nucleic Acids Res*, **32**, W394–399 (2004).

131. H.R. Bigelow, D.S. Petrey, J. Liu, D. Przybylski, and B. Rost, Predicting transmembrane beta-barrels in proteomes, *Nucleic Acids Res*, **32**, 2566–2577 (2004).

132. W.T. Doerrler, Lipid trafficking to the outer membrane of gram-negative bacteria, *Mol. Microbiol.*, **60**, 542–552 (2006).

4

Template Based Prediction of Three-dimensional Protein Structures: Fold Recognition and Comparative Modeling

Jan Kosiński, Karolina L. Tkaczuk, Joanna M. Kasprzak and Janusz M. Bujnicki

4.1 Introduction

The protein structure is intrinsically connected to the protein function. In particular, a catalytic activity of enzymes is determined by a specific orientation of appropriate amino acid residues in the three-dimensional space. Likewise, a substrate specificity of the enzymes and transport proteins depends on how their structures can accommodate to geometrical and electrostatic features of the substrates and ligands. In the case of structural proteins, their three-dimensional shape directly determines the way in which they can be used in building the cellular components. Therefore, it is undisputable that the knowledge of a protein structure can help in determining and understanding the protein function.

With the three-dimensional structure in hand a scientist can often infer what activity the protein has and what substrate it processes or which ligand it must bind to perform its function. The efforts aimed at revealing the structures of proteins started about 1935, when the first X-ray diffraction pattern of a protein crystal was reported.[1] Twenty-two years later the first protein structure has been solved (the structure of a myoglobin at 6Å resolution[2]). Today, after 50 years from the release of that first protein structure, about 45,000 structures are known. However, this set corresponds only to about 16,000 unique proteins, because

Prediction of Protein Structures, Functions, and Interactions   Edited by Janusz Bujnicki
© 2009 John Wiley & Sons, Ltd

multiple structures are often available for the same protein. At the same time, there are over 6,600,000 of known and predicted proteins in the UniProtKB database (as of October 2008) and sequences of millions of other predicted proteins are already deposited in other public databases.

This increasingly large gap between the number of structures and protein sequences results from limitations of present structure determination techniques. The structure determination by X-ray crystallography is impeded by the requirement of the high-level expression, efficient purification and successful crystallization of the target proteins. NMR techniques are able to overcome some of these difficulties, but are still limited with respect to the size of protein that can be tackled on a routine basis. Moreover, both X-ray crystallography and NMR methods are very expensive, time-consuming and require highly specialized equipment.

The alternative approach assumes that protein structures can be modeled using computer programs. This idea is based on Anfinsen's hypothesis[3] that protein structure in a given environment is determined by protein sequence and the protein folding is a process of assuming the conformation of the lowest free energy that we should be able to simulate with accurate knowledge of laws of physics. However, mainly due to limited computational power of current computers and the high cost of precise free energy calculations, the protein folding problem has not been solved yet. Fortunately, it was found that evolutionary related proteins tend to have similar structures,[4] which suggested that many protein structures can be modeled based on already known structures of related proteins (Figure 4.1). This technique, referred to as comparative or homology modeling, has become the most successful protein structure modeling method, which often leads to accurate models suitable for detailed functional analyses. The term '*comparative modeling*' refers to all techniques that model protein structures based on the coordinates of known structure under the assumption that the structure of the *target* (i.e. the protein we want to model) is similar to the structure of the *template* (i.e. the structure that will serve a framework for modeling the target). The specific case of comparative modeling is '*homology modeling*', which is technically the same as comparative modeling but additionally implies that target and the template are evolutionary related (i.e. homologous).

The prerequisite to obtain a comparative model of high accuracy is the availability of a known structure of a structurally similar protein that can be used as a modeling template. This is not guaranteed, because the number of solved structures is still limited, and thus far we have not yet obtained structural representatives for all protein families. The paucity of structures was one of the major driving forces for emergence of a new field: structural genomics. The main goal of the structural genomics initiative is to provide an exhaustive structural coverage of diverse protein structure universe to enable modeling any other structure using comparative modeling tools. Therefore, structural genomics consortia aim at selecting the most representative proteins from structurally uncharacterized sequence families for experimental structure determination. Other proteins may then be modeled with computer programs based on the already solved structures. It is expected that with the advance of structural genomics projects, the coverage of protein structure universe will get more complete, the number of proteins unrelated to any protein with known structure will decrease significantly and the comparative modeling will become easier and more accurate, thus becoming fully complementary to experimental structure determination techniques.

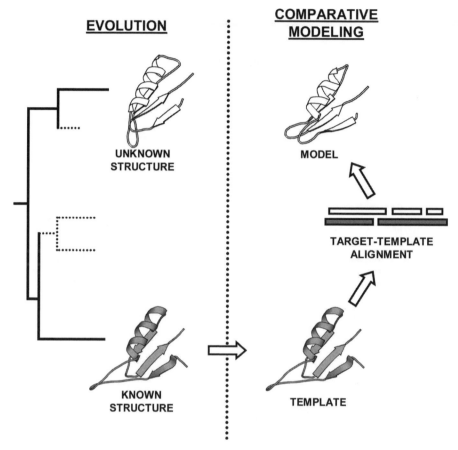

EVOLUTION

COMPARATIVE MODELING

UNKNOWN STRUCTURE

MODEL

TARGET-TEMPLATE ALIGNMENT

KNOWN STRUCTURE

TEMPLATE

Figure 4.1   *Evolutionary principles of comparative modeling*

4.2   Steps of Comparative Modeling

Comparative modeling is a multi-step procedure involving a variety of programs and alternative approaches (Figure 4.2). It starts from a prediction of the target fold and identification of the best template structure(s) bearing this fold. A single template or multiple templates (if available) are then used as a source of coordinates for modeling the target protein. Modeling is usually an iterative process including model building, model refinement and model evaluation. The following sections review in detail these steps, starting from protein fold recognition and ending with model quality assessment.

4.2.1   Fold Recognition

Fold recognition (FR) is the first step of the comparative modeling process. It is defined as the prediction of a structural fold of the target protein from its amino acid sequence by detecting of proteins of known three-dimensional structure that are likely to have a similar fold to the target. FR is accomplished by searching a database of known protein

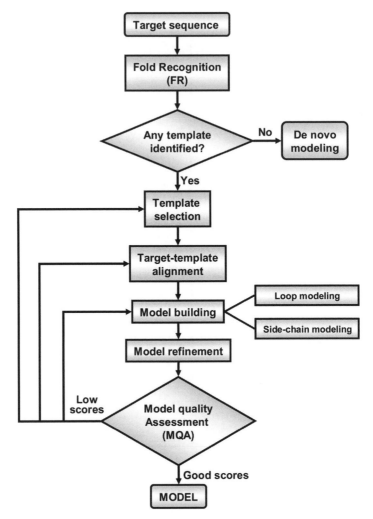

Figure 4.2   *Steps of comparative modeling*

structures (PDB or its non-redundant subset) for structures that are the most 'compatible' with the target sequence. This sequence-structure compatibility can be evaluated by various means depending on the method (see below). Typically, FR methods produce rankings of templates and provide target-template alignments and various scores describing the level of sequence-structure compatibility. This information can be directly used for template selection and comparative modeling (see sections below). FR methods can be generally divided into two general categories: (1) sequence-only based methods and (2) methods that use both sequence and structural information. Publicly available FR programs are listed in Table 4.1 along with description of their characteristic features.

Sequence-based methods for protein fold recognition rely on traditional approaches for remote homology detection that are comprehensively reviewed in Chapter 1 in this

Table 4.1 Summary of publicly available programs for protein Fold Recognition. Brief descriptions of methods are included. In each section programs are listed in the alphabetical order. The description of methodology used for construction of profiles or for scoring has been provided where applicable

Method	Search strategy	Description
Sequence-based methods:		
FFAS[15] ffas.ljcrf.edu/	Profile–profile	Profile Construction: Profiles are calculated from alignments returned by PSI-BLAST,[5] sequences are weighted based on their dissimilarity with respect to the whole family. Comparison Score: a dot product of corresponding vectors.
FORTE[16] www.cbrc.jp/forte/	Profile–profile	Profile Construction: A profile is a re-formatted PSSM from a PSI-BLAST[5] search Comparison Score: correlation coefficient between each two profile columns (vectors)
PDB-BLAST E.g. at genesilico.pl/meta	Profile–sequence	Target sequence is used as a seed to construct a sequence profile by a PSI-BLAST[5] search against the protein sequence database. Then, the profile is used to search against the sequences from the PDB database to identify possible templates
SUPERFAMILY[81] supfam.org/SUPERFAMILY/	Scoring with HMMs	Scores the target sequence with HMMs from the Superfamily library of HMMs built for structurally characterized superfamilies derived from SCOP[82]
Sequence/Structure-based methods:		
FUGUE[28] www-cryst.bioc.cam.ac.uk/~fugue	Sequence/profile – profile	Profile Construction: A profile for the target is calculated from a user-provided or PSI-BLAST-derived MSA (sequences weighted using the VA scheme[83]). A profile for the template contains two matrices: a scoring matrix derived from environment-specific substitution tables and a gap penalty matrix derived from structural alignments from HOMSTRAD[84] Comparison Score: score from scoring matrix pre-calculated for each template family, gap penalties from the gap penalty matrix
HHPRED[24] toolkit.tuebingen.mpg.de/hhpred	HMM–HMM	Target sequence or MSA is used for building a HMM, which is aligned with all HMMs representing annotated proteins or domains with known structure. HMMs can include SS information (experimentally determined or predicted)

(continued overleaf)

Table 4.1 (Continued)

Method	Search strategy	Description
INUB (bioinbgu)[26] inub.cse.buffalo.edu/query.html	Sequence/profile – profile	INUB is new version of bioinbgu. Bioinbgu selects consensus prediction from five components that use the same alignment algorithm but with different sequence similarity scoring functions. Profile Construction: uses MSAs or PSSMs from PSI-BLAST[5] searches Comparison Score: a linear combination of SS similarity score and one of five sequence similarity scores (derived from bi-directional sequence-sequence, sequence-PSSM, sequence-MSA and MSA-PSSM comparisons)
Meta-BASIC[85] meta.bioinfo.pl	Profile-profile	Runs several profile-profile comparison methods that use different parameters of profile construction and methods for calculations of alignment scores. Profile Construction: Meta-profile derived from PSI-BLAST[5] searches with different numbers of iterations, profiles include SS predictions by PSIPRED[86]. Comparison Score: dot product of corresponding profile vectors or multiplication of the first vector by BLOSUM32 matrix and then by the second vector
mGenThreader[19] bioinf.cs.ucl.ac.uk/psipred/	Profile-profile	Evaluates profile-profile alignments using structure based scoring function (incl. SS, pairwise contacts statistical potentials and solvation terms). Profile Construction: PSI-BLAST search against sequence database, uses PSSMs directly from PSI-BLAST. Comparison Score: dot product of corresponding vectors (but using only positive scores from target PSSM)[87]
nFOLD2[87] www.biocentre.rdg.ac.uk/ bioinformatics/nFOLD/	Profile-profile	An extended version of mGenThreader[19] that takes models generated by mGenThreader and scores them with several MQAPs, a secondary structure element alignment score and functional site detection score
PHYRE[25] www.sbg.bio.ic.ac.uk/~phyre/	Profile-profile/metaserver	Runs several sequence, profile and SS comparison methods to generate a set of models, from which best models are selected using the 3D-Colony method (the 3D-Jury algorithm with models weighted according to scores provided by the servers used to generate them)

Method (URL)	Approach	Description
PROSPECTOR/TASSER-Lite[88,89] cssb.biology.gatech.edu/skolnick/ webservice/tasserlite/	Profile-sequence/threading	Multi-pass procedure starting from a sequence profile for the target. Scores templates with this profile to get target-template alignments, which are evaluated by various pairwise potentials in the next passes. Profile Construction: two FASTA[90] searches to generate close and distant profiles (the latter containing more distant homologs), can include SS predictions Comparison Score: Score from the target profile
RAPTOR[91] www.bioinformaticssolutions.com/ raptoronline/	Threading	Performs threading using efficient linear programming methodology. The threading scoring function consists of scores that take into account residue solvent accessibility, secondary structure, amino acid substitution matrix and pairwise interaction score (based on statistical potential). Models are ranked using SVM
SAM-T06[27] www.soe.ucsc.edu/research/ compbio/SAM_T06/T06-query.html	Scoring with HMMs	HMMs are constructed for both the target and templates. Then both the target and templates are scored using target- and template-derived HMMs to give combined scores for templates. HMMs include SS and other local structure properties incl. burial (observed for the templates, predicted for the target)
SP4[18] sparks.informatics.iupui.edu/SP4/	Profile-profile	Profile Construction: PSSMs from PSI-BLAST for target and template, residue-depth dependent, structure-derived sequence profile for template, SS profiles and SA profiles Comparison Score: linear combination of weighted dot products of corresponding profile vectors and weighted scoring functions for SS and SA matches, gap penalties are SS-dependent
THREADER[20,23] bioinf.cs.ucl.ac.uk/threader/	Threading	Aligns target and template sequences using double dynamic programming algorithm that can take into account pairwise contact potentials and solvation energies as well as sequence and SS similarity
WURST[92] www.zbh.uni-hamburg.de/wurst/	Profile-sequence	Profile construction: profile from PSI-BLAST search. Comparison Score: linear combination of fragment-based sequence-to-structure compatibility score and sequence similarity score based on a substitution matrix derived from superpositions of known structures. Models based on different templates are additionally scored using a pseudo-energy function

Table 4.1 (continued)

Method	Search strategy	Description
Metaservers (with associated Meta-predictors):		
Bionfo.PL[30] meta.bioinfo.pl	Metaserver	Provides: Unified view of outputs from different servers; FR servers; SS servers; full atom models; interactive scoring of alignments with Verify3D
3D-JURY[93] bioinfo.pl/meta	Consensus method	Interactive meta predictor that takes any set of models as an input and compares them all against all. 3D-JURY scores best models that appear to contain the largest recurrent subset of common coordinates
GeneSilico[31] genesilico.pl/meta	Metaserver	Provides: Unified view of outputs from different servers; FR servers; SS servers; solvent accessibility prediction; Disorder prediction; transmembrane helix prediction; full atom models
PCONS5[94] www.sbc.su.se/~bjorn/Pcons5	Consensus method	A local copy of PCONS5 server run on models generated by Genesilico Metaserver
3D CONSENS genesilico.pl/meta	Consensus method	A local implementation of the 3D-Jury algorithm
LOMETS[95] zhang.bioinformatics.ku.edu/LOMETS/	Metaserver	Provides: Unified presentation of outputs from different servers; FR servers only; No consensus method; full atom models; Additional restraint files in the output
MetaPP[96] www.predictprotein.org/meta.php	Metaserver	Provides: FR servers; SS servers; Homology modeling servers; contact prediction; transmembrane helix prediction; signal peptides; posttranslational modifications; No consensus method
Pcons.net[32] pcons.net/	Metaserver	Provides: Unified presentation of outputs from different servers; FR servers only; full atom models
PCONS5[94] www.sbc.su.se/~bjorn/Pcons5	Consensus method	Makes superposition of top models produced by different FR servers and ranks the models according to they average similarity to the entire ensemble of the models
Pmodeller[97] pcons.net/	Consensus method	Runs PCONS but additionally builds models with MODELLER and scores them with ProQ method[98]. The final ranking is made based on the score combining PCONS and ProQ scores
@TOME[99] bioserv.cbs.cnrs.fr/HTML_BIO/frame_meta.html	Metaserver	Provides: Unified presentation of outputs from different servers; FR servers; single SS server; full atom models; No consensus method

Meta-predictors/fragment assembly methods:

3D-SHOTGUN[100] http://inub.cse.buffalo.edu/	Meta-predictor/fragment assembly	For each FR model in the input set, a hybrid model is created by merging it with the most frequently recurring fragments from other at least partially superimposable models. Then, the best models is selected based on the original scores of initial models and a score reflecting similarity between that model and other hybrid models from the ensemble. 3D-SHOTGUN is now a part of the INUB server
FRankenstein's Monster[37] 1. genesilico.pl/toolkit/unimod? method=FrankensteinWeb OptimizeAlignment 2. genesilico.pl/meta	Meta-predictor/fragment assembly	A hybrid model is constructed by merging the consensus and best scoring (according to MQAP) fragments of models from FR servers. This hybrid model is superimposed on the templates to generate a target-template alignment, which is then used to build a new model. If this model contains poorly scoring regions, sampling of alternative target-template alignments is carried out and additional generation of models are built and recombined
I-TASSER[39] http://zhang.bioinformatics.ku.edu/ I-TASSER/	Meta-predictor/fragment assembly	Runs several profile–profile alignment methods that differ in the way of profile construction and alignment algorithm. Aligned regions are used as fragments, from which a pool of models is assembled with missing fragment added de novo. The models are then clustered and representative structures undergo second round of re-assembly/refinement. Conformation of the lowest energy is selected as the final model
PROTINFO[101] protinfo.compbio.washington.edu/ protinfo_abcmfr/	Meta-predictor/fragment assembly	Takes a set of FR models and performs exhaustive recombination of the models using a graph-theoretic clique finding (CF) approach and all-atom conditional probability discriminatory function (RAPDF)[102] for selecting best combinations
ROBETTA[43] robetta.bakerlab.org	Meta-predictor/fragment assembly	Selects best templates indicated by PSI-BLAST[5] and 3D-Jury[93] and generates a pool of target-template alignments using K*Sync[43]. The ensemble of models is then generated based on these alignments with missing fragments added de novo. The final models are selected based on energy scoring and structure clustering

volume (Kaminska *et al.*), therefore here they will be discussed only briefly. The simplest sequence-only based FR approach uses a pairwise sequence alignment algorithm (for instance BLAST[5]) to search a database of protein sequences from PDB for potential template proteins that exhibit significant sequence similarity to the target protein. However, in this approach distantly related templates are often missed due to low sensitivity of simple sequence searches.[5–8] More sensitive methodology includes construction of a sequence profile representing a whole protein family and using this profile to search against the sequence database that includes sequences of proteins with known structure (as implemented in PSI-BLAST[5]). However, sequence profiles may miss templates significantly similar to few family representatives but not significantly similar to the entire family as a whole. This problem may be solved by engaging the intermediate sequence search method (ISS) strategy,[9–12] where sequences of homologs identified in one round of a search are used as queries for new searches. If a sequence of a protein with known structure is detected during one of these subsequent searches, it can be used as a template for modeling the target protein. Although the ISS approach can reveal distant similarities, it is computationally demanding and usually returns many false positive hits.

Sequence-sequence and sequence-profile comparisons perform well when target-template sequence identity is above 30–40%. When the sequence identity drops below 30%, methods based on profile-profile and HMM-HMM comparisons become significantly more accurate and more sensitive.[13,14] Profile-profile approach is in use to detect homology by comparing profiles calculated both for the target and the template. There are many programs for remote homology detection and alignment using profile-profile comparisons (see Chapter 1 by Kaminska *et al.* about protein sequence analyses) and some of them, for example FFAS[15] and FORTE,[16] have been actually designed as FR servers.

To further increase sensitivity and specificity of template searches, one can use algorithms relying on Hidden Markov Models (HMMs). HMMs are conceptually very similar to profiles, but they represent a protein family in a specific probabilistic model that includes not only probabilities of transition of one amino acid to another at a given position of protein sequence, but also position-specific probabilities of deletions and insertions. Owing to this feature, FR methods based on HMMs are capable of constructing very accurate alignments.[14,17]

Profiles and HMMs have the advantage of incorporating not only sequence information but also structural information. The template profile or HMM can incorporate structural information derived from the three-dimensional structure. The target profile or HMM can be enriched with the structural features predicted from the sequence only (see Chapter 1 by Kaminska *et al.* about protein sequence analyses). Profiles can easily include such information as they are represented as matrices and adding more information constitutes of adding new vectors or dimensions to the sequence-only derived matrix. HMMs can include structural information for each residue as additional states in HMM. Comparisons of performance of similar methods with and without structural information indicated that structural data increases both sensitivity and accuracy.[14,18,19]

FR methods that use both sequence and structural information can be grouped according to the amount of structural information they use. Historically, the first type of methods relied almost exclusively on structural information. In the so-called threading approach, the target sequence is first aligned to each structure from a database using a double dynamic programming algorithm and an empirical energy function that assesses contacts between residues of the target, mounted on the backbone of the potential template.[20–22] Then, the resulting crude models (with the backbone conformation taken from the template, but with amino acid residues of the target) are scored according to the energy function, and ranked. Energy functions used in threading are typically based on pairwise residue-residue contact potentials derived from statistical analysis of known structures, and may also include a separate solvation potential to account for interactions of residues with the solvent. The best known threading program is THREADER[20,23] that currently, in addition to statistical contact and solvation potentials, can use secondary structure predictions and target-template sequence similarity.

Another variety of methods that turned out to outperform the threading approach use sequence profiles or HHMs combined with only one structural feature – secondary structure (e.g. HHPRED,[24] PHYRE[25] and INUB (new version of BIOINBGU[26])). More advanced methods (e.g. SP4[18] (a successor of SPARKS and SP4) and SAM-T06 (a successor of SAM-T04[27])) include features like residue accessibility, residue depth and various protein backbone properties. Generally, the utility of a structural feature (like secondary structure state or solvent accessibility) for the success of a FR protocol depends on whether this feature can be confidently predicted for the target. One way to overcome this limitation is to use structural data derived from known three-dimensional structures to construct structure-derived substitution matrices (e.g. FUGUE[28]) or to employ generic statistical potentials that are applicable to all proteins (e.g. mGenThreader[19]). Structural substitution matrices and/or statistical potentials can be used as a scoring function within the alignment construction algorithm (FUGUE) or to score alignments calculated by a separate, faster sequence-based algorithm (mGenThreader).

More recently it was found that the best approach to recognize a protein fold is the so called meta-prediction, i.e. running multiple FR methods, comparing their results and selecting the most promising model.[29] In general, meta-predictors look for predictions that are most common among the FR results. This can be further enhanced by incorporation of model quality assessment methods (see Chapter 6 by Wallner and Elofsson in this volume) to improve the ranking of models. Meta-predictors are often available as web servers that provide a single submission facility to query multiple 'primary' FR servers and retrieve their predictions. Additionally, meta-servers often display results of other predictions, including secondary structure or disordered regions. The most popular FR meta-servers include Bioinfo.pl MetaServer at http://bioinfo.pl/meta,[30] GeneSilico MetaServer at http://genesilico.pl/meta,[31] and Pcons.net at pcons.net/.[32]

4.2.2 Template Identification and Selection

Once the structural fold of the target protein is predicted using FR programs, the potentially best modeling template(s) must be selected. Selection of the optimal template is of fundamental importance for the quality of a comparative model because different structures with the same fold may differ from each other as much as they differ from the (unknown) structure of the protein being modeled. For instance, two similar templates, despite sharing the common structural core and topologies of secondary structures, can differ in relative orientation of the secondary structure elements, conformation of loops or even relative orientation of structural domains. Moreover, templates can contain structurally different elaborations of the common core that may or may not be present in the target. Thus, the objective is to select, from several alternatives, a template that is likely to be most structurally similar to the (unknown) structure of the target. Usually, if one template is much better than others, most FR servers select this template as the top match in their ranking list. Therefore, if there is a consensus template among FR predictions, it should be used as the main (or the only) template. However, if there is no clearly preferred template, a careful template selection must be performed. The available methods for selecting templates can be grouped into four categories: sequence-based, evolutionary, structural and knowledge-based.

Sequence-based methods rely on the assumption that the template with the highest sequence similarity to the target should also exhibit the highest structural similarity. In practice, the sequence of the target and potential templates are compared by building pairwise alignments or by running BLAST or PSI-BLAST[5] against the database that includes sequences of the templates. Matches to sequences of templates are ranked according to E-values. BLAST and PSI-BLAST are a method of choice for selecting templates, when target-template sequence identity is high (above 40% identity).[33,34] More sensitive sequence-based methods for template selection include comparison of profiles or HHMs built for sequence families of the target and all alternative templates (e.g. HHPRED[24]). These methods outperform BLAST programs in template selection when target-template sequence identity drops below 40%.[34]

Evolutionary methods rely on the assumption that the template that is closest to the target on the phylogenetic tree should exhibit the highest structural similarity. This approach requires calculation of a sequence alignment and a phylogenetic tree for a group of related protein sequences, including the target and all templates under considerations.[35] For closely related sequences, a phylogenetic inference using current evolutionary models and maximum likelihood or Bayesian methodologies is a much better estimator of evolutionary distances than similarity scores from pairwise sequence comparison. Therefore, constructing a phylogenetic tree might be useful especially when there are no significant differences in sequence similarities of the target to alternative templates. However, phylogenetic calculations are seldom reliable for distantly related sequences; therefore evolutionary methods are less universal and can be recommended only when sequence similarity between the target and alternative templates is high.

Structural methods perform an estimation of a 'fit' of the target sequence to each alternative template. The fit can be assessed based on the Z-scores from FR programs. However, more accurate template selection may be obtained after building several alternative models, each based on a different template, and scoring them by Model Quality Assessment Programs (MQAPs, see Chapter 6 by Wallner and Elofsson). Structural methods for template selection are recommended for cases of no significant sequence similarity between the target and the templates.

A knowledge-based approach constitutes a set of rules that should be taken into consideration during template selection, in particular to discriminate between structures of the same protein solved under different experimental conditions. The most important rule states that the structure of a template should be solved under similar conditions (in a similar environment) to the conditions desired for a model. For instance, for modeling of a target protein in a ligand-bound conformation, the template structure solved with a ligand should be used rather than a ligand-free structure. This rule should be followed especially when ligand-bound and free forms differ substantially in their conformations. Also, if the modeled protein in the biologically active form is an oligomer and different templates exhibit different quaternary structures, an oligomeric template with the same number of subunits and symmetry as the target should be used, and the model should be also built and evaluated as an oligomer, not as a monomer. Other rules include a preference for template structures solved by X-ray crystallography rather than NMR, structures with higher resolution and better R-factor, and structures with lower B-factor or without missing coordinates due to intrinsic disorder.

If it is not possible to unequivocally select a single best template from a set of alternatives, a model can be built based on multiple templates. This is accomplished either by averaging the coordinates of superposed templates or by modeling different regions of the target based on different templates. Selection of more than one template was shown to be very effective and accurate[36–39] and provided fundaments for several most successful protein structure prediction methodologies (e.g. FRankenstein's monster approach[37] or I-TASSER,[39] see Section 4.2.5, 'Model Refinement', below).

4.2.3 Target-Template Alignment

After selecting a template or several templates predicted to be optimal for comparative modeling of the target structure, a target-template sequence alignment must be generated to specify which residues of the target are to be modeled based on which residues of the template. A correct alignment is a prerequisite for successful modeling – in the case of misalignment residues are placed in wrong positions of the structure. Alignment shift of a single residue along the polypeptide chain corresponds to placement of this residue's C-alpha atom at ∼3.8Å away from its correct location – an error that cannot be automatically corrected by optimization procedures implemented in most model-building programs.

The FR analysis includes both calculation and evaluation of alignments; therefore all FR methods provide target-template alignments as an output. It is important to emphasize that these alignments are usually suboptimal and should not be used to construct a final model, but rather serve as a starting point of the refinement process aiming at localizing and correcting all potentially misaligned regions. In a traditional approach, the refinement process starts from discrimination between regions aligned with high confidence and all other regions. It is not straightforward to identify confident parts of the alignment. As a rule of thumb, one should take as confident regions where target-template matches returned by different FR servers are the same and, preferably, where target-template sequence similarity is high, e.g. common motifs are present.[40] Another rule of thumb is that alignments should generally preserve the localization of hydrophobic residues in the protein core while keeping charged residues exposed to the solvent. Besides, all insertions and deletions (indels) should be placed in solvent-exposed regions, preferably outside secondary structure elements, in loops that show variability between different homologous structures and/or conformational flexibility in the selected template(s). Indels also should not be introduced within a protein core or in predicted binding sites (e.g. in an active site or at site of protein–protein interactions if the protein is an oligomer). These rules rely on the observation that protein core, secondary structure elements and functional sites usually accumulate fewer changes than solvent exposed loops. Moreover, they ensure that the overall structure of the model will not be greatly distorted by the model building program.

During the optimization of target-template alignments, confident parts of the target-template alignment are kept as granted (at least initially) and regions with uncertain alignment are modified. Subsequently, new models are built for the modified alignments, and evaluated. Those variants of alignments that yield locally best evaluated structures are retained and regions with unsatisfactory scores may be rebuilt and evaluated again. This iterative target-template alignment optimization may be done manually, with the usage additional information derived from the literature (e.g. about localization of functional regions, mutagenesis studies etc.) and several experienced modelers proved that it can lead to very accurate predictions.[36,41,42]

Because refining the target-template alignment by hand is laborious and time-consuming and often requires subjective knowledge of an expert structural biologist, several methods have been developed to automate this process and make it more objective. K*SYNC program[43] generates a large ensemble of alternative alignments by varying the parameters of alignment algorithm (e.g. using various gap penalties or secondary structure predictions). HMM-KALIGN[44] generates so called sub-optimal alignments using generalized Viterbi algorithm for HMM alignment. Both methods produce large sets of alignments that then can be evaluated by various means. The final alignment is usually identified as the one that produces the best-scoring model. A number of model evaluation functions have been implemented in different programs that take into account the different accuracy of models.[37–39,45,46] The alternative models (or the corresponding alignments) may be also spliced and recombined to yield new model (or alignment) variants. A progressive automatic refinement procedure has been implemented e.g. in the MOULDER program,[47] which uses a genetic algorithm to recombine and mutate models built from preliminary alignments. Details of model recombination methods are covered in Section 4.2.5, 'Model Refinement'.

4.2.4   Model Building

In comparative modeling, model building is a procedure of creating a three-dimensional structural model based on a template structure and a target-template alignment. The target-template alignment is treated as a specification which residues of the target should be modeled based on particular residues of the template. Model building based on an alignment is usually fully automated. Model building methods can be grouped into three classes: rigid-body assembly, segment matching, or coordinate reconstruction and modeling by satisfaction of spatial restraints[48–50](see Table 4.2 for detailed description of individual model building programs).

Rigid body assembly methods build a structural model from fragments of templates aligned with the target sequence, supplemented with additional fragments derived from a structural database for regions where the target is not aligned to any template. There are various implementations of this procedure including 3D-JIGSAW,[51] BUILDER,[52,53] COMPOSER[54] and SWISS-MODEL[55] but the general procedure is similar. These methods first create a framework by selecting or averaging coordinates of superposed templates, then select best local segments from templates and superpose them onto the framework. Second, loops and insertions are added as fragments derived from a database of known protein structures. Finally, the fully assembled model is energy-minimized to optimize connections between the fragments and to reduce their potential steric conflicts. The

Table 4.2 Summary of publicly available programs for comparative modeling

Method	Description
	Rigid body assembly programs
SWISS-MODEL[55] swissmodel.expasy.org/SWISS-MODEL.html	Model is constructed by: (i) building the framework structure (Cα coordinates averaged over all templates); (ii) adding loops (de novo or from a database) and sidechains (iso-sterically with sidechains in templates and using rotamer libraries); (iii) reconstructing complete backbone and energy minimization of the whole structure
3D-JIGSAW[51] www.bmm.icnet.uk/servers/3djigsaw/	Model is constructed from SS fragments derived from templates and, for insertions, from fragments derived from a protein structure database. Side-chains are built using rotamers from templates or from rotamer libraries. The full structure is built using a self-consistent mean field method[53] followed by energy minimization
NEST[57] wiki.c2b2.columbia.edu/honiglab_public/index.php/Software:nest	Applies artificial evolution algorithm – changes the template into the model of the target by applying stepwise mutations followed by energy minimization
	Segment matching programs
SEGMOD/ENCAD[61] csb.stanford.edu/levitt/segmod/	Builds a framework of Cα coordinates based on the template and searches a structural database for segments matching the framework. The final model is an energy minimized mean model obtained from averaging alternative models generated in the segment matching step
	Modeling by satisfaction of spatial restraints programs
MODELLER[59] www.salilab.org/modeller/	Based on the target-template alignment derives structural restraints from the templates and builds a model that satisfies these restraints by minimizing a function representing these restraints and stereochemical constraints. The program supports also de novo loops modeling, optimization of various models and incorporation of secondary structure predictions and user-defined restraints.

above-mentioned programs differ in the way of selecting segments from the templates and structural database and in side-chain building and model refinement methodology. Another program usually classified together with rigid body assembly methods[48,56] is NEST,[57] whose special feature is the implementation of artificial evolution in its modeling scheme.

Modeling by satisfaction of spatial restraints is a process of constructing the structural model that optimally satisfies restraints derived from the template structure. This approach is based on an assumption that conformations of homologous residues in both structures are similar. The modeling methodology is based on procedures used for structure determination using restraints derived from NMR measurements.[58] The most widely used modeling program in this class is MODELLER.[59] In this method, various restraints are derived from the template and mapped onto the corresponding residues of the target based on a user-defined target-template alignment. Restraints include bond lengths and angles, planarity of peptide groups and side-chain rings, chiralities, van der Waals contact distances, bond angles, dihedral angles etc. and are encoded in the form of conditional probability density functions (PDFs), which also take into account correlations obtained from a database of superpositions of known structures and a physical force field. The model is then obtained by minimization of target function combining the PDFs, using methods of conjugate gradients and molecular dynamics with simulated annealing. Users may optionally add their own restraints, such as secondary structure elements or distances between particular residues, or to include ligands.

Segment matching comprises three steps: (1) generation of a guiding structure of the target based on the sequence alignment to the template; (2) identification in a database of known structures short segments that match the guiding structure (based on energetic and geometric criteria); and (3) assembly of the full atom model from these segments using atoms of the guiding structure as anchor points. The method is based on the approach used in X-ray crystallography where segment derived from known structures are used for building a polypeptide chain based on electron density.[60] The segment matching method has been implemented for comparative modeling in program SEGMOD[61] that, despite its age (not substantially further developed since 1992), still exhibits performance comparable to state-of-the-art programs.[48,56]

According to benchmarks,[48,56] analyzing the performance of most commonly used comparative modeling programs (including MODELLER, SEGMOD/ENCAD, SWISS-MODEL, 3D-JIGSAW, NEST and BUILDER), there is no program that clearly outperforms the others. Nevertheless, MODELLER, NEST, and SEGMOD/ENCAD perform better than the others and MODELLER performs best on the average. Taking advantage of the fact that all these programs are fast and generate models in a time up to a couple of minutes, it is recommended to use several programs and either select the best model or build the final model by splicing best parts of models generated by different methods.[37,38] In the experience of the authors of this chapter, MODELLER seems to be best suited for generating models for protein function prediction, as it offers a possibility for including ligands and user-defined restraints.

4.2.5   Model Refinement

Comparative modeling, albeit very successful, has severe limitation: it rarely leads to models that are closer to the native structure than the best single template structure.[62]

Therefore, a considerable effort is undertaken to develop refinement methods that could decrease a deviation of comparative models from the native structures.

Methods for model refinement either try to improve the overall quality of the model or focus on particular parts of structure. Refinement of loops and side-chains is discussed in separate sections of this chapter (Sections 4.2.6 ('Loop Modeling') and 4.2.7 ('Side-chain Modeling'). Methods for refinement of full models aim at generating a model close to the native structure, starting from comparative models or *de novo* models. They can be grouped into two main classes, depending on whether they use physical or knowledge-based statistical potentials. Methods that use physical potentials usually include refinement with molecular dynamics or energy minimization. In the methodology developed by Chen and Brooks,[63] replica exchange molecular dynamics with Generalized Born implicit solvent model is performed, starting from the comparative model. Constraints are introduced to restrict conformational sampling during simulation, under the assumption that the starting model is close to the native structure. A method by Summa and Levitt[64] performs energy minimization in vacuum of models close to native structure by using physical or knowledge-based potentials. According to these authors, knowledge-based potentials perform better than the physical ones. Knowledge-based potentials are also used by the ROSETTA program[65] that can perform full-atom refinement with a Monte Carlo search procedure.[66]

A specific type of model refinement consists of selection of fragments from multiple templates. It is based on the notion that template coordinates optimal for modeling different regions of the target are often present in different protein structures of the same fold. Therefore, modeling on these optimal fragments selected from alternative templates should lead to a model that is closer to the native structure than any model based on a single template. In the FRankenstein's monster approach[36,37] a set of preliminary models (based on alignments from various FR servers) is evaluated using model quality assessment programs and a hybrid model is constructed by merging consensus fragments (i.e. fragments found in most of the models) and best scoring non-consensus fragments. This hybrid model is then superimposed onto the templates, from the superposition a new target-template alignment is derived, and used to construct a new model. If the new model contains poorly scoring fragments, this procedure can be further iterated (e.g. by building additional generations of models based on alternative alignments). In the I-TASSER program,[39] a set of preliminary models from FR servers is used as a source of fragments, from which the model is assembled in the course of a Monte Carlo simulation. The assembly procedure is done in an iterative way. In the first iteration, the ensemble of models is generated from aligned fragments, while unaligned regions are modeled *de novo*. Then, the representative structures are selected from the ensemble by means of clustering, the cluster centroids undergo second round of fragment re-assembly/refinement and the lowest energy structures are chosen as final models.

4.2.6   Loop Modeling

One of the difficulties that a modeler may encounter is the reconstruction of the three-dimensional structure in regions of low or no similarity to the template(s). These evolutionarily variable regions often encompass insertions, deletions or unstructured regions that may play a crucial role for biological function of a protein (i.e. constitute a binding site).

Therefore, not only the overall protein structure has to be modeled reliably for accurate function prediction, but also its variable regions such as loops.

There are two main approaches to loop modeling: (1) database search methods and (2) *de novo* methods. There are also hybrid methods that combine both of them. State-of-the-art programs for loop modeling are listed in Table 4.3.

Database search methods build loops based on the conformations of loops derived from a database of known structures. Usually, a database of loops is created first and then, during modeling loops of the target, it is searched for conformations that fit best to the loops being modeled. This has been implemented in model building programs BUILDER[52,53] and SEGMOD[61] (see Section 4.2.4, 'Model Building'). The ARCHPRED server,[67] which is devoted specifically to loop modeling, searches a pre-existing loops library for loops that are similar in sequence to the query loop. These candidate loops are attached to stem regions of the model as to minimize steric clashes with the rest of the structure and maximize geometry matching propensities for secondary structure and main chain dihedral angles at the point of loop attachment. The prerequisite for successful loop modeling using database search is a sufficient coverage of conformations of loops in a database, which can be fulfilled only for loops of length up to about 10 residues.[67] The number of possible conformations grows exponentially with the loop length.[68] and becomes astronomical for loops longer than 10 residues. For shorter loops that are well represented in the database, the limitation of database search approaches is however in a scoring function.

De novo **methods** aim to predict conformations of loops using various conformational search methods using physical potentials, often in combination with knowledge-based potentials (the fundamentals of *de novo* modeling of entire proteins is covered in another Chapter 5 in this book by Gront *et al.*). Most *de novo* methods for loop modeling first generate a large number of alternative initial loop conformations randomly (e.g. MODLOOP[69] or LOOPY[70] or by exhaustive systematical sampling (e.g. PLOP[71]). These conformations are then optimized in physical or pseudo-physical potentials and the lowest energy conformations or representatives of largest clusters of similar conformations are selected and combined with the rest of the model. In theory, there is no limitation on the length of the loop to be modeled by *de novo* methods. However, with the increasing loop length, the number of possible conformations increases rapidly, making the modeling computationally prohibitive.

Hybrid methods that combine database search and *de novo* approaches typically use separate algorithms depending on the situation (e.g. database search for longer loops and *de novo* method for short ones (e.g. SWISS-MODEL[55])) or generate two populations of conformations from database searches and *de novo* modeling (e.g. CODA[72]). One of the most successful approaches for structure prediction, ROSETTA,[65] first builds loops based on fragments derived from database of fragments derived from known structures and then performs a conformational search and Monte Carlo minimization.

4.2.7 Side-Chain Modeling

For a structural model to be useful for functional analyses such as predicting a catalytic site or ligand binding residues, it is important to accurately predict conformation of individual side-chains. Side-chains can be built by most comparative modeling programs at the same time when the backbone is modeled or they can be rebuilt or even added anew to the naked backbone of the final model. Comparative modeling programs use information about torsion

Table 4.3 *Summary of publicly available programs for loop modeling*

Database search approaches

ARCHPRED[67]
www.fiserlab.org/servers/archpred

Performs fragment-search and selects loop fragments by matching the length, the types of bracing SS elements and by satisfying the geometrical restraints imposed by the stem residues

De novo methods

LOOPY[70]
wiki.c2b2.columbia.edu/honiglab_public/index.php/Software:Loopy

Builds multiple random initial conformations using an ab-initio method, makes fast energy minimization and uses colony energy to sort out the best predictions

PLOP[71]
francisco.compbio.ucsf.edu/~jacobson/plop_manual/plop_overview.htm

Performs exhaustive conformational sampling followed by optimization and clustering of solutions and energy minimization and iterative refinement of loop subsections

MODLOOP[69]
salilab.org/modloop

Builds multiple random initial conformations and optimizes them by energy minimization combined with molecular dynamics. Selects best loops according to pseudo-energy

Hybrid methods

CODA[72]
www-cryst.bioc.cam.ac.uk/coda/

Generates alternative conformations using two methods: knowledge-based and ab initio, clusters all conformations and select best clusters using knowledge-based rules

ROSETTA LOOPS[103]
depts.washington.edu/ventures/UW_Technology/Express_Licenses/Rosetta/

Combines assembly of database-derived fragments and Monte Carlo minimization, using knowledge-based energy function.

angles of side-chains of the template for modeling conserved residues and/or utilize rotamer libraries for modeling other residues. The majority of programs designed specifically for side-chains prediction are based on rotamer libraries that contain information about torsion angles derived from statistical analysis of known structures. Widely used are backbone-dependent rotamer libraries that contain information on side-chains conformations as a function of backbone dihedral angels. Various methods have been developed to explore the spectrum of conformation in an efficient way, from simple combinatorial searches using rotamer libraries (e.g. SCWRL[73]), to molecular dynamics optimization of entire models[74,75] (see Table 4.4 for detailed description of some of these methods).

The most successful programs for side-chain modeling are those based on rotamer libraries.[48,49,76] Nevertheless, when the target-template sequence identity is high (>50%) the best approach is to use comparative modeling programs to model side-chains mainly based on the information from templates (i.e. to retain the conformation of conserved residues as close to that in the template as possible). Only when the sequence identity drops down, it is advisable to use programs that try conformations from rotamer libraries for non-conserved residues or, in the case of very low sequence identities (below 30%) – for all residues.[48] In general, it is advisable to keep the conformation of conserved residues as close as possible to the conformation in the template(s) and assign the conformation of the remaining residues using backbone-dependent rotamer libraries.

4.2.8 Model Quality Assessment

Model quality assessment (MQA) constitutes an inseparable part of all structure prediction methods. The goal of model quality assessment programs (MQAPs) is to assess the overall accuracy of the final model and/or a local accuracy of individual fragments of the model. A variety of MQAPs exist that implement a wide spectrum of methodologies from physical potentials to knowledge-based scoring functions. A detailed survey of MQAPs is covered in Chapter 6 in this book by Wallner and Elofsson.

Measures of local structural accuracy are of the particular importance, as they allow for identifying regions that score poorly and should be remodeled. Remodeling can be accomplished either by going back to the stage of target-template alignment generation and trying different versions of target-template alignments, or by refining the poorly scoring region (with e.g. *de novo* refinement methods). If the poorly scored region cannot be improved by any means, it may be taken as indication that this part of the final model is unreliable and should be taken with caution during functional interpretation of the structure.

4.3 Accuracy of Comparative Models

4.3.1 Model Accuracy

The resolution and local accuracy of structures determined by X-ray crystallography or NMR can be estimated by analyzing parameters that are not optimized in the process of model building, but are correlated with the model quality (e.g. Φ and Ψ angles on the Ramachandran plot). Alternatively, a part of the data collected during the structure determination experiment may be set aside and not used for model building, but only later, for model verification. On the other hand, the comparative modeling procedure does not

Table 4.4 *Summary of publicly available programs for side-chain modeling*

SCWRL[73] dunbrack.fccc.edu/SCWRL3.php	Represents side-chains as vertices in a graph and utilizes tools from graph theory to find the set of conformations corresponding to the global minimum using a new Dunbrack backbone-dependent rotamer library[73]
SCCOMP[104] www.weizmann.ac.il/cgi-bin/sgedg/sccomp1.cgi	Performs either iterative modeling of side-chains starting from buried ones or stochastic modeling where side-chains are selected at random with buried residues being selected more often. Uses an optimized scoring function and a modified Dunbrack rotamer library[105]
SCAP[106] wiki.c2b2.columbia.edu/honiglab_public/index.php/Software:Scap	Selects the lowest energy set of rotamers from an ensemble of side-chain conformations generated by energy minimization using different starting conditions and by selecting rotamers from a detailed rotamer library (represented in Cartesian coordinates and observed rather than idealized bond lengths and angles)
SMOL[107] Contact Nick V. Grishin: Nick.Grishin@UTSouthwestern.edu	Performs Monte Carlo search starting from random placement of all side-chains followed by rotamer substitutions using the Dunbrack rotamer library[105] and optimized scoring function

rely on experimental measurements, but on copying information from the templates (e.g. the Ramachandran plot of the model reflects the quality of the template rather the accuracy of the modeling procedure) and usually there is no external data to test the model validity.

The overall accuracy of the model can be roughly estimated based on the notion that models built based on the template with sequence identity to the target above certain threshold are very close to the native structure (exhibit high accuracy), while models based on templates with sequence identity below that threshold become less similar to the native structures (and their accuracy is usually lower). Currently, models based on templates with at least 40% sequence identity are expected to be highly accurate (approaching the quality of NMR models or low resolution X-ray structures) and models from the zone below 40% are expected to exhibit decreasingly low accuracy[49]. Global and local accuracy of the model can be also assessed by statistical methods (MQAPs) that compare the features of the model to those generally observed in protein structures. Currently, some MQAPs (e.g. MetaMQAP[76a] and ProQres[77]) make it possible to predict deviation of individual residues from their counterparts in the (unknown) native structures, and these values can be used to calculate the global accuracy of the model (e.g. the predicted root mean square deviation from the native structure).

4.3.2   Errors in Comparative Models

It must be emphasized that comparative models almost always contain errors. The type and magnitude of these errors are different from those seen sometimes in crystallographic or NMR structures[78] and usually depend on the modeling software and the assumptions of the modeling process. Obviously, these errors may or may not affect the applicability of the model to further biological investigations, depending on the research question being asked. Thus, while numerical assessments of model accuracy is essential for understanding the 'resolution' of the model, the awareness of the type and magnitude of errors that most commonly appear in computational models is very important for proper functional interpretation of the modeled structures.

Wrong fold. The most severe error in comparative modeling may be matching the target sequence to a wrong fold. The fold assignment depends on the template selected for comparative modeling during the FR step. If the template has a different fold than the target, the resulting model will be definitely erroneous, because the fold cannot be changed during the model optimization step. In a model with a wrong fold, all conserved functional residues will almost certainly have biologically irrelevant arrangement in space, thus precluding any functional inference from the model. The uncertainty in the fold assignment may be inferred from the lack of consensus among FR servers. Thus, all comparative models based on uncertain FR results should be taken with extreme caution.

Misalignments. Misalignments to a template with a correct fold lead to the second most severe type of errors in comparative modeling, namely misthreading of residues along the protein backbone. Thus, target-template alignment used for model building is, after template selection, the second most important factor influencing the model correctness. In the case of a misalignment, residues are placed in wrong positions of the structure, at a minimum of 3.8 Å (the distance between consecutive residues) from their correct localization. This may lead to misplacement of functionally important residues along the backbone and disruption of residue clusters that form functional sites. Misalignments are common at the low target-template identity (below 30–40%)[49] and are often difficult to detect. Some misalignments in structural elements located on the protein surface lead to the placement of hydrophilic residues in a hydrophobic environment and vice versa, and can be identified by MQAPs, but others (e.g. small shifts in buried β-strands composed entirely of hydrophobic residues) are unlikely to cause structural disturbance. One solution to overcome the problem with misalignments is to consider many alternative alignments during the functional analysis, for instance by generation and detailed inspection of corresponding alternative models for possible problems in packing in protein core or orientation of surface-exposed side chains that may be important for protein function.

Displacements of local structures. Another class of common errors in comparative models results from reorientation of secondary structure elements (e.g. minor rotations, translations, and/or bending), a trend in protein evolution that is not uncommon among distant homologs. However, this can happen also for elements with conserved sequence, even when the local residue identity is above 50%.[49] In such cases, correct target-template alignments are usually obtained, but the assumption that sequence-conserved regions do not change their structure, is violated. A displacement of a structural element can significantly change the position of its residues with respect to other parts of the protein. Such errors may be identified by MQAPs, but it is typically impossible to ameliorate them by changing the alignment. Thus, an inability to obtain a reasonable MQAP score for a given segment despite trying many alternative alignments may indicate that structures of the target and the template are locally different. One way to deal with such structural changes is an additional refinement of the model, in particular within the suspected region (see Section 4.2.5, 'Model Refinement', above).

Misfolded loops and insertions. Loops and insertions with no counterpart in the template are the major source of errors in all comparative models. Insertions may have completely wrong conformation or despite having correct conformation they may be positioned in a wrong orientation relatively to the rest of the structure. The modeling of insertions may fail due to limitations of loop modeling programs and/or due to the insufficient accuracy of regions that flank the loop in the model. Generally, the conformation of long loops in comparative models should be regarded as unreliable during the functional analysis, unless there are particular reasons to believe that they have been modeled correctly. This is especially the case if the loop belongs to the functional site and its structure is influenced by the ligand. For instance, modeling of a loop in the absence of a ligand may place the loop in a conformation that may be incompatible with ligand binding. Errors in loops might appear regardless the target-template identity, but in general their frequency and magnitude increases with decreasing overall target-template sequence similarity.

Wrong side-chain conformations. Errors in side-chain conformations are very common in comparative models. It is generally difficult to predict side-chain conformations correctly, as small inaccuracies in the conformation of the protein backbone greatly affects the distribution of allowed side-chain rotamers. When the target-template sequence identity is high (above 50%) the conformations of side-chains conserved in both target and template tend to be similar.[48] At lower target-template identities, the preservation of side-chain conformations is restricted to conserved functional sites and, only to some extent, the protein core. Therefore, the orientation of side-chains should be considered in functional investigations only if the target-template sequence identity in a given region is high or if there are other constraints on side-chain conservation (e.g. binding of the same ligand).

4.4   Protein Structure Modeling and Function Prediction

The model of protein structure is built usually to help in addressing specific biological questions. These questions usually do not only aim at guessing 'what is the structure of my protein' but also 'what is the function of my protein' and 'how my protein performs its function'. Ultimately, the amount of information that can be derived from a model depends on the accuracy of this model.

The assessments of model accuracy, although very approximate, roughly correlate with a number of expected errors in the model. Therefore, they can be used as convenient guidelines for determining what kind of information can be derived from the model.

4.4.1   High Accuracy Models

The accuracy of answers the model can provide decreases together with target-template sequence similarity. Above 50% sequence identity it is generally feasible to obtain a structurally accurate and functionally interpretable model, as the fold is usually correct, alignment errors are rare, backbone conformation is typically close to the native one, and conformations of many side-chains are modeled correctly.[35,49] In such case, the functional characterization of the target can be carried out by comparing it to the template, if the latter is already functionally characterized. For instance, residues belonging to the catalytic or ligand binding site can be predicted based on their correspondence to residues of the template that have been already attributed to this activity. On the other hand, in the structural genomics era, hundreds of structures are solved without any functional annotations and without any ligands, thus many potential templates offer little functional information. Nevertheless, these structures and models of closely related proteins built using 'functionally uncharacterized' structures as templates, can serve as good platforms for detailed bioinformatic function predictions (see Chapter 7 by Kinoshita *et al.* in this volume).

4.4.2   Low Accuracy Models

When the target-template sequence similarity is low (<40%), the model usually contains regions of low accuracy that may or may not have been identified in the process of model quality assessment. Functional analysis of such models becomes more difficult because functional regions might be built up by e.g. misplaced secondary structure elements with wrong side-chains conformations. Fortunately, functional sites accumulate mutations, and therefore structural changes, slower than the rest of the protein, thus if the target and template bear similar catalytic activity, the catalytic sites often can be modeled with high accuracy even for targets with 10–20% sequence identity to the template. A good example can be found among nucleases with the PD-(D/E)XK fold where catalytic residues can be usually predicted and modeled with high accuracy even if target-template sequence similarity is very low (even below 10% sequence identity).[79,80] Importantly, although these catalytic residues are correctly placed on the protein backbone, the exact conformation of the side-chains is often not accurately predicted (precluding for instance an inference of the exact geometry of the active site).

In models of low accuracy (e.g. those based on templates of less than 40% sequence identity) it is also often possible to roughly predict functional regions such as DNA-binding sites[79] or protein–protein interaction sites. For instance, for predicting a DNA binding site it might be sufficient to map sequence conservation and electrostatic potential onto the surface of the protein model. The presence of conserved and positively-charged surface patches, in particular comprising deep clefts, can suggest nucleic acid binding sites. Based on this prediction it is possible to rationally design a limited set of experiments, with which to verify the hypothesis about the DNA binding site (for instance by site-directed mutagenesis of putative binding residues).

4.4.3 Summary

Summarizing, the accuracy of the answer the model can give depends on the accuracy of the model. In particular, models that are of expected low accuracy should not be used to address high-resolution questions, e.g. about detailed conformation of residues in the active site. Therefore, a biologist interested in using a model should carefully assess its potential accuracy, both globally and locally, and adjust the questions to the estimated accuracy of the model and its different parts. If the predicted accuracy of the model appears sufficient to address the desired question, numerous bioinformatics tools for structure analysis and function prediction (reviewed in other chapters of this book) may be used to formulate new hypotheses and guide experimental studies of the protein and the biological system it is involved in.

Acknowledgements

We thank present and former members of the Bujnicki lab in IIMCB and at the UAM for stimulating discussions and contribution of ideas and information to this article. The authors acknowledge the support from past and current grants for the development of bioinformatics methods from Polish Ministry of Science, Framework Programme of the EU, EMBO, and HHMI. JK and KLT have worked on this article while being supported by scholarships from the Foundation for Polish Science.

References

1. J.D. Bernal, and D. Crowfoot, X-ray photographs of crystalline pepsin, *Nature*, **133**, 794–795 (1934).
2. J.C. Kendrew, G. Bodo, H.M. Dintzis, R.G. Parrish, H. Wyckoff, and D.C. Phillips, A three-dimensional model of the myoglobin molecule obtained by x-ray analysis, *Nature*, **181**, 662–666 (1958).
3. C.B. Anfinsen, E. Haber, M. Sela, and F.H. White, Jr., The kinetics of formation of native ribonuclease during oxidation of the reduced polypeptide chain, *Proc Natl Acad Sci U S A*, **47**, 1309–1314 (1961).

4. C. Chothia, and A.M. Lesk, The relation between the divergence of sequence and structure in proteins, *EMBO J.*, **5**, 823-826 (1986).

5. S.F. Altschul, T.L. Madden, A.A. Schaffer, *et al.*, Gapped BLAST and PSI-BLAST: a new generation of protein database search programs, *Nucleic Acids Res*, **25**, 3389–3402 (1997).

6. S.E. Brenner, C. Chothia, and T.J. Hubbard, Assessing sequence comparison methods with reliable structurally identified distant evolutionary relationships, *Proc Natl Acad Sci USA*, **95**, 6073–6078 (1998).

7. Z. Chen, Assessing sequence comparison methods with the average precision criterion, *Bioinformatics*, **19**, 2456–2460 (2003).

8. A.J. Reid, C. Yeats, and C.A. Orengo, Methods of remote homology detection can be combined to increase coverage by 10% in the midnight zone, *Bioinformatics*, **23**, 2353–2360 (2007).

9. B. John, and A. Sali, Detection of homologous proteins by an intermediate sequence search, *Protein Sci*, **13**, 54–62 (2004).

10. L.N. Kinch, K. Ginalski, L. Rychlewski, and N.V. Grishin, Identification of novel restriction endonuclease-like fold families among hypothetical proteins, *Nucleic Acids Res*, **33**, 3598–3605 (2005).

11. J. Pei, and N.V. Grishin, The P5 protein from bacteriophage phi-6 is a distant homolog of lytic transglycosylases, *Protein Sci*, **14**, 1370–1374 (2005).

12. W. Li, F. Pio, K. Pawlowski, and A. Godzik, Saturated BLAST: an automated multiple intermediate sequence search used to detect distant homology, *Bioinformatics*, **16**, 1105–1110 (2000).

13. T. Ohlson, B. Wallner, and A. Elofsson, Profile-profile methods provide improved fold-recognition: a study of different profile-profile alignment methods, *Proteins*, **57**, 188–197 (2004).

14. J. Soding, Protein homology detection by HMM-HMM comparison, *Bioinformatics*, **21**, 951–960 (2005).

15. L. Jaroszewski, L. Rychlewski, Z. Li, W. Li, and A. Godzik, FFAS03: a server for profile–profile sequence alignments, *Nucleic Acids Res*, **33**, W284–288 (2005).

16. K. Tomii, and Y. Akiyama, FORTE: a profile-profile comparison tool for protein fold recognition, *Bioinformatics*, **20**, 594–595 (2004).

17. S.R. Eddy, Profile hidden Markov models, *Bioinformatics*, **14**, 755–763 (1998).

18. S. Liu, C. Zhang, S. Liang, and Y. Zhou, Fold recognition by concurrent use of solvent accessibility and residue depth, *Proteins*, **68**, 636–645 (2007).

19. L.J. McGuffin, and D.T. Jones, Improvement of the GenTHREADER method for genomic fold recognition, *Bioinformatics*, **19**, 874–881 (2003).

20. D.T. Jones, W.R. Taylor, and J.M. Thornton, A new approach to protein fold recognition, *Nature*, **358**, 86–89 (1992).

21. M.J. Sippl, and S. Weitckus, Detection of native-like models for amino acid sequences of unknown three-dimensional structure in a data base of known protein conformations, *Proteins*, **13**, 258–271 (1992).

22. S.H. Bryant, and C.E. Lawrence, An empirical energy function for threading protein sequence through the folding motif, *Proteins*, **16**, 92–112 (1993).

23. R.T. Miller, D.T. Jones, and J.M. Thornton, Protein fold recognition by sequence threading: tools and assessment techniques, *Faseb J*, **10**, 171–178 (1996).

24. J. Soding, A. Biegert, and A.N. Lupas, The HHpred interactive server for protein homology detection and structure prediction, *Nucleic Acids Res*, **33**, W244–248 (2005).

25. R.M. Bennett-Lovsey, A.D. Herbert, M.J. Sternberg, and L.A. Kelley, Exploring the extremes of sequence/structure space with ensemble fold recognition in the program Phyre, *Proteins* (2007).

26. D. Fischer, Hybrid fold recognition: combining sequence derived properties with evolutionary information, *Pacific Symp. Biocomp.*, 119–130 (2000).

27. K. Karplus, S. Katzman, G. Shackleford, *et al.*, SAM-T04: what is new in protein-structure prediction for CASP6, *Proteins*, **61 Suppl 7**, 135–142 (2005).

28. J. Shi, T.L. Blundell, and K. Mizuguchi, FUGUE: sequence-structure homology recognition using environment-specific substitution tables and structure-dependent gap penalties, *J Mol Biol*, **310**, 243–257 (2001).

29. L. Rychlewski, and D. Fischer, LiveBench-8: the large-scale, continuous assessment of automated protein structure prediction, *Protein Sci*, **14**, 240–245 (2005).
30. J.M. Bujnicki, A. Elofsson, D. Fischer, and L. Rychlewski, Structure prediction meta server, *Bioinformatics*, **17**, 750–751 (2001).
31. M.A. Kurowski, and J.M. Bujnicki, GeneSilico protein structure prediction meta-server, *Nucleic Acids Res*, **31**, 3305–3307 (2003).
32. B. Wallner, P. Larsson, and A. Elofsson, Pcons.net: protein structure prediction meta server, *Nucleic Acids Res*, **35**, W369–374 (2007).
33. B. Contreras-Moreira, P.W. Fitzjohn, and P.A. Bates, In silico protein recombination: enhancing template and sequence alignment selection for comparative protein modelling, *J Mol Biol*, **328**, 593–608 (2003).
34. M.I. Sadowski, and D.T. Jones, Benchmarking template selection and model quality assessment for high-resolution comparative modeling, *Proteins*, **69**, 476–485 (2007).
35. M.A. Marti-Renom, A.C. Stuart, A. Fiser, R. Sanchez, F. Melo, and A. Sali, Comparative protein structure modeling of genes and genomes, *Annu Rev Biophys Biomol Struct*, **29**, 291–325 (2000).
36. J. Kosinski, I.A. Cymerman, M. Feder, M.A. Kurowski, J.M. Sasin, and J.M. Bujnicki, A 'FRankenstein's monster' approach to comparative modeling: merging the finest fragments of Fold-Recognition models and iterative model refinement aided by 3D structure evaluation, *Proteins*, **53 Suppl 6**, 369–379 (2003).
37. J. Kosinski, M.J. Gajda, I.A. Cymerman, *et al.*, FRankenstein becomes a cyborg: the automatic recombination and realignment of fold recognition models in CASP6, *Proteins*, **61 Suppl 7**, 106–113 (2005).
38. A. Kolinski, and J.M. Bujnicki, Generalized protein structure prediction based on combination of fold-recognition with de novo folding and evaluation of models, *Proteins*, **61 Suppl 7**, 84–90 (2005).
39. Y. Zhang, Template-based modeling and free modeling by I-TASSER in CASP7, *Proteins*, **69**, 108–117 (2007).
40. M.L. Tress, D. Jones, and A. Valencia, Predicting reliable regions in protein alignments from sequence profiles, *J Mol Biol*, **330**, 705–718 (2003).
41. K. Ginalski, and L. Rychlewski, Protein structure prediction of CASP5 comparative modeling and fold recognition targets using consensus alignment approach and 3D assessment, *Proteins*, **53 Suppl 6**, 410–417 (2003).
42. C. Venclovas, and M. Margelevicius, Comparative modeling in CASP6 using consensus approach to template selection, sequence-structure alignment, and structure assessment, *Proteins*, **61 Suppl 7**, 99–105 (2005).
43. D. Chivian, and D. Baker, Homology modeling using parametric alignment ensemble generation with consensus and energy-based model selection, *Nucleic Acids Res*, **34**, e112 (2006).
44. E. Becker, A. Cotillard, V. Meyer, H. Madaoui, and R. Guerois, HMM-Kalign: a tool for generating sub-optimal HMM alignments, *Bioinformatics* (2007).
45. D. Chivian, D.E. Kim, L. Malmstrom, *et al.*, Automated prediction of CASP-5 structures using the Robetta server, *Proteins*, **53 Suppl 6**, 524–533 (2003).
46. D.E. Kim, D. Chivian, and D. Baker, Protein structure prediction and analysis using the Robetta server, *Nucleic Acids Res*, **32**, W526–531 (2004).
47. B. John, and A. Sali, Comparative protein structure modeling by iterative alignment, model building and model assessment, *Nucleic Acids Res*, **31**, 3982–3992 (2003).
48. B. Wallner, and A. Elofsson, All are not equal: a benchmark of different homology modeling programs, *Protein Sci*, **14**, 1315–1327 (2005).
49. M.S. Madhusudhan, M.A. Marti-Renom, N. Eswar, *et al.*, Comparative protein structure modeling, *in The Proteomics Protocols Handbook*, J.M. Walker (ed.), Humana Press, Totowa, N.J., 2005.
50. A. Fiser, Protein structure modeling in the proteomics era, *Expert Rev Proteomics*, **1**, 97–110 (2004).
51. P.A. Bates, L.A. Kelley, R.M. MacCallum, and M.J. Sternberg, Enhancement of protein modeling by human intervention in applying the automatic programs 3D-JIGSAW and 3D-PSSM, *Proteins*, **Suppl 5**, 39–46 (2001).

52. P. Koehl, and M. Delarue, Application of a self-consistent mean field theory to predict protein side-chains conformation and estimate their conformational entropy, *J.Mol.Biol.*, **239**, 249–275 (1994).
53. P. Koehl, and M. Delarue, A self consistent mean field approach to simultaneous gap closure and side-chain positioning in homology modelling, *Nat.Struct.Biol.*, **2**, 163–170 (1995).
54. M.J. Sutcliffe, I. Haneef, D. Carney, and T.L. Blundell, Knowledge based modelling of homologous proteins, Part I: Three-dimensional frameworks derived from the simultaneous superposition of multiple structures, *Protein Eng*, **1**, 377–384 (1987).
55. T. Schwede, J. Kopp, N. Guex, and M.C. Peitsch, SWISS-MODEL: An automated protein homology-modeling server, *Nucleic Acids Res*, **31**, 3381–3385 (2003).
56. J.A. Dalton, and R.M. Jackson, An evaluation of automated homology modelling methods at low target template sequence similarity, *Bioinformatics*, **23**, 1901–1908 (2007).
57. D. Petrey, Z. Xiang, C.L. Tang, *et al.*, Using multiple structure alignments, fast model building, and energetic analysis in fold recognition and homology modeling, *Proteins*, **53 Suppl 6**, 430–435 (2003).
58. T.F. Havel, and M.E. Snow, A new method for building protein conformations from sequence alignments with homologues of known structure, *J Mol Biol*, **217**, 1–7 (1991).
59. A. Sali, and T.L. Blundell, Comparative protein modelling by satisfaction of spatial restraints, *J Mol Biol*, **234**, 779–815 (1993).
60. T.A. Jones, and S. Thirup, Using known substructures in protein model building and crystallography, *Embo J*, **5**, 819–822 (1986).
61. M. Levitt, Accurate modeling of protein conformation by automatic segment matching, *J Mol Biol*, **226**, 507–533 (1992).
62. M. Tress, I. Ezkurdia, O. Grana, G. Lopez, and A. Valencia, Assessment of predictions submitted for the CASP6 comparative modeling category, *Proteins*, **61 Suppl 7**, 27–45 (2005).
63. J. Chen, and C.L. Brooks, 3rd, Can molecular dynamics simulations provide high-resolution refinement of protein structure?, *Proteins*, **67**, 922–930 (2007).
64. C.M. Summa, and M. Levitt, Near-native structure refinement using in vacuo energy minimization, *Proc Natl Acad Sci U S A*, **104**, 3177–3182 (2007).
65. C.A. Rohl, C.E. Strauss, K.M. Misura, and D. Baker, Protein structure prediction using Rosetta, *Methods Enzymol*, **383**, 66–93 (2004).
66. P. Bradley, K.M. Misura, and D. Baker, Toward high-resolution de novo structure prediction for small proteins, *Science*, **309**, 1868–1871 (2005).
67. N. Fernandez-Fuentes, J. Zhai, and A. Fiser, ArchPRED: a template based loop structure prediction server, *Nucleic Acids Res*, **34**, W173–176 (2006).
68. K. Fidelis, P.S. Stern, D. Bacon, and J. Moult, Comparison of systematic search and database methods for constructing segments of protein structure, *Protein Eng*, **7**, 953–960 (1994).
69. A. Fiser, and A. Sali, ModLoop: automated modeling of loops in protein structures, *Bioinformatics*, **19**, 2500–2501 (2003).
70. Z. Xiang, C.S. Soto, and B. Honig, Evaluating conformational free energies: the colony energy and its application to the problem of loop prediction, *Proc Natl Acad Sci U S A*, **99**, 7432–7437 (2002).
71. K. Zhu, D.L. Pincus, S. Zhao, and R.A. Friesner, Long loop prediction using the protein local optimization program, *Proteins*, **65**, 438–452 (2006).
72. C.M. Deane, and T.L. Blundell, CODA: a combined algorithm for predicting the structurally variable regions of protein models, *Protein Sci*, **10**, 599–612 (2001).
73. A.A. Canutescu, A.A. Shelenkov, and R.L. Dunbrack, Jr., A graph-theory algorithm for rapid protein side-chain prediction, *Protein Sci*, **12**, 2001–2014 (2003).
74. M.R. Lee, D. Baker, and P.A. Kollman, 2.1 and 1.8 A average C(alpha) RMSD structure predictions on two small proteins, HP-36 and s15, *J Am Chem Soc*, **123**, 1040–1046 (2001).
75. H. Lu, and J. Skolnick, Application of statistical potentials to protein structure refinement from low resolution ab initio models, *Biopolymers*, **70**, 575–584 (2003).
76. Z. Xiang, P.J. Steinbach, M.P. Jacobson, R.A. Friesner, and B. Honig, Prediction of side-chain conformations on protein surfaces, *Proteins*, **66**, 814–823 (2007).

76a. M. Pawlowski, M.J. Gajda, R. Matlak, and J.M. Bujnicki, MetaMQAP: a meta-server for the quality assessment of protein models, *BMC Bioinformatics*, **9**, 403 (2008).

77. B. Wallner, and A. Elofsson, Identification of correct regions in protein models using structural, alignment, and consensus information, *Protein Sci*, **15**, 900–913 (2006).

78. J.M. Bujnicki, M. Feder, L. Rychlewski, and D. Fischer, Errors in the *D. radiodurans* large ribosomal subunit structure detected by protein fold-recognition and structure validation tools, *FEBS Lett*, **525**, 174–175 (2002).

79. J. Kosinski, E. Kubareva, and J.M. Bujnicki, A model of restriction endonuclease MvaI in complex with DNA: a template for interpretation of experimental data and a guide for specificity engineering, *Proteins*, **68**, 324–336 (2007).

80. J. Orlowski, M. Boniecki, and J.M. Bujnicki, I-Ssp6803I: the first homing endonuclease from the PD-(D/E)XK superfamily exhibits an unusual mode of DNA recognition, *Bioinformatics*, **23**, 527–530 (2007).

81. J. Gough, K. Karplus, R. Hughey, and C. Chothia, Assignment of homology to genome sequences using a library of hidden Markov models that represent all proteins of known structure, *J Mol Biol*, **313**, 903–919 (2001).

82. A.G. Murzin, S.E. Brenner, T. Hubbard, and C. Chothia, SCOP: a structural classification of proteins database for the investigation of sequences and structures, *J.Mol.Biol.*, **247**, 536–540 (1995).

83. M. Vingron, and P. Argos, A fast and sensitive multiple sequence alignment algorithm, *Comput Appl Biosci*, **5**, 115–121 (1989).

84. K. Mizuguchi, C.M. Deane, T.L. Blundell, and J.P. Overington, HOMSTRAD: a database of protein structure alignments for homologous families, *Protein Sci*, **7**, 2469–2471 (1998).

85. K. Ginalski, M. von Grotthuss, N.V. Grishin, and L. Rychlewski, Detecting distant homology with Meta-BASIC, *Nucleic Acids Res*, **32**, W576–581 (2004).

86. D.T. Jones, Protein secondary structure prediction based on position-specific scoring matrices, *J Mol Biol*, **292**, 195–202 (1999).

87. D.T. Jones, K. Bryson, A. Coleman, *et al.*, Prediction of novel and analogous folds using fragment assembly and fold recognition, *Proteins*, **61 Suppl 7**, 143-151 (2005).

88. J. Skolnick, D. Kihara, and Y. Zhang, Development and large scale benchmark testing of the PROSPECTOR_3 threading algorithm, *Proteins*, **56**, 502–518 (2004).

89. S.B. Pandit, Y. Zhang, and J. Skolnick, TASSER-Lite: An automated tool for protein comparative modeling, *Biophys J*, **91**, 4180–190 (2006).

90. W.R. Pearson, Empirical statistical estimates for sequence similarity searches, *J Mol Biol*, **276**, 71–84 (1998).

91. J. Xu, M. Li, G. Lin, D. Kim, and Y. Xu, Protein threading by linear programming, *Pac Symp Biocomput*, 264–275 (2003).

92. A.E. Torda, J.B. Procter, and T. Huber, Wurst: a protein threading server with a structural scoring function, sequence profiles and optimized substitution matrices, *Nucleic Acids Res*, **32**, W532–535 (2004).

93. K. Ginalski, A. Elofsson, D. Fischer, and L. Rychlewski, 3D-Jury: a simple approach to improve protein structure predictions, *Bioinformatics*, **19**, 1015–1018 (2003).

94. B. Wallner, and A. Elofsson, Pcons5: combining consensus, structural evaluation and fold recognition scores, *Bioinformatics*, **21**, 4248–4254 (2005).

95. S. Wu, and Y. Zhang, LOMETS: a local meta-threading-server for protein structure prediction, *Nucleic Acids Res*, **35**, 3375–3382 (2007).

96. V.A. Eyrich, and B. Rost, META-PP: single interface to crucial prediction servers, *Nucleic Acids Res*, **31**, 3308–3310 (2003).

97. B. Wallner, H. Fang, and A. Elofsson, Automatic consensus-based fold recognition using Pcons, ProQ, and Pmodeller, *Proteins*, **53 Suppl 6**, 534–541 (2003).

98. B. Wallner, and A. Elofsson, Can correct protein models be identified?, *Protein Sci*, **12**, 1073–1086 (2003).

99. D. Douguet, and G. Labesse, Easier threading through web-based comparisons and cross-validations, *Bioinformatics*, **17**, 752–753 (2001).

100. D. Fischer, 3D-SHOTGUN: a novel, cooperative, fold-recognition meta-predictor, *Proteins*, **51**, 434–441 (2003).
101. L.H. Hung, S.C. Ngan, T. Liu, and R. Samudrala, PROTINFO: new algorithms for enhanced protein structure predictions, *Nucleic Acids Res*, **33**, W77–80 (2005).
102. R. Samudrala, and J. Moult, An all-atom distance-dependent conditional probability discriminatory function for protein structure prediction, *J Mol Biol*, **275**, 895–916 (1998).
103. C.A. Rohl, C.E. Strauss, D. Chivian, and D. Baker, Modeling structurally variable regions in homologous proteins with ROSETTA, *Proteins*, **55**, 656–677 (2004).
104. E. Eyal, R. Najmanovich, B.J. McConkey, M. Edelman, and V. Sobolev, Importance of solvent accessibility and contact surfaces in modeling side-chain conformations in proteins, *J Comput Chem*, **25**, 712–724 (2004).
105. R.L. Dunbrack, Jr., and F.E. Cohen, Bayesian statistical analysis of protein side-chain rotamer preferences, *Protein Sci*, **6**, 1661–1681 (1997).
106. Z. Xiang, and B. Honig, Extending the accuracy limits of prediction for side-chain conformations, *J Mol Biol*, **311**, 421–430 (2001).
107. S. Liang, and N.V. Grishin, Side-chain modeling with an optimized scoring function, *Protein Sci*, **11**, 322–331 (2002).

5

Template-free Predictions of Three-dimensional Protein Structures: From First Principles to Knowledge-based Potentials

Dominik Gront, Dorota Latek, Mateusz Kurcinski and Andrzej Kolinski

5.1 Introduction

The number of known protein sequences is much larger (by a factor of a thousand) than the number of experimentally solved protein structures. Depending on genome, for about half of newly determined sequences protein structure can be predicted by means of comparative modeling methods.[1] For the remaining sequences there is no structurally similar protein in the databases of protein structures, or the appropriate template cannot be identified by means of the existing fold recognition methods. In such cases molecular models need to be constructed using protein sequence as the only target-specific information. This is the case of *de novo* protein modeling. *De novo* protein structure prediction has been the 'holy grail' of computational biology for the last forty years, and toward that goal there has been steady progress. New methods have been developed and the enormous increase of computational resources have enabled new applications. Nevertheless, *de novo* (or template free) *in silico* protein structure prediction is still limited to relatively small and topologically simple structures. It should be pointed out that development of *de novo* methods is important not only for protein structure prediction, but also for prediction of protein folding pathways and protein–protein interactions. In this chapter we outline the most typical and the most successful approaches to *de novo in silico* protein folding.

Prediction of Protein Structures, Functions, and Interactions   Edited by Janusz Bujnicki
© 2009 John Wiley & Sons, Ltd

5.2   Force Fields and Scoring Functions

5.2.1   Quantum Mechanics Applications in de Novo Modeling of Proteins

The mathematical theories of quantum mechanics predated the first computers by many years. Accordingly, its early applications were restricted to very simple systems: single atoms, ions, or small and highly symmetrical molecules. Advances in computational techniques made it possible to apply quantum mechanics to more demanding problems. Calculations of geometrical features, electric multipoles and reaction pathways for medium-size molecular systems have recently become standard procedures. Quantum-mechanical computation is also a reliable source of various data, of quality comparable to those obtained by experimental techniques. However, long-scale evolution of the time-dependant Schrödinger's equation for even a small protein is still far beyond the computing powers of present-day supercomputers. A molecule of a small protein consists of thousands of nuclei and tens of thousands of electrons. Adding solvent molecules to the system approximately double these numbers. The quantum description of such a large molecular system is only feasible with some serious simplifications.

One modeling approach proposed by Car and Parrinello treats the system classically, save that after every simulation step the distribution of charges is recalculated using quantum mechanics methods.[2] A different approach is presented by McCammon and Lesyng.[3] In their methodology the whole system is divided into three regions: the active site of the proteins is subject to quantum mechanical description, the rest of the protein is treated classically, and water molecules are approximated by a continuous solvent model. There are a number of other hybrid quantum-classical theories which could be used for protein modeling, but their applications are still rather limited due to high computing costs, even despite serious simplifications. The most significant contribution of quantum mechanics to protein modeling comes from its most basic approximations. Adiabatic and Born-Oppenheimer approximations separate the movement of nuclei from electrons and lead straightforwardly to the idea of the force field.

5.2.2   Empirical Force Fields Models

In 1967 Bixon and Lifson published a work entitled 'Potential functions and conformations in cycloalkanes',[4] where the idea of the force field was first introduced. This work contains an analysis of free energy variations with respect to conformation changes in several cykloalkanes, which was actually a derivation of a simple force field. From the mathematical point of view, a force field is a function, where the argument is a multidimensional vector describing the conformation of the given system and the resulting value is the system's energy. As most of the known force fields contain components parameterized by various factors, the definition of a particular force field should also include a parameter set intended for a particular problem.

There are two basic types of molecular force fields: statistical (or knowledge-based) and empirical. The first one is derived from observed regularities in certain datasets (i.e. the protein structure database) and will be discussed in detail in the following sections. The idea of an empirical force field comes strictly from the Born-Oppenheimer approximation, where the total wave function is written in the form of a product of nuclear and electronic factors. In a brute-force approach, one would solve the electronic Schrödinger's equation for every possible nuclei conformation ($\mathbf{R}$), which would give the $E(\mathbf{R})$ function, or in other words the desired force field. However, this approach is impossible for large systems, due to the enormous computing costs required. In the potentials obtained in this way, general and predictable tendencies could be observed, like the attraction of oppositely charged atoms.

The first fundamental theorem of empirical force fields is the simple assumption that easily measured structural properties observed in one molecule (such as the C-C bond length in ethane) can be used to predict similar features in other molecules (the C-C bond length in polyethylene). Thanks to quickly-developing experimental techniques (especially spectroscopy), as well as quantum calculations, we are presently capable of measuring the distances and angles between atoms and bonds with high accuracy. This information can be easily incorporated in very simple formulas describing bond stretching and bending. The second fundamental theorem of empirical force field construction assumes that the total energy of the system is a sum of independent contributions from all atoms, bonds, etc. Although there is no one exact way to describe certain molecular properties, most widely used empirical force fields consist of very similar elements, differing only by sets of parameters. In the next few paragraphs, brief descriptions of the most common components of empirical force field components are presented.

Bond stretching. The energy curve of a typical bond has one minimum corresponding to the equilibrium bond length,* from which point a change in the bond length (stretching or compressing) causes an energy increase. A good analytical approximation of such an energy curve is the Morse potential in the following form:

$$E_{bond}(r) = D_e \left[1 - \exp\left\{ -\sqrt{\frac{k}{2D_e}}(r - r_0) \right\} \right]^2 \tag{5.1}$$

D_e is the depth of the minimum, r_0 is the reference (equilibrium) bond length and k is the stretching constant of the bond. This form, however, is rarely used to reproduce the bond stretching effect, since it requires the calculation of both three parameters for each bond (D_e, k, r_0), and the relatively expensive exp() function. The most elementary approach is to use Hook's law instead, which approximates the bond energy curve by a parabola.

$$E_{bond}(r) = \frac{k}{2}(r - r_0)^2 \tag{5.2}$$

This form has one major inadequacy in that the bond described by it cannot be broken, but chemical reactions are rarely modeled with empirical force fields.

Angle bending. Variations of angles from their reference values are usually described in a similar fashion to bond stretching:

$$E_{angle}(\alpha) = \frac{k}{2}(\alpha - \alpha_0)^2 \tag{5.3}$$

where α_0 is a reference angle value and k is the bending constant.

---

* The bond equilibrium (or reference) length is set when all other terms in potential function are set to zero. The same condition applies to the reference angle between bonds, dihedral angle, etc.

Electrostatic interactions. Interactions between charged atoms may be considered at various levels of accuracy. To a crude approximation, attraction and repulsion of point charges may be described by Coulomb's law:

$$E_{electr}(q_i, q_j, r_{ij}) = \frac{q_i q_j}{4\pi \varepsilon_0 r_{ij}} \tag{5.4}$$

$q_{i,j}$ are the charges of atoms i and j and r_{ij} is the distance between these atoms. This very simple model is not sufficient for many tasks that require a more accurate description of the charge distribution. The most widely used are partial charges and/or central multipole expansion. According to premises taken from experimental data, one may assign partial charges to particular atoms. Partial charges can also be determined using quantum mechanical calculations. Such an approach renders the charge distribution in a more realistic way, while electrostatic interactions may still be modeled simply by Coulomb's formula. More sophisticated models incorporate the central multipole expansion to include not only charges, but also the higher multipole moments – dipoles, quadrupoles, etc. – in the interactions. Moreover, in order to account for the mutual fitting of the electrostatic potential and the molecular conformation the charge distribution may be updated after a certain number of simulation steps.

Non-bonded interactions. The most important forces dictating the properties of biomolecules are the non-bonded interactions. Van der Waals forces are most commonly described by the Lennard-Jones (LJ) potential, which approximates very well both the strong repulsion at short distances, reflecting the excluded volume effect, and the weak dispersive attraction when interacting atoms are far apart. The following form of the Lennard-Jones potential is also called the 12-6 potential, where 12 and 6 denote the powers of the repulsive and attractive terms, respectively:

$$E_{vdW} = 4\varepsilon \left[\left(\frac{r_0}{r}\right)^{12} - \left(\frac{r_0}{r}\right)^6 \right] \tag{5.5}$$

ε is the value of the Lennard-Jones potential at minimum (potential well depth) and r_0 is the collision parameter (at r_0 Lennard-Jones potential is equal to zero). Other forms of m-n LJ potentials besides 12-6 are also used, however the 12-6 potential is most favorable due to its computing efficiency. r^{-12} can be quickly obtained by squaring the r^{-6} term and thus can be obtained without time-consuming square root calculations. There are no other particular arguments for using this form of potential, except for the fact that it produces reasonable results. This property also relates to the other terms in empirical force fields, as suggested by the name 'empirical'.

Dihedral angles. The following formula describes the energy change as a function of dihedral (torsional) angle as a single bond is rotated:

$$E_{torsion} = \sum_{n=0}^{N} \frac{E_n}{2} [1 + \cos(n\theta - \theta_0)] \tag{5.6}$$

Every set of four bonded atoms A-B-C-D contributes a dihedral term to the total energy. $E_{torsion}$ changes as atoms or groups of atoms AB are rotated around bond B-C. E_n is the barrier height, n is the number of minima on the potential curve as θ ranges from $0°$ to

$360°$, and θ_0 is the phase factor indicating the minimum value of the torsional potential. Although this effect can be modeled using only the bond length, angle, and non-bonded interaction terms described above, most force fields also include dihedral angle terms, producing more accurate results.

As mentioned before, most of the currently used force fields assume that the total energy of a system is a sum of independent contributions from all atoms, bonds, etc. Thus, the complete formula used to calculate the total energy could be written as follows:

$$E_{total}(\mathbf{R}) = \sum_i E_{bond}(r_i) + \sum_i E_{angle}(\alpha_i) + \sum_i \sum_j E_{electr}(q_i, q_j, r_{ij}) +$$
$$+ \sum_i \sum_j E_{vdW}(r_{ij}) + \sum_i E_{torsion}(\theta_i) \tag{5.7}$$

The terms contributing to the energy in this equation are widely used in the most common biomolecular force fields, including CHARMM,[5] AMBER,[6] GROMOS,[7] and others. Beyond this equation some other additional or alternative terms may also be used to cover both internal and external features of the force field, including higher-order treatments of bond bending and stretching, cross-terms reflecting the dependence of one vibration on another, and others.

5.2.3 Exploring the Potential Energy Surface in Classical Molecular Mechanics

Cartographic maps present altitude with respect to geographical coordinates. In the same way molecular force fields describe variations of a system's total potential energy as a function of its atomic coordinates, which is usually referred as the potential energy surface. But unlike geographic maps, which represent a function of just two coordinates (latitude and longitude), the potential energy surface (PES) function requires 3N variables to calculate the energy of a molecular system, where N is the number of atoms. It means that to model a small protein surrounded by solvent molecules, one requires a function of many thousands of variables. The fact that we cannot draw a map of such a surface may be the least important problem.

In molecular modeling we are interested in finding the minima of the PES, especially the global minimum, which according to Anfinsen's theorem corresponds to the native structure of a protein.[8] For such a highly dimensional function, one cannot simply apply classical differential analysis to locate minima or saddle points, which correspond to low-energy conformations and energy barriers, respectively. Due to the enormous computing cost required, an exhaustive brute-force search of the conformational space is also practically impossible. The only reasonable method to explore a potential energy surface is to start from an initial conformation of a protein molecule and to modify it according to certain rules, which should lead to conformations producing a lower the potential energy value.

Energy minimization (molecular mechanics). One way to locate a minimum would be the so-called 'downhill walking' method – changing the conformation of a molecule by moving always towards structures of a lower energy. Using such an approach one finds only the nearest minimum, which is not necessarily the global one. This is the main idea of molecular mechanics (MM) minimization, which is often used to improve the quality of near-native models. Energy minimization algorithms may be divided into three groups:

the non-derivative (e.g. the simplex method), first-order derivative (the steepest decent and conjugate gradient methods) and second-order derivative (the Newton-Raphson method). No single algorithm has proven most suitable for all purposes – one should choose the one most appropriate for a given task. The process of inverting the Hessian matrix, required for the Newton-Raphson method, is very simple for small systems, while it remains a very time-consuming task when the number of atoms is large. On contrary, for the simplex method no matrix inversion is required, but the total energy must be calculated for a large number of structures. This requirement restricts the application of the simplex algorithm only to cases where the energy function is not too complicated and may be quickly evaluated.

Molecular dynamics. While molecular mechanics minimization methods explore the potential energy surface only in the closest vicinity of the starting conformation, molecular dynamics (MD) can probe a much wider area of the PES. The most basic idea of MD is to generate successive configurations of the molecule by integrating Newton's laws of motion. The negative gradient of the energy function is used to calculate the net force acting on every atom, which enables calculation of velocities. Assuming a finite time step, the new positions of the atoms are determined and subsequently used for calculation of the potential energy of the system. Repeating this scheme produces a trajectory of consecutive structures which can be further analyzed.

5.2.4   Knowledge-based Force Fields

The advantage of the empirical potentials is that, in principle, they are derived from the laws of physics. The disadvantage is that the calculation of the free energy is very difficult because the computation should include atomic level descriptions of both the protein and the surrounding solvent. Currently this type of computation is too expensive for protein folding simulations because of too many degrees of freedom treated explicitly (see Table 5.1). It is also quite complicated to include subtle, but important, multi-body contributions to the energy function. Usually such effects are modeled by means of cross-terms that describe correlations between the torsional and planar angles[9, 10] and explicit polarization models.[11] These approaches (especially the latter) can substantially increase the expense of the simulation.

Knowledge-based scoring functions are based on the Anfinsen hypothesis: that proteins in their native conformation are in thermodynamic equilibrium with their environment.[8] Thus the main goal of the parameterization of a knowledge-based force field is to minimize the energy of a native structure with respect to the misfolded structures, also known as decoys. In general, the methods for knowledge-based force field derivation can be roughly divided into two groups.

In the first group are methods based on statistical mechanics or Bayesian statistics. These methods were first proposed by Tanaka and Scheraga[12] and later refined by Miyazawa and Jernigan[13] and Sippl.[14, 15] An energy value E_j assigned to an interaction of type j is given by the Boltzmann-inversion formula:

$$E_j = -kT \ln \left(\frac{n_j^{native}}{n_j^{reference}} \right) \qquad (5.8)$$

Table 5.1 Comparison of time scales of protein folding processes and applicability of modeling methods

PROCESS	TIME SCALE	APPLICABLE MODELING METHOD	PROCESS	TIME SCALE	APPLICABLE MODELING METHOD
Side-chain rotation	$10^{-13} - 10^{-11}$ s	Quantum Mechanics Molecular Dynamics	β-hairpin folding	$10^{-6} - 10^{-3}$ s	Molecular Dynamics Reduced Models
Loop closure	$10^{-9} - 10^{-7}$ s	Molecular Dynamics Reduced Models	Domain folding	$10^{-4} - 10^{3}$ s	Reduced Models
α-helix formation	$10^{-8} - 10^{-4}$ s	Molecular Dynamics Reduced Models	Protein aggregation	$10^{-3} - 10^{8}$ s	Molecular Docking

Quantum Mechanics: Self Consistent Filed (SCF), Density Functional Theory (DFT); **Molecular Dynamics:** Amber, Gromacs, Sybil; **Reduced Models:** Rosetta, Unres, Refiner, CABS; **Molecular Docking:** 3DDOCK, ZDOCK, CABS.

where n_j^{native} is the number of interactions observed in native structures and $n_j^{reference}$ is the number of similar interactions in a reference population. The same result can be also derived from Bayesian principles.[16]

Methods that belong to the second group search for a set of force field parameters where the native conformation has the lowest energy value among a large set of decoys. Numerous variants of this approach differ by the criteria for optimization, e.g. achieving the largest energy gap between the native state and the best of decoy structures,[17] maximization of the average probability of successful prediction, minimization of the free energy of the native state,[18] z-score optimization[19] (the energy difference between the average decoy structure and the native structure in units of standard deviation) and many others.

Xia and Levitt[20] recently proposed a formalism that unifies the approaches outlined above. The authors assumed that the total energy of a given protein conformation may be calculated as a linear combination of all the energy components involved.

5.3   Reduced Models: Representation and Conformational Sampling

The choice of representation of a protein molecule often depends upon the sampling methodology and force field employed in structure prediction, ranging from empirical potentials for coarse-grained models sampled with Monte Carlo and genetic algorithm methods to complex, atom-based, potentials that directly approximate the physical interactions in the system.

A protein molecule can be represented in a computer program in many different ways. One of the most important factors is how many degrees of freedom are treated explicitly. In the most detailed representation, each atom of a macromolecule is treated as an interaction center. Additionally a macromolecule may be immersed in solvent molecules. Such a model, usually sampled by means of molecular dynamics, is computationally too expensive for routine applications. Therefore mesoscopic models are widely employed for protein structure calculations.

In their classical work Levitt and Warshel[21] introduced a protein model based only on two interaction centers per residue: Cα and a united atom describing the side chain. To the present day, many mesoscopic protein model approaches have been proposed. In many cases researchers follow the work of Levitt and Warshel, reducing the protein backbone to a trace of Cα atoms. The side chain moieties may be represented by single united atoms, sets of several atoms or ellipsoids. In the Rosetta approach, the backbone is represented explicitly while the side chains are represented by united atoms. A survey of several mesoscopic protein models is given in Table 5.2.

The straightforward advantage of a mesoscopic model versus an all-atom model is that a single simulation step takes less CPU time because the number of degrees of freedom is smaller. As a consequence, fewer energy terms need to be calculated. Because in mesoscopic models high-frequency motions are neglected, the time step in MD simulations can be larger. Another advantage of the coarse-grained mesoscopic approach is that the force field derived for the united atoms lead to a much smoother energy surface than that for the all-atom energy function. In other words, coarse-graining removes many local energy minima in which the system could become trapped during the simulation.

Table 5.2 Overview of the most successful protein structure prediction algorithms

Model/name of a group participating in CASP6	Representation of the conformational space	Coordinates system	Sampling method	Protein representation	CASP6 results			Availability
					Number of first models in the new fold category with GDT-TS > 25	Average GDT-TS of all first models in the new fold category		
						All new folds targets	All sent new folds targets	
UNRES/ Scheraga	Continuous space model	Cartesian coordinates and torsion angles of the backbone	Molecular dynamics, Monte Carlo		2	16.65	24.97	UNRES: http://www.chem.univ.gda.pl/~adam/local/docs/index.htm as a part of PROTARCH: http://cbsu.tc.cornell.edu/software/protarch
ROSETTA/ Baker	Continuous space model	torsion angles of the backbone	Monte Carlo		3	27.38	27.38	Rosetta: http://www.rosettacommons.org Automatic predictions (Robetta): http://robetta.bakerlab.org
CABS/ Kolinski-Bujnicki	Lattice model (grid=0.61 Å)	Cartesian coordinates	Replica Exchange Monte Carlo		4	25.18	25.18	http://www.biocomp.chem.uw.edu.pl/services.php
REFINER/ boniaki_pred	Continuous space model	Cartesian coordinates	Replica Exchange Monte Carlo		2	17.46	19.64	Unavailable
FRAGFOLD/ Jones-UCL	Continuous space model	Cartesian coordinates	simulated annealing		4	24.43	27.49	Unavailable

(continued overleaf).

Table 5.2 *(continued)*

CAS/ Skolnick-Zhang	Lattice model (grid=0.87 Å)	Cartesian coordinates	Replica Exchange Monte Carlo		3	23.97	23.97	As a part of TASSERLite – a tool for comparative modeling and threading; http://cssb.biology.gatech.edu/skolnick/files/TASSERLite
PROTINFO/ Samudrala AB	Continuous space model	torsion angles of the backbone	Monte Carlo simulated annealing		2	14.02	25.24	http://protinfo.compbio.washington.edu/protinfo_abcmfr
Simfold/ Rokko	Continuous space model	torsion angles of the backbone	Reversible fragment assembly with multi-canonical ensemble Monte Carlo		2	24.65	24.65	Unavailable

Note: The protein representation for each algorithm is presented graphically (UNRES: Cα, SC, PB; Rosetta: backbone atoms, SC; CABS: Cα, Cβ, SC, PB; REFINER: Cα, SC₁, SC₂; FRAGFOLD: all-atom; CAS: Cα, SC; PROTINFO: all-atom; Simfold: backbone atoms, SC). The following abbreviations are used: Cα – alpha carbon atom, Cβ – beta carbon atom, SC – a pseudoatom which is located in a side-chain center of mass and represents side-chain atoms, SG – a pseudoatom which is located in a side-chain center of mass and represents all side-chain atoms except the C-beta atom, PB – a pseudoatom which is located in a geometrical center of a pseudobond between two consecutive C-alpha atoms and represents a peptide bond. SC₁ and SC₂ stand for two united atoms representing two fragments of a side chain in REFINER. CASP6 (the sixth edition of the Critical Assessment of Techniques for Protein Structure Prediction) is a worldwide biennial competition in which different groups predict structural models of typically more than one hundred proteins for which only the sequence is published. To describe the CASP6 results we used the GDT-TS (Global Distance Test Total Score) measuring the average percentage of residues for which all distances between the model and the target are shorter than the chosen cutoff distances. The differences in the last two columns reflect the fact that some groups attempted predictions of a selected subset of new fold targets.

Once a protein model representation has been selected, and a suitable energy function defined, one may try to find the global energy minimum for a given amino acid sequence. Such an attempt relies on the Anfinsen hypothesis and on the assumptions that the simplifications introduced to the model (e.g. neglecting posttranslational modification, using a simplified representation, applying a continuous solvent etc.) do not alter the native state drastically.

Unfortunately, to find the lowest energy conformation one has to overcome the multiple minima problem. In the few past decades of extensive research many methods have been devised for the global optimization of protein structures, clusters or crystals.[22] Among the most successful approaches are Monte Carlo plus Minimization (MCM),[23] Conformational Space Annealing[24] and the Diffusion Equation Method.[25] Genetic algorithms have been also employed in *de novo* protein structure prediction.[26–28]

In many approaches minimization is performed by wandering around on the potential energy surface. This is especially important for protein structure prediction because the simplified (and thus computationally tractable) force fields are only approximations of the true energy function of biomolecules. Therefore it is crucial to locate multiple energy minima. For example, in multi-scale approaches low resolution models are expected to visit as many low energy states as possible. Then a more precise energy evaluation is performed for only these states. A number of methods based on Monte Carlo dynamics have been proposed for this purpose.[29] Parallel Tempering (PT),[30–32] known also as Replica Exchange Monte Carlo, is probably the most widely used method. In this approach several independent copies of the same system are simulated in parallel. The simulations are usually conducted by means of an isothermal Metropolis algorithm, although a molecular-dynamics-based PT method has also been proposed.[33] Each copy of the modeled system is simulated at a different temperature. For adjacent replicas configurations can be exchanged (swapped) between temperatures. The high temperature systems are generally able to sample large volumes of the energy surface, whereas energy minimization is performed by the low-temperature replicas. The exchange mechanism prevents the low-temperature replicas from becoming trapped in local energy minima. The PT method has proven to be general and very robust. One provides only a set of temperatures as the controlling parameters of the simulations. Several methods have been proposed to choose optimal temperature sets, e.g. [34–36].

5.4   Successful Approaches to *de Novo* Protein Modeling

Continuous-space models based on all-atom representations lack the ability to find a global free energy minimum of a typical protein in a tractable period of time. The most straightforward solution to this problem is to reduce the large number of degrees of freedom by employing a reduced representation of a protein chain or/and simplified interaction scheme.

5.4.1   Continuous Space Reduced Models

The classical work by Levitt and Warshel[21] assumed a two-center approximation of a single residue of a polypeptide chain: the Cα atom and a spherical united atom representing the side group. The only degrees of freedom were the torsion angles defined by four consecutive Cα atoms. This representation of a protein is also used with some alterations in the definition of side-chains in UNRES, a recently developed continuous-space reduced model.[37] UNRES assumes the form of ellipsoids for the side chains. Additionally, peptide bonds are represented as united atoms located in the middle of two consecutive Cα atoms. The free energy function accounts explicitly for the interactions of only two types of united atoms: the united side chain atoms and the peptide groups. The Cα atoms are used only for precise definition of the chain geometry during the simulation. The virtual Cα-Cα bonds are of a fixed length of 3.8 Å. The only degrees of freedom in continuous space are the bond and torsion pseudoangles defined in the Cα-based local coordinate systems. The free energy function of a given conformation includes hydrophobic (or hydrophilic) interactions between the side chains, steric repulsion between side chains groups and peptide groups, and electrostatic interactions between peptide groups. Local conformational propensities of a polypeptide are described by torsional and angle-bending potentials. The multibody correlations between local interactions and electrostatics, which are the most important for reproducing regular secondary structure elements, are described by higher order terms. It should be pointed out that the weighting factors of the free energy terms of the UNRES force field were optimized using a sophisticated method based on a hierarchical classification of decoys according to their content of native-like elements.[38] The conformational space is sampled by a variety of methods, including simple Monte Carlo schemes, CSA (Conformational Space Annealing – a combination of a genetic algorithm and local minimization), variants of molecular dynamics (such as MREMD – Multiplexing Replica Exchange Molecular Dynamics), and other elaborate techniques of global minimization. The UNRES approach to protein structure prediction proved its high modeling efficiency in several Critical Assessment of Techniques for Protein Structure Prediction (CASP) competitions (see Table 5.2), providing (for example) the best model for the T0215 target (the fold-recognition analogy category) in CASP6. CASP is a biannual competition, in which competitors from all over the world predict models for sequences of proteins with unpublished structures. After the competition is closed the native structures are released and all previously sent models are assessed.

Another recently developed protein model is REFINER, which increases the resolution of side chain representation (as compared to UNRES) by adding another united atom (so that the side chains of most residues are represented by two united atoms, excepting glycine and alanine).[39] The radii of both united atoms are correlated with the chemical structure of the side-chain. The energy terms include short-range potentials, long-range contact-type potentials dependent on the mutual orientation of interacting side-chains, and the main chain hydrogen-bonding potential. The force field was derived similarly to the CABS model (described below), using structural statistical regularities extracted from known protein structures. The REFINER approach performed well not only in CASP (see Table 5.2), but also in simulations of autocatalytic misfolding followed by aggregation of the model peptides. These computational experiments provided a simplified molecular explanation of 'conformational' diseases.[40]

A different approach to the reduced representation of proteins has been employed in the Rosetta model developed by Baker and co-workers.[41] The Rosetta representation of a protein structure includes all non-hydrogen backbone atoms but reduces side chains to single united atoms located in the side chain centers of mass. For higher-resolution modeling, discrete all-atom side chain conformations may be added from a rotamer library,[42] using Monte Carlo simulated annealing protocols. All bonds lengths and angles in the backbone are set to the ideal values for a polyalanine peptide chain. The only degrees of freedom are the backbone torsion angles, which change continuously. The Rosetta strategy differs from one protein-modeling application to another, but the main idea remains the same. Namely, short fragments of known protein structures are assembled using the Monte Carlo method, and from those fragments native-like protein conformations are usually obtained. This approach mimics real protein folding by sampling fluctuations of different local structures accessible for a given sequence that finally lead to compact global conformations, which is consistent with the model of local and non-local interactions. Usually, a Monte Carlo conformational search starts from an extended conformation, where nine-residue windows from the query sequence are randomly selected. By using a fragment library derived from a non-redundant database of protein structures, the best fitting structural fragments to the given sequence windows are determined. Fragment assembly is divided into several steps, including comparison of PSI-BLAST-constructed profiles of the database fragments and the query sequence windows, and calculations of similarity between the DSSP-derived secondary structure of the assembled fragments and the predicted secondary structure for the query sequence. After a random selection of the best fitting fragments, all torsion angles in the windows of the query protein chain are replaced with these from the selected fragments. Such a change of the torsion angles is called a 'move'. Alternatively, the move may be carried out by random perturbation of the torsion angles. After the move, the energy of the entire chain is evaluated and the standard Metropolis criterion is applied. For applications in *de novo* modeling, a simplified energy function has been designed using Bayes' theorem. The simplified interaction scheme includes several sequence-dependent and sequence-independent terms consistent with the coarse-grained representation. The sequence-specific terms correspond to solvation and electrostatic effects. Electrostatics and disulfide bonding are evaluated by a distance-dependent residue-based pair potential. The sequence-independent energy terms include the van der Waals attraction and steric repulsion of the backbone atoms and the side-chain centroids. The global arrangement of the secondary structure elements is moderated by both strand-strand and helix-strand packing potentials. In simplified versions of the Rosetta model, hydrogen bonding is accounted for only implicitly as a knowledge-based strand-pairing potential. Except for the van der Waals interactions, all the remaining force field components were derived via an elaborate statistical analysis of known protein structures. The Rosetta strategy proved to be very successful in all editions of CASP (see an example from CASP6 in Table 5.2).

Since Rosetta was published, many other *de novo* methods based on fragment assembly has been developed,[43] many of which are available as web servers (see Table 5.2). For example, the FRAGFOLD method developed by Jones[44] forms a protein conformation from both the best fitted super-secondary structures (two or three sequential secondary structure elements) and short tri-, tetra- and pentapeptides from a library during the simulated annealing process. It is worth noticing that simulations based on conventional fragment assembly lack reversibility and detailed balance. Consequently, not all efficient sampling methods can be used in such simulations. SimFold, developed by Takada *et al.*,[45] deals with this problem by a reversible fragment assembly approach combined with a powerful conformational sampling method – Multicanonical Ensemble Monte Carlo. A different approach to fragment assembly is used by Samudrala in PROTINFO.[46] Instead of copying angles or coordinates directly from the best fitting fragments from known protein structures, these fragments are used to generate the binned distribution of backbone torsion angles for the given sequence. During the simulated annealing, every change of torsion angles is based on this distribution.

A detailed representation of the main chain and a reduced description of the side-chains, which is characteristic for the Rosetta method and other fragment assembly methods, is also employed by Betancourt in his protein model.[47] For the definition of a knowledge-based potential based on dihedral Φ and Ψ angles, which reflects correlations of local conformations of consecutive residues, an all-atom representation of the backbone was used. For the definition of other knowledge-based potentials, e.g. the pairwise potentials that depend on distance and mutual positions along the sequence, a reduced representation of the backbone and the side-chains is used. The coarse-grained representation of a residue is defined by one to three pseudoatoms (depending on a residue type). Backbone atom positions, and side chain sizes and orientations are used for the definition of interaction centers. Side chain atoms of a residue are grouped into united atoms according to their radial distances from the average position of the backbone atoms. Namely, the first pseudoatom is formed by the backbone atoms and the side chain α and β atoms (for example Cα or Cβ), the second approximates γ and δ atoms, and the third pseudoatom replaces distal (ε, ζ and η) atoms. Positions of pseudoatoms are defined by spherical coordinates using the backbone geometry as a reference system with Cα atom at the origin. Controlled by the Monte Carlo scheme, the conformation is updated by changes in a continuous space of torsion Φ and Ψ angles of a fragment of a protein (so-called 'pivot' moves and 'fixed end' moves). The described model has been used mainly in threading applications and protein folding with the Go contact potential, but its innovative approach could be very successful in *de novo* structure prediction.

5.4.2   High Resolution Lattice Models

High resolution lattice models have proven to be very effective not only in protein structure prediction,[48] but also in prediction of protein interactions[49, 50] and protein folding mechanisms.[51, 52] The CABS (Cα, Cβ, Side chain) model is probably the most representative example of such approach. In this model the main chain is reduced to an alpha carbon trace restricted to a cubic lattice with a grid spacing equal to 0.61 Å. By slightly fluctuating around the equilibrium length of 3.8 Å, Cα-Cα virtual bonds can adapt 800 different orientations. The Cα-trace is used as a reference frame for the definition of the beta carbon and the centers of the remaining portions of the side chains, both of which have coordinates off the lattice. A two-rotamer approximation (two possible rotational isomers of a side chain) of the conformations of the side chains is used to speed up computations. The force field is entirely statistical and knowledge-based, and contains short-range generic and sequence-specific terms. These terms provide conformational biases for 4-residue fragments, a model of the main-chain hydrogen bonds (defined by the local Cα-trace geometry of the interacting fragments), and context-specific long-range interactions between the side chains. The side chain interactions account in an implicit fashion for the solvent effects and complex multibody correlations. Sampling of the CABS energy surface is performed by means of various Monte Carlo techniques, including simulated annealing, parallel tempering, and other multi-copy techniques. Due to 'prefabricated' local moves stored in large reference tables, pre-computed interactions for various local geometries, and the smoothed energy surface due to the lattice approximation, the CABS simulations are much more efficient than an otherwise equivalent continuous space model. Starting from an earlier version of CABS, Zhang and coworkers[53–55] developed a very efficient hierarchical method for automated modeling of proteins based on fragmentary and structurally remote templates.

A unique approach to a reduced representation of conformational space is represented by the SICHO (SIde CHain Only) model. In SICHO,[56] only the centers of mass of side chains are modeled explicitly, and are restricted to a cubic lattice with a grid spacing equal to 1.45 Å. The distribution of distances between consecutive side chains is amino-acid-specific and mimics the distribution seen in known protein structures. The virtual bonds between subsequent side chains belong to a set of 646 lattice vectors. The positions of the alpha carbons (used for the definition of the main chain hydrogen bond patterns) are approximated from the positions of the side chains, using a set of simple knowledge-based rules. The force field of SICHO is similar in spirit to that of CABS. Although somewhat less accurate, the SICHO model is computationally more efficient than CABS. The cost of simulations for very small proteins is the same as for CABS model, although the scaling of the computing time as a function of chain length is much better. For this reason the SICHO algorithm is more suitable for modeling of large proteins and macromolecular assemblies. The predictive strength of SICHO was successfully tested during the earlier CASP competitions.[57, 58] The SICHO and CABS models could easily be combined into a single hierarchical, multiscale modeling pipeline.

5.5 Multiscale Modeling

Multiscale approaches to protein structure modeling employ a few (at least two) distinct models of protein chains. A low resolution representation is used to effectively search low-energy regions in the conformational space. A higher-resolution model, usually an all-atom representation, is devoted to accurate assessment of the best protein conformations generated in the low-resolution search. The first high-accuracy multiscale modeling was conducted for the GCN4 leucine zipper dimmer.[59] Low resolution simulations were performed by means of a very simplified lattice model called '310 Hybrid'. The simulations started from random conformations of the two spatially separated monomers. The only prior knowledge introduced into the simulation was the amino acid sequence and a secondary structure assignment providing a weak bias towards the known helix-like geometry. From low resolution simulations the five models of lowest energy were selected. The coordinate root-mean-square distances after the best superimposition (CRMSD) between the models and the native structure ranged from 2.3 Å to 3.7 Å. In the second stage of the multiscale modeling, an all-atom CHARMM[5] force field with explicit solvent was used to refine and assess the models. After energy minimization, the resulting models had CRMSDs of 0.81–2.29 Å from the native structure, and the best structures could be easily selected based on objective geometrical and energetic criteria.

A very interesting multiscale procedure has been proposed by Levitt and coworkers.[20, 60] First a large number of compact, diamond lattice polymers are generated, assuming that a single lattice site may correspond to more than one residue of the modeled protein chain.

Then the obtained lattice scaffolds were used for construction of detailed models from small protein fragments excised from the structural database. The large set of models generated was scored with a simplified energy function and then refined using molecular mechanics. Frequently quite reasonable models were generated by this hierarchical approach.

In another example of multiscale modeling,[61] *de novo* models were computed for 16 small globular proteins. For each of these sequences, 20,000 to 30,000 trial conformations were generated with the Rosetta modeling tool.[41] Then all-atom representations were built for each of the models and the all-atom energy was minimized. The final models were selected by means of structural clustering. The method employed in this work demanded a lot of computational resources. The authors presented results for protein sequences having 69 amino acids on average. For each of these sequences, the calculations took approximately 150 CPU days. The authors identified conformational sampling as the computational bottleneck of their approach.

Yet another successful approach to all-atom refinement of mesoscopic models was presented by Kmiecik *et al.*[51] Contrary to the method by Bradley *et al.*,[61] in this work the authors optimized all non-hydrogen atoms of a model except for the $C\alpha$ trace, which remained frozen. This made the minimization step much shorter. Another consequence of keeping the $C\alpha$ atoms restricted to their initial positions is that the conformational space is explored only in the nearest neighborhood of the energy hyper-surface. A global search was performed with the CABS modeling tool.

The approaches described above utilize an all-atom protein model only at the final stage of modeling as a tool for structure refinement and highly accurate energy evaluation. In the approach proposed by Brooks *et al.*[62] a multiscale simulation is conducted at two levels of accuracy. The mesoscopic lattice-based model SICHO is used for efficient sampling of conformational space. After a short SICHO simulation, the resulting structures are subjected to all-atom energy minimization with the CHARMM force field. The optimized conformations are accepted or rejected according to the Metropolis criterion. Both simulation layers may utilize parallel tempering for further sampling enhancement. This modeling procedure is a part of the MMTSB (Multiscale Modeling Tools for Structural Biology) protein modeling package, consisting of several simulation programs and Perl scripts for maintaining the pipeline. The programs are publicly available from the MMTSB webpage: http://mmtsb.scripps.edu/software/mmtsbToolSet.html.

A necessary component of multiscale modeling is a tool for fast and reliable reconstruction of an all-atom protein representation of a mesoscopic model. For example, in the MMTSB protocol described above, such a reconstruction routine must be called for every Monte Carlo step of the high-resolution simulation layer. Usually the all-atom reconstruction is performed in two steps. Firstly, positions of non-hydrogen atoms of protein backbone are retrieved. The accuracy of the first step is important because the rebuilding of side chains is usually based on the φ and ψ dihedral angles, calculated from the reconstructed backbone. Several algorithms have been proposed for the backbone reconstruction. They can be roughly divided in two groups: a rigid fitting of short peptides and statistics-based approaches. Methods from the first group[63–65] utilize short (usually 3 to 10 residue long) fragments (or 'spare parts') extracted from high-resolution protein structures. These methods can be very accurate, as is the algorithm by Claessens *et al.*,[63] although recently conducted large-scale tests[66] have shown that they fail to reconstruct a fraction of the proteins from the test set. Methods that rely on structural statistics extracted from the

Table 5.3 *Summary of the results of the reconstruction of backbone in a set of 81 native protein structures*

Method	% of successfully rebuilt structures	Average results for 35 proteins rebuilt by various methods			
		Average CRMSD on backbone	Φ correlation	Ψ correlation	Average running time per protein [s]
MaxSprout	46.25%	0.47 Å	0.75	0.82	1.71 s
BB	100%	0.64 Å	0.52	0.65	56.98 s
Pulchra	100%	0.59 Å	0.65	0.78	1.06 s
Sybyl	91.25%	0.39 Å	0.77	0.86	172.6 s
BBQ	**100%**	**0.42 Å**	**0.81**	**0.84**	**0.37 s**

PDB[62, 66–70] are usually more robust. A comparison of the performance of several methods for main chain reconstruction is given in Table 5.3.

In the second stage of all-atom reconstruction, i.e. side chain rebuilding, backbone-dependent rotamer libraries are commonly used.[71, 72] Typically, *de novo* protein structure calculations generate a large number of decoys. Correct identification of the native-like structures among the decoys still remains a challenging task. The force fields used in protein simulations usually reflect only the energy terms. For estimation of the entropic effects a clustering algorithm may be employed.[73–76] Clustering procedures combine elements into groups according to a defined distance measure. A size of a given group of similar structures combined with their average conformational energy can provide valuable hints for the final model selection.

5.6 Experimental and Predicted Restraints in Guided *de Novo* Modeling

De novo methods provide theoretical models of reasonable resolution for proteins of a maximum length of 150 residues. This barrier of protein length on a reachable resolution could be overcome with the use of some sparse and non-specific experimental data in the folding simulations, which is much easier to obtain in comparison to the time- and effort-consuming experimental determination of complete protein structures. If we combine *de novo* methods with such simple experiments we could accelerate the process of solving a protein structure without decreasing the resolution of the resulting molecular models.

5.6.1 Structure Prediction Supported by Sparse Experimental Data

As discussed above, successful *de novo* methods can employ experiment-based restraints in their algorithms. The Rosetta method was enhanced with scoring potentials satisfying various experimental constraints, and the new server Rosetta-NMR was developed.[77] Rosetta-NMR takes advantage of the data from NMR measurements which are provided additionally by the user. It is possible to supply chemical shifts (CS), residual dipolar couplings (RDC) and NOEs. Briefly, chemical shifts are used in the Rosetta model as a source of local information about torsion angles Φ and Ψ obtained by a modified version of the TALOS program,[78] while NOEs are used in the form of simple distance constraints. RDCs, which define the degenerated orientations of internuclear vectors (typically ^{15}N-^{1}H) with respect to a specific reference frame (a molecule alignment tensor), are used in the same way as CSs and NOEs in the scoring function, which is used in the selection of fragments from the database.

It is worth mentioning that most standard NMR structure determination programs can deal with ambiguous NOE signals, but the degeneracy of the RDC data is still an impediment for many programs, requiring at least enough data to reduce or eliminate the ambiguities.[77] Consequently, RDCs are used commonly in structure refinement but rarely in the beginning stages of structure determination. For *de novo* structure prediction protocols such as Rosetta-NMR, which do not act in a deterministic way, the effects of degeneracy of the RDCs could be reduced by the fact that the model force field could discriminate against false or unsuitable constraints, thereby allowing the incorporation of such data in the early stages of *in silico* folding. Moreover, due to its frequent degeneracy and non-specificity, sparse experimental data should be used during the modeling in a complementary (with respect to the force field employed) fashion, as it is implemented in the Rosetta-NMR or CABS models. What is perhaps more important for experimentalists (potentially saving them time), is the fact that Rosetta-NMR is also capable of providing reasonable protein models even in cases when the NMR signals have not been assigned yet to specific residues.[79]

Chemical shift data were also exploited in the CABS *de novo* method in the form of secondary structure bias and angular constraint potentials as supplementary terms in the energy function. Even such local information on the protein conformation enables a significant reduction of the explored conformational space, and therefore accelerates the prediction procedure without decreasing the resolution of the final models.[80]

5.6.2   Combining Experimental and Theoretical Restraints in *de Novo* Modeling

De novo modeling can also be enhanced with some additional data from other theoretical methods. The most successful approaches in *de novo* modeling are based on combinations of different kinds of data and methods to obtain the most probable consensus solution. In CASP6 one of the most successful methods in the 'new fold' category was a combined approach involving two diametrically different methods: the CABS model by Kolinski and the Frankenstein3D model by Bujnicki.[48] The latter method was treated as a source of spatial constraints which guided *de novo* protein folding carried out by the former. Frankenstein3D, originally developed for comparative modeling or fold-recognition applications, uses an optimized procedure of selecting and scoring pairwise target-template alignments to build a hybrid model. The hybrid model is obtained from the best fitted fragments from preliminary models which are assessed by statistical potentials, structural comparison and an external method (previously Verify3D and currently MetaMQAP[81]).

Multiple sequence alignments provide various data which can be used not only in comparative or fold-recognition modeling, but also in *de novo* approaches. The various methods that supply information from sequence alignments most commonly employ neural networks, Support Vector Machines (SVM) or Hidden-Markov Models (HMM,). These methods change input or optimize parameters according to the type of information they would finally provide, yielding valuable sets of 'fuzzy' structural constraints of both local and global nature.

Information about a local conformation of a protein can be obtained much more precisely than about global arrangements. For example, predictions of secondary structure, which are based on the analysis of position-specific sequence profiles, are 70–80% accurate.[82] Secondary structure prediction methods are used most commonly as a bias in folding simulations.[83] Various methods are applied for secondary structure prediction,[84] e.g. PSIPRED and PROFsec employ neural networks, and SAM-T06 uses an HMM based method.[85, 86]

Multiple sequence alignments are also a source of information about solvent accessibility, which is especially useful in structure-based prediction of protein function, because residues in active sites are typically exposed to solvent. Usually, residue hydrophobicity is also taken into account in accessibility predictions.[87, 88] Separate predictions of solvent accessibility from servers are rarely used in *de novo* methods, because corresponding statistical potential terms are already introduced into many force fields.[41, 89] Such burial potentials, implicitly describing interactions with solvent, typically include distributions of the number of inter-residue contacts for a given amino acid type.

Before applying any *de novo* methods for structure prediction it is important to find out if a query protein is globular, because most of these methods are not suited for predictions of the structure of proteins with transmembrane, disordered or coiled-coil regions. The first step in *de novo* protocols should be to predict if the query protein contains any of these regions. Such prediction is also based on analysis of multiple sequence alignments together with the distributions of hydrophobic and of positively charged amino acids (the 'positive-inside-rule' for prediction of transmembrane regions),[87, 90] or together with the distributions of hydrophobic and hydrophilic residues (for coiled-coil predictions).[91] In the prediction of disordered regions (which are frequently located at the N- and C-termini and typically have a biased composition of amino acids), PSI-BLAST profiles, secondary structure predictions, average hydrophobicity, net charges, particular amino acid frequencies and sequence complexity are involved.[92] In CASP7 the most successful prediction methods employing this information were VSL2.[93] POODLE[94] and DISOPRED[95] servers.

Most *de novo* methods are used for the prediction of single domain proteins. Consequently, reliable division of a query sequence into separate domains is crucial. Typically, each domain is folded by *de novo* methods separately, and the final orientation of the domains (an extremely difficult task), is predicted by other methods, often including evolutionary data.[48] Domain boundaries are usually detected based on multiple sequence alignment analysis,[96] or by using only the statistics of the appearance of amino acids at the boundaries, which could be especially useful for new folds.[97]

Apart from the local information obtained from multiple sequence alignments, some spatial constraints of much lower accuracy can be derived from the prediction of Cβ-Cβ contacts (defined as a contact when two Cβ atoms are closer in space than 8 Å). Such probable contacts are derived from multiple alignments, solvent accessibility and secondary structure correlations. In CASP6 the methods for contact prediction provided data high above random (20%) distributions and were quite useful in modeling new folds in the cases where the meta-predictors did not find any reliable templates. However, excepting rare examples,[98] most *de novo* structure prediction protocols still lack regular implementation of predicted contacts in the modeling pipelines. Information about residues which are close in space, if correct, is crucial for successful *de novo* methods mimicking protein folding. Particularly, in the folding of proteins the distance between cysteine residues plays the most important role in whether disulfide bonds form. Information about the disulfide bridges can be obtained either from NMR and mass spectroscopy experiments, or from theoretical predictions involving SVM or neural networks with profiles, often including solvent accessibility and secondary structure as additional inputs.[99] Disulfide bonds data is typically used in *de novo* modeling in the form of distance constraints between cysteine residues. Some *de novo* methods employ statistical contact-type potentials which favor interactions between every two cysteines,[89] but more precise information about possible disulfide bridges, as predicted by other methods, could be critical.

5.7 Perspectives

Although quite successful in application to small, single domain proteins, routine *de novo* protein structure prediction remains one of the most challenging unsolved problems of computational biology. It is clear that in the near future the problem will not be solved by the brute force of rapidly increasing computer power. The most promising methods appear to be various multiscale, hierarchical procedures, where crude models are built by means of mesoscopic algorithms and then refined by all-atom molecular mechanics. If needed, good quality *de novo* models could be bootstrapped with quantum mechanical simulations. Cascades of mesoscopic models at various resolution could be very useful, especially when applied to large systems. It is also clear that successful procedures need to be divided into two (or three) partially separated tasks: fold generation, fold selection and structure refinement. In this context, a very promising approach could be atomic-level knowledge-based statistical scoring functions. Finally, what is very important for molecular biology is to move a few steps beyond single protein structure prediction: designing efficient computational tools for protein folding pathway prediction and for modeling protein-protein/nucleic acid/lipid/peptide interactions. Progress in these areas will be extremely beneficial for understanding metabolic and signaling pathways, mechanisms of molecular transport, and other processes taking place in a living cell. It is also worth to note that even low resolution models are often sufficient for functional annotation by means of the fold assignment, enable finding binding/active sites by methods of three-dimensional motif searching, can be very helpful in refining/guiding the results of protein NMR experiments and in supporting site-directed mutagenesis experiments.[100] All of these aspects of structure prediction are extremely important for high-level computer-aided drug design and for modern biotechnology.

References

1. D. Baker, and A. Sali, Protein structure prediction and structural genomics, *Science*, **294**, 93–96 (2001).
2. R. Car, and M. Parrinello, Unified approach for molecular dynamics and density-functional theory, *Phys Rev Lett*, **55**, 2471–2474 (1985).
3. B. Lesyng, and J.A. McCammon, Molecular modeling methods. Basic techniques and challenging problems, *Pharmacol Ther*, **60**, 149–167 (1993).

4. M. Bixon, and S. Lifson, Potential functions and conformations in cycloalkanes', *Tetrahedron*, **23**, 769 (1967).
5. B. Brooks, R. Bruccoleri, B. Olafson, D. States, S. Swaminathan, and M. Karplus, CHARMM: A program for macromolecular energy, minimization, and dynamics calculations, *Journal of Computational Chemistry*, **4**, 187–217 (1983).
6. W. Cornell, P. Cieplak, C. Bayly, I. Gould, K. Merz, D. Ferguson, D. Spellmeyer, T. Fox, J. Caldwell, and P. Kollman, A Second Generation Force Field for the Simulation of Proteins, Nucleic Acids, and Organic Molecules, *J. Am. Chem. Soc.*, **117**, 5179–5197 (1995).
7. H.J.C.B. Wilfred and F. van Gunsteren. 'Computer Simulation of Molecular Dynamics: Methodology, Applications, and Perspectives in Chemistry.' 992–1023, 1990.
8. C. Anfinsen, Principles that govern the folding of protein chains, *Science*, **181**, 223–230 (1973).
9. A. Mackerell, M. Feig, and C. Brooks, Extending the treatment of backbone energetics in protein force fields: Limitations of gas-phase quantum mechanics in reproducing protein conformational distributions in molecular dynamics simulations, *Journal of Computational Chemistry*, **25**, 1400–1415 (2004).
10. A.D. Mackerell, Jr., Empirical force fields for biological macromolecules: overview and issues, *J Comput Chem*, **25**, 1584–1604 (2004).
11. H. Yu, and W. van Gunsteren, Accounting for polarization in molecular simulation, *Computer Physics Communications*, **172**, 69–85 (2005).
12. S. Tanaka, and H.A. Scheraga, Medium- and long-range interaction parameters between amino acids for predicting three-dimensional structures of proteins, *Macromolecules*, **9**, 945–950 (1976).
13. S. Miyazawa, and R.L. Jernigan, Estimation of effective inter-residue contact energies from protein crystal structures: quasi-chemical approximation, *Macromolecules*, **18**, 534–552 (1985).
14. M.J. Sippl, Calculation of conformational ensembles from potentials of mean force. An approach to the knowledge-based prediction of local structures in globular proteins, *J Mol Biol*, **213**, 859–883 (1990).
15. M.J. Sippl, Boltzmann's principle, knowledge-based mean fields and protein folding. An approach to the computational determination of protein structures, *J Comput Aided Mol Des*, 7, 473–501 (1993).
16. K.T. Simons, C. Kooperberg, E. Huang, and D. Baker, Assembly of protein tertiary structures from fragments with similar local sequences using simulated annealing and Bayesian scoring functions, *J Mol Biol*, **268**, 209–225 (1997).
17. T.L. Chiu, and R.A. Goldstein, Optimizing energy potentials for success in protein tertiary structure prediction, *Fold Des*, **3**, 223–228 (1998).
18. M.H. Hao, and H.A. Scheraga, How optimization of potential functions affects protein folding, *Proc Natl Acad Sci U S A*, **93**, 4984–4989 (1996).
19. L.A. Mirny, and E.I. Shakhnovich, How to derive a protein folding potential? A new approach to an old problem, *J Mol Biol*, **264**, 1164–1179 (1996).
20. Y. Xia, E.S. Huang, M. Levitt, and R. Samudrala, Ab initio construction of protein tertiary structures using a hierarchical approach, *J Mol Biol*, **300**, 171–185 (2000).
21. M. Levitt, and A. Warshel, Computer simulation of protein folding, *Nature*, **253**, 694–698 (1975).
22. D.J. Wales, and H.A. Scheraga, Global optimization of clusters, crystals, and biomolecules, *Science*, **285**, 1368–1372 (1999).
23. Z. Li, and H.A. Scheraga, Monte Carlo-minimization approach to the multiple-minima problem in protein folding, *Proc Natl Acad Sci U S A*, **84**, 6611–6615 (1987).
24. J. Lee, H. Scheraga, and S. Rackovsky, New optimization method for conformational energy calculations on polypeptides: Conformational space annealing, *Journal of Computational Chemistry*, **18**, 1222–1232 (1997).
25. L. Piela, J. Kostrowicki, and H. Scheraga, On the multiple-minima problem in the conformational analysis of molecules: deformation of the potential energy hypersurface by the diffusion equation method, *Journal of Physical Chemistry*, **93**, 3339–3346 (1989).

26. T. Dandekar, and P. Argos, Folding the main chain of small proteins with the genetic algorithm, *J Mol Biol*, **236**, 844–861 (1994).
27. S. Sun, Reduced representation model of protein structure prediction: statistical potential and genetic algorithms, *Protein Sci*, **2**, 762–785 (1993).
28. R. Unger, and J. Moult, Genetic algorithms for protein folding simulations, *J Mol Biol*, **231**, 75–81 (1993).
29. U.H. Hansmann, and Y. Okamoto, New Monte Carlo algorithms for protein folding, *Curr Opin Struct Biol*, **9**, 177–183 (1999).
30. U. Hansmann, Parallel Tempering Algorithm for Conformational Studies of Biological Molecules, *Chem. Phys. Lett.*, **281**, 140–150 (1997).
31. D. Gront, A. Kolinski, and J. Skolnick, Comparison of three Monte Carlo conformational search strategies for a proteinlike homopolymer model: Folding thermodynamics and identification of low-energy structures, *The Journal of Chemical Physics*, **113**, 5065–5071 (2000).
32. D. Earl, and M. Deem. 'Parallel Tempering: Theory, Applications, and New Perspectives.' 3910–3916, 2005.
33. Y. Sugita, and Y. Okamoto, Replica-exchange molecular dynamics method for protein folding, *Chemical Physics Letters*, **314**, 141–151 (1999).
34. D. Gront, and A. Kolinski, Efficient scheme for optimization of parallel tempering Monte Carlo method, *Journal of Physics: Condensed Matter*, **19**, 036225–036234 (2007).
35. H. Katzgraber, S. Trebst, D. Huse, and M. Troyer, Feedback-optimized parallel tempering Monte Carlo, *Journal of Statistical Mechanics: Theory and Experiment*, **2006**, P03018 (2006).
36. N. Rathore, M. Chopra, and J.J. de Pablo, Optimal allocation of replicas in parallel tempering simulations, *J Chem Phys*, **122**, 024111 (2005).
37. A. Liwo, P. Arlukowicz, S. Oldziej, C. Czaplewski, M. Makowski, and H.A. Scheraga, Optimization of the UNRES Force Field by Hierarchical Design of the Potential-Energy Landscape. 1. Tests of the Approach Using Simple Lattice Protein Models, *J. Phys. Chem. B*, **108**, 16918–16933 (2004).
38. A. Liwo, P. Arlukowicz, C. Czaplewski, S. Oldziej, J. Pillardy, and H.A. Scheraga, A method for optimizing potential-energy functions by a hierarchical design of the potential-energy landscape: application to the UNRES force field, *Proc Natl Acad Sci U S A*, **99**, 1937–1942 (2002).
39. M. Boniecki, P. Rotkiewicz, J. Skolnick, and A. Kolinski, Protein fragment reconstruction using various modeling techniques, *J Comput Aided Mol Des*, **17**, 725–738 (2003).
40. E. Malolepsza, M. Boniecki, A. Kolinski, and L. Piela, Theoretical model of prion propagation: a misfolded protein induces misfolding, *Proc Natl Acad Sci U S A*, **102**, 7835–7840 (2005).
41. C.A. Rohl, C.E. Strauss, K.M. Misura, and D. Baker, Protein structure prediction using Rosetta, *Methods Enzymol*, **383**, 66–93 (2004).
42. R.L. Dunbrack, Jr., and F.E. Cohen, Bayesian statistical analysis of protein side-chain rotamer preferences, *Protein Sci*, **6**, 1661–1681 (1997).
43. J.M. Bujnicki, Protein-structure prediction by recombination of fragments, *Chembiochem*, **7**, 19–27 (2006).
44. D.T. Jones, K. Bryson, A. Coleman, *et al.*, Prediction of novel and analogous folds using fragment assembly and fold recognition, *Proteins*, **61 Suppl 7**, 143–151 (2005).
45. G.F. Chikenji, Y.; Takada, S., A reversible fragment assembly method for de novo protein structure prediction, **119**, 6895–6903 (2003).
46. L.H. Hung, S.C. Ngan, T. Liu, and R. Samudrala, PROTINFO: new algorithms for enhanced protein structure predictions, *Nucleic Acids Res*, **33**, W77–80 (2005).
47. M.R. Betancourt, A reduced protein model with accurate native-structure identification ability, *Proteins*, **53**, 889–907 (2003).
48. A. Kolinski, and J. Bujnicki, Generalized protein structure prediction based on combination of fold-recognition with de novo folding and evaluation of models, *Proteins*, **61**, 84–90 (2005).
49. M. Kurcinski, and A. Kolinski, Hierarchical modeling of protein interactions, *J Mol Model*, **13**, 691–698 (2007).
50. M. Kurcinski, and A. Kolinski, Steps towards flexible docking: modeling of three-dimensional structures of the nuclear receptors bound with peptide ligands mimicking co-activators' sequences, *J Steroid Biochem Mol Biol*, **103**, 357–360 (2007).

51. S. Kmiecik, D. Gront, and A. Kolinski, Towards the high-resolution protein structure prediction. Fast refinement of reduced models with all-atom force field, *BMC Struct Biol*, **7**, 43
(2007).
52. S. Kmiecik, and A. Kolinski, Characterization of protein-folding pathways by reduced-space
modeling, *Proc Natl Acad Sci U S A*, **104**, 12330–12335 (2007).
53. Y. Zhang, A. Kolinski, and J. Skolnick, TOUCHSTONE II: a new approach to ab initio protein
structure prediction, *Biophys J*, **85**, 1145–1164 (2003).
54. Y. Zhang, A. Arakaki, and J. Skolnick, TASSER: An automated method for the prediction of
protein tertiary structures in CASP6, *Proteins*, **61 Suppl 7**, 91–98 (2005).
55. Y. Zhang, I-TASSER server for protein 3D structure prediction, *BMC Bioinformatics*, **9**, 40
(2008).
56. A. Kolinski, and J. Skolnick, Assembly of protein structure from sparse experimental data: an
efficient Monte Carlo model, *Proteins*, **32**, 475–494 (1998).
57. A. Kolinski, P. Rotkiewicz, B. Ilkowski, and J. Skolnick, A method for the improvement of
threading-based protein models, *Proteins*, **37**, 592–610 (1999).
58. J. Skolnick, A. Kolinski, and A.R. Ortiz, MONSSTER: a method for folding globular proteins
with a small number of distance restraints, *J Mol Biol*, **265**, 217–241 (1997).
59. M. Vieth, A. Kolinski, C.L. Brooks, 3rd, and J. Skolnick, Prediction of the folding pathways
and structure of the GCN4 leucine zipper, *J Mol Biol*, **237**, 361–367 (1994).
60. E. Huang, R. Samudrala, and J. Ponder, Ab Initio fold prediction of small helical proteins using
distance geometry and knowledge-based scoring functions, *Journal of Molecular Biology*, **290**,
267–281 (1999).
61. P. Bradley, K. Misura, and D. Baker, Toward High-Resolution deNovo Structure Prediction for
Small Proteins, *Science*, **309**, 1868–1871 (2005).
62. M. Feig, P. Rotkiewicz, A. Kolinski, J. Skolnick, and C. Brooks, Accurate reconstruction of all-
atom protein representations from side-chain-based low-resolution models, *Proteins: Structure,
Function, and Genetics*, **41**, 86–97 (2000).
63. M. Claessens, E. van Cutsem, I. Lasters, and S. Wodak, Modelling the polypeptide
backbone with 'spare parts' from known protein structures, *Protein Eng.*, **2**, 335–345
(1989).
64. L. Holm, and C. Sander, Database algorithm for generating protein backbone and side-chain
co-ordinates from a C alpha trace application to model building and detection of co-ordinate
errors, *J Mol Biol*, **218**, 183–194 (1991).
65. T.A. Jones, and S. Thirup, Using known substructures in protein model building and crystallography, *EMBO J*, **5**, 819–822 (1986).
66. D. Gront, S. Kmiecik, and A. Kolinski, Backbone building from quadrilaterals: A fast and accurate algorithm for protein backbone reconstruction from alpha carbon coordinates, *J Comput
Chem*, **28**, 1593–1597 (2007).
67. E.O. Purisima, and H.A. Scheraga, Conversion from a virtual-bond chain to a complete polypeptide backbone chain, *Biopolymers*, **23**, 1207–1224 (1984).
68. R. Kazmierkiewicz, A. Liwo, and H. Scheraga, Energy-based reconstruction of a protein backbone from its Alpha-carbon trace by a Monte-Carlo method, *Journal of Computational Chemistry*, **23**, 715–723 (2002).
69. S.A. Adcock, Peptide backbone reconstruction using dead-end elimination and a knowledge-
based forcefield, *J Comput Chem*, **25**, 16–27 (2004).
70. M. Milik, A. Kolinski, and J. Skolnick, Algorithm for rapid reconstruction of protein
backbone from alpha carbon coordinates, *Journal of Computational Chemistry*, **18**, 80–85
(1996).
71. R. Shetty, P. de Bakker, M. Depristo, and T. Blundell, Advantages of fine-grained side chain
conformer libraries, *Protein Eng.*, **16**, 963–969 (2003).
72. Dunbrack, Jr., and M. Karplus, Backbone-dependent Rotamer Library for Proteins Application
to Side-chain Prediction, *Journal of Molecular Biology*, **230**, 543–574 (1993).
73. D. Shortle, K.T. Simons, and D. Baker, Clustering of low-energy conformations near
the native structures of small proteins, *Proc Natl Acad Sci U S A*, **95**, 11158–11162
(1998).

74. M. Betancourt, and J. Skolnick, Finding the needle in a haystack: educing native folds from ambiguous ab initio protein structure predictions, *Journal of Computational Chemistry*, **22**, 339–353 (2001).
75. D. Gront, and A. Kolinski, A new approach to prediction of short-range conformational propensities in proteins, *Bioinformatics*, **21**, 981–987 (2005).
76. D. Gront, and A. Kolinski, HCPM–program for hierarchical clustering of protein models, *Bioinformatics*, **21**, 3179–3180 (2005).
77. C.A. Rohl, and D. Baker, De novo determination of protein backbone structure from residual dipolar couplings using Rosetta, *J Am Chem Soc*, **124**, 2723–2729 (2002).
78. G. Cornilescu, F. Delaglio, and A. Bax, Protein backbone angle restraints from searching a database for chemical shift and sequence homology, *J Biomol NMR*, **13**, 289–302 (1999).
79. J. Meiler, and D. Baker, Rapid protein fold determination using unassigned NMR data, *Proc Natl Acad Sci U S A*, **100**, 15404–15409 (2003).
80. D. Latek, D. Ekonomiuk, and A. Kolinski, Protein structure prediction: combining de novo modeling with sparse experimental data, *J Comput Chem*, **28**, 1668–1676 (2007).
81. J. Kosinski, I. Cymerman, M. Feder, M. Kurowski, J. Sasin, and J. Bujnicki, A 'FRankenstein's monster' approach to comparative modeling: Merging the finest fragments of Fold-Recognition models and iterative model refinement aided by 3D structure evaluation, *Proteins: Structure, Function, and Genetics*, **53**, 369–379 (2003).
82. K. Bryson, L.J. McGuffin, R.L. Marsden, J.J. Ward, J.S. Sodhi, and D.T. Jones, Protein structure prediction servers at University College London, *Nucleic Acids Res*, **33**, W36–38 (2005).
83. A.R. Ortiz, A. Kolinski, P. Rotkiewicz, B. Ilkowski, and J. Skolnick, Ab initio folding of proteins using restraints derived from evolutionary information, *Proteins*, **Suppl 3**, 177–185 (1999).
84. B. Rost, Review: protein secondary structure prediction continues to rise, *J Struct Biol*, **134**, 204–218 (2001).
85. D.T. Jones, Protein secondary structure prediction based on position-specific scoring matrices, *J Mol Biol*, **292**, 195–202 (1999).
86. K. Karplus, S. Katzman, G. Shackleford, *et al.*, SAM-T04: what is new in protein-structure prediction for CASP6, *Proteins*, **61 Suppl 7**, 135–142 (2005).
87. B. Rost, G. Yachdav, and J. Liu, The PredictProtein server, *Nucleic Acids Res*, **32**, W321–326 (2004).
88. B. Rost, Prediction in 1D: secondary structure, membrane helices, and accessibility, *Methods Biochem Anal*, **44**, 559–587 (2003).
89. A. Kolinski, Protein modeling and structure prediction with a reduced representation, *Acta Biochim Pol*, **51**, 349–371 (2004).
90. D.T. Jones, Improving the accuracy of transmembrane protein topology prediction using evolutionary information, *Bioinformatics*, **23**, 538–544 (2007).
91. A. Lupas, M. Van Dyke, and J. Stock, Predicting coiled coils from protein sequences, *Science*, **252**, 1162–1164 (1991).
92. F. Ferron, S. Longhi, B. Canard, and D. Karlin, A practical overview of protein disorder prediction methods, *Proteins*, **65**, 1–14 (2006).
93. K. Peng, P. Radivojac, S. Vucetic, A.K. Dunker, and Z. Obradovic, Length-dependent prediction of protein intrinsic disorder, *BMC Bioinformatics*, **7**, 208 (2006).
94. K. Shimizu, Y. Muraoka, S. Hirose, and T. Noguchi, Feature Selection Based on Physicochemical Properties of Redefined N-term Region and Cterm Regions for Predicting Disorder., *Proceedings of 2005 IEEE Symposium on Computational Intelligence in Bioinformatics and Computational Biology*, **337**, 635–645 (2005).
95. J.J. Ward, L.J. McGuffin, K. Bryson, B.F. Buxton, and D.T. Jones, The DISOPRED server for the prediction of protein disorder, *Bioinformatics*, **20**, 2138–2139 (2004).
96. D.E. Kim, D. Chivian, L. Malmstrom, and D. Baker, Automated prediction of domain boundaries in CASP6 targets using Ginzu and RosettaDOM, *Proteins*, **61 Suppl 7**, 193–200 (2005).
97. N.V. Dovidchenko, M.Y. Lobanov, and O.V. Galzitskaya, Prediction of number and position of domain boundaries in multi-domain proteins by use of amino acid sequence alone, *Curr Protein Pept Sci*, **8**, 189–195 (2007).

98. D. Kihara, H. Lu, A. Kolinski, and J. Skolnick, TOUCHSTONE: an ab initio protein structure prediction method that uses threading-based tertiary restraints, *Proc Natl Acad Sci U S A*, **98**, 10125–10130 (2001).
99. J. Cheng, H. Saigo, and P. Baldi, Large-scale prediction of disulphide bridges using kernel methods, two-dimensional recursive neural networks, and weighted graph matching, *Proteins*, **62**, 617–629 (2006).
100. R. Sanchez, U. Pieper, F. Melo, *et al.*, Protein structure modeling for structural genomics, *Nat Struct Biol*, **7 Suppl**, 986–990 (2000).

6

Quality Assessment of Protein Models

Björn Wallner and Arne Elofsson

6.1 Introduction

When building a protein model, with or without the aid of experimental information, it is often necessary to use an independent measure to evaluate the correctness of the model. This is the role of Model Quality Assessment programs (MQAPs). Different types of MQAPs have been developed during the last decades. The goal of all these methods is to assess the quality of protein models. However, the definition of quality differs depending on the problem, thus it is always important to consider the specific problem to be solved when using an MQAP.

Traditionally MQAPs are methods that evaluate the quality of a protein model. However, during recent years other types of MQAPs, including consensus based MQAP, has increased in importance. Until recently most work has been focused on the development of methods aimed at detecting the native structures and to separate these from incorrect models. However, today one of the most important uses of MQAPs is to select the best out of a set of models built by homology or by other methods. Although these two problems clearly are related, there is no guarantee that a method that works well on the first problem works well on the other, in particular when all of the plausible models are of low quality.

In this chapter we will first discuss how MQAPs have been used in the past and how they are used today and finally we will present an analysis of how MQAPs performed in CASP7.[1]

The first use of MQAPs was to detect erroneous models from X-ray crystallography. X-ray based models might be wrong, in particular when the resolution of the diffraction data is quite low, but other types of errors, including tracing a chain backwards through the electron density also occurs. Here, the number of residues in disallowed regions in the

Prediction of Protein Structures, Functions, and Interactions   Edited by Janusz Bujnicki
© 2009 John Wiley & Sons, Ltd

Ramachandran plot can often be used as an indicator for the quality of the X-ray model. This so called 'stereochemical correctness' could for instance have detected the wrongly built small subunit of Rubisco by Eisenberg and coworkers.[2] In the early 1990s a number of MQAPs were developed to identify wrongly built models using Ramachandran plots and other measured of 'stereochemical correctness' as the main source of information. The best known of these methods are PROCHECK[3] and WHATCHECK.[4] Today with improved refinement methods the 'stereochemical correctness' of X-ray models is almost always very good. Also, with the introduction of the R-free method[5] where a fraction of the data is only used for testing the need for MQAPs to validate protein models built from crystallographic constraints has decreased. However, the recent discovery that a number of globular[6] and membrane[7] protein models were wrong indicated that their days are not completely over.

Besides the 'stereochemical' MQAP a number of 'statistical' MQAPs were also developed in the 1990s, e.g. Verify3D[8] and ProsaII.[9]

Several of these methods were developed with a dual purpose, the identification of erroneous structures and the identification of proteins that share the same fold. These methods were based on statistical features of correct protein structures therefore, they are therefore also often referred to as knowledge-based energy functions. The general idea is that features that commonly seen in known structures will yield good scores and unusual features will yield bad scores. Different methods utilize different features, Sippl pioneered the use of pairwise contacts both for fold recognition and as an MQAP, while Eisenberg and co-workers developed similar methods using 3D-1D-profiles, i.e. calculations on the probability to find a certain amino-acid type in a particular structural environment.

The second use of MQAPs is in molecular modeling. Protein modeling by homology used to be a time-consuming art, where hours of CPU and user time were used to create a single model. However, with the computing power of today it is possible to generate hundreds of alternative models for a given target with just a single click on a meta-server, such as Pcons.net,[10] bioinfo.pl[11] or genesilico.pl.[12] Here, with the increasing number of models it has become increasingly more important to use an MQAP that can identify the most accurate model.

Here, the ultimate goal of a MQAP is to select the most accurate model from a set of many alternative models and also to provide a reasonable estimate of the accuracy of the models. In the discussion part of this chapter we will review how well different MQAPs performed on this task during CASP7.

6.2   Short Historical Overview

6.2.1   Verify3D

The 3D-1D-profile method was primarily developed to detect distantly related proteins.[13] However, it was later also used to identify incorrect protein models,[8] for this purpose it was later renamed to Verify3D. In retrospect, the Verify3D MQAP has most likely been more successful than 3D-1D-profile method. Even today Verify3D quite well even compared with the best MQAPs. While, methods using multiple sequence alignments have been shown to perform significantly better than 3D-1D-profiles for remote homology detection.

The basic idea of Verify3D (and 3D-1D-profiles) is to assign an environmental class to each residue in a protein. In Verify3D the environments are divided into 18 classes based on the secondary structure, area buried and the fraction of polar contacts. Secondly, the probability for each amino acid type to be in each type of environment is calculated. When a model is evaluated the sum of probabilities over a window, or the entire proteins, is calculated and if the probability is low it is likely that the model is incorrect.

One feature that should be noted in the original 3D-1D-profiles is the use of 'area buried' and not 'fraction buried'. This causes for instance a glycine, which can not have an area buried larger than the area of its side chain, to be restricted to a few of the 18 environments, while a large amino acids such as tryptophan could be in all environments.

6.2.2   ProsaII

In parallel with the development of the 3D-1D-profiles several groups developed fold recognition, or threading, methods based on the contact probability between pairs of residues.[9, 14] Although these methods are not straightforward to use in fold recognition, as they break the conditions for standard dynamic programming, they showed some great success in the early rounds of CASPs. However, today these methods have also been outrun by the rapid increase in sequence database sizes and methods that utilize this information better. But as for 3D-1D-profiles, the MQAP ProsaII is still very useful.

The basis for ProsaII, and for Threader, is the probability for two residues to be at a specific distance from each other. In the simplest methods of this type[15] only the probability to be a contact or not is included, while in more sophisticated methods, such as ProsaII, the amino acid types, the distance as well as the sequence separations are used.[16] The distances are normally calculated from the $C\alpha$ or $C\beta$ atoms of a residues but the closest distance from any atom in the side chain can also be used. A typical problem in these methods is normalization, e.g. how should the fact that hydrophobic residues are more frequent at the interior of the protein or that side chain sized differs be utilized.

6.2.3   ERRAT

Physical based energy functions are almost always built on the potential of atomic inter-action energies, while the knowledge-based energy functions discussed so far are based on residue properties. However, there also exist some knowledge-based energy-functions that utilize atomistic properties. Colovos *et al.* (1993) used the distribution of atom-atom contacts to develop the ERRAT method,[17] Melo *et al.* developed a mean force potential at the atomic level[18] while others have used a distant-dependent atomic potential.[19, 20]

ERRAT is based on the probability that two atoms of a particular type are in contact. The major difference in comparison with the residues contact based methods is that ERRAT does not base it on a sum of probabilities that two atoms are in contact, instead in ERRAT the fraction of all contacts that is of a particular type is used.

6.2.4   PROCHECK and WHATCHECK

At the same time as the development of statistical methods to evaluate the correctness of a protein model, methods to evaluate the stereochemical correctness of these models were developed.[3, 21] In these methods the number of residues in disallowed Ramachandran plots, the number of strange bond lengths or angels etc. are measured. A clear correlation between the accuracy of a protein structure and these criteria was found. However, these criteria are not very useful to evaluate theoretically constructed models as they are easy to fulfill and nowadays virtually all modeling programs are able to build models with very good stereochemistry even using alignments that are completely wrong.

6.3 MQAPs Developed to Predict Quality of Models

The methods discussed above have been developed with the main focus to distinguish between native and non-native structures. However, in even the best protein models for difficult modeling targets are sometimes not very native-like. Therefore, another objective of MQAP can be to find the best possible model. A task that is far more difficult than to simply distinguish between native and non-native structures.

6.3.1 ProQ

ProQ[22] was one of the first methods that used protein models with different similarity as a target function. In ProQ various properties from a model were calculated and used to train a set of neural networks. Each structure was described by a set of structural features such as: atom-atom contacts, residue-residue contacts, surface area exposure and secondary structure agreement and the neural network was trained to predicted the protein model quality.

6.3.2 Victor/FRST

Victor/FRST[23] was one of the top performing MQAPs in CAFASP4 MQAP. It is a statistical potential with four energy terms representing pairwise, solvation, hydrogen bond, and torsion angle potentials, combined with a linear weighting function.

6.3.3 ABIpro-h

ABIpro-h is one of the more advanced methods developed by Baldi and coworkers.[24] It combines predicted structural features such as secondary structure, relative solvent accessibility and residue-residue contacts, with physical energy terms for hydrogen bonding, van der Waals interactions and electrostatics. In addition it also includes statistical terms for residue solvent environment and local structure residue pairing.

6.3.4 Circle-QA

Circle-QA uses 3D-1D-profile combined with predicted secondary structure.[25] The 3D-1D-profiles are similar to the profiles used in Verify3D and consist of three parameters: fraction of buried area, fraction of polar area and secondary structure. The main differences are that the 3D-1D-profiles and the combined scores are difficult dependent. For difficult targets a larger weight is given to the secondary structure information in the 3D-1D-profile. The rational for having difficult dependent scores is that different features are important to distinguish the best models among high quality models compared to among low quality models. When the model quality is low giving higher scores to models where the secondary

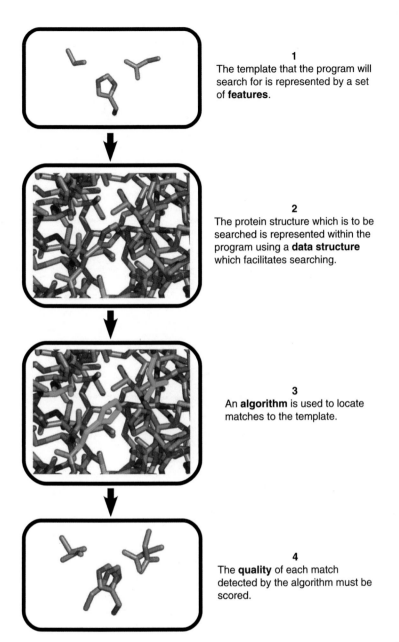

1
The template that the program will search for is represented by a set of **features**.

2
The protein structure which is to be searched is represented within the program using a **data structure** which facilitates searching.

3
An **algorithm** is used to locate matches to the template.

4
The **quality** of each match detected by the algorithm must be scored.

Figure 8.1 *Components of a template matching method. This figure was prepared using Pymol (www.pymol.org)*

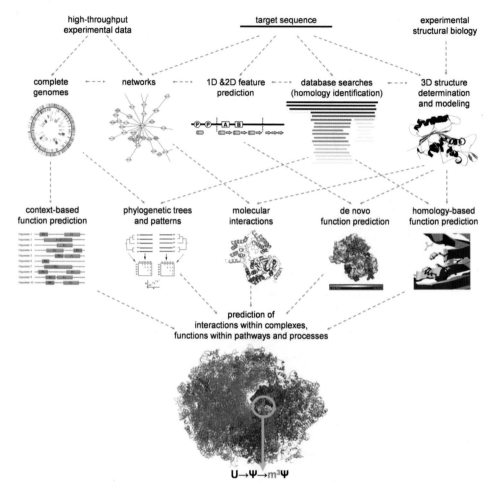

high-throughput
experimental data

target sequence

experimental
structural biology

complete
genomes

networks

1D &2D feature
prediction

database searches
(homology identification)

3D structure
determination
and modeling

context-based
function prediction

phylogenetic trees
and patterns

molecular
interactions

de novo
function prediction

homology-based
function prediction

prediction of
interactions within complexes,
functions within pathways and processes

$U \rightarrow \Psi \rightarrow m^3\Psi$

Figure 11.3 *Integration of different prediction approaches and experimental data. Genomic context and network analyses can be used to identify various types of interactions and suggesting cellular pathways the protein participates in, rather than by suggesting a specific biochemical activity. On the other hand, homology-based predictions can be used to make complementary predictions about biochemical details, starting from identification of putative active sites of enzymes and suggesting potential reaction types to be catalyzed. De novo methods are capable of identifying structural and functional analogies between unknown systems and previously studied systems, regardless of the presence or absence of evolutionary relationships. Combination of context-based methods with homology-based methods and de novo predictions (e.g. macromolecular docking of models guided by restraints) can be used to predict not only which proteins interact, but also how, e.g. what is the structure of the complex, allowing further inferences to be made about the spatial and temporal aspects of biological processes*

structure agrees with the predicted is probably a good idea. For high quality models the secondary structure is most likely already correct and the exact side chain packing becomes increasingly important.

6.3.5   Meta Methods

Lately a number of methods that combine the output from several other MQAPs have been used. Two of these are described below.

MetaMQAP. MetaMQAP uses an SVN to predict the CA-CA deviation per residue based on scores returned by five primary MQAPs. An updated version of the program, MetaMQAPII uses a regression model instead of an SVN and eight primary MQAPs: Verify3D,[8] ProsaII,[9] BALA,[26] ANOLEA,[18] PROVE,[27] TUNE,[28] REFINER,[29] and ProQ.[22] This method was not available when the benchmark was conducted and all references to the MetaMQAP are to the old method.

QA-ModFOLD. QA-ModFOLD[30] is another meta-MQAP that uses artificial neural networks and the output scores from MODCHECK,[31] ProQ[22] and ModSSEA to predict the protein model quality. ModSSEA is a simple MQAP that was develop together with QA-ModFOLD. It is based on secondary structure element alignments (SSEA) between predicted and actual secondary structure assignment.

6.3.6   Consensus Based MQAPs

A completely different set of MQAPs was introduced a few years ago.[32] These consensus MQAPs do not try to evaluate the quality of a model, instead they use the similarity between a model and many other models. Certainly this type of evaluation is only useful when a large set of models for the same target. However, nowadays this is frequently the case when building models using homology.

Pcons. Pcons was the first consensus method. Pcons and other consensus methods utilize a set of alternative protein models as their input.[10, 32, 33] In Pcons a structural superposition algorithm, Lgscore,[34] is used to search for recurring structural patterns in the whole set of models. Pcons predicts the quality of all models, by assigning a score to each model reflecting the average similarity to the entire ensemble of models. The idea being that recurring patterns are more likely to be correct as compared to patterns than only occur in one or a few models. In addition, Pcons also contains terms from predicted model quality (ProQ) and any server score attached to the models.

6.3.7   3D-Jury

The 3D-Jury method introduced by Rychlewski and co-workers is a simplified version of Pcons.[35] Here only the similarity between models is used. At the bioinfo.pl meta server 3D-Jury can be used to evaluate different sets of models. In several benchmarks it has been shown that 3D-Jury, despite its simplicity, performs on par with Pcons.

6.4   Local versus Global Predictors

Several of the methods described above predict a single global quality measure for each protein model. One drawback of such a measure is that it does not recognize correct and incorrect regions in a protein model. High scoring models might contain regions that can be improved or combined to produce a more accurate model. Thus, an obvious extension of the global measure is to analyze the details of various structural regions and predict a local quality score for each individual residue. This knowledge can in turn be used as a guide during the refinement process to provide confidence measures for what parts of protein model to trust. ProsaII[9] and Verify3D[8, 36] are perhaps the most utilized MQAPs for assessing local quality correctness and they have been used successfully in CASP to select well- and poorly-folded fragments.[37, 38]

6.4.1   ProQprof, ProQres and ProQlocal

Three recently developed local quality predictors are ProQres, ProQprof and ProQlocal.[10, 39] ProQres and ProQprof predict the local quality of a residue using different types of information. ProQres evaluates the structure, while ProQprof base the prediction on alignment information. These two approaches provide complementary information that is combined in ProQlocal, e.g. the structure might be OK while the sequence similarity is low or vice versa. In a recent benchmark, the combined approach was found to be slightly better than any of the individual methods.[39]

The structural features used in ProQres are identical to the ones in ProQ,[40] i.e. atom-atom contacts, residue-residue contacts, solvent accessibility surfaces and secondary structure information. However, in order to achieve a localized quality prediction the environment around each residue is described by calculating the structural features for a sliding window around the central residue. Hereby, the quality of the central residue is predicted not only by its own features, but also by the environment of the whole window, i.e. by all features (contacts, solvent accessibility and secondary structure) involving the residue in the window and their contacts.

ProQprof utilize the target-template alignment to achieve its prediction. This type of analysis was first performed by Tress *et al.* (2003) where a profile-derived alignment score were used to predict reliable regions in protein alignments.[41] They concluded that positions in the models with a high profile alignment score were more likely to be correct compared to positions with lower scores. In ProQprof this idea is extended to a window of profile-profile scores that are used as input to a neural network trained to predict the local quality for the central residue in the window.

6.4.2   Pcons – local version

A simple approach to predict the local quality based on *consensus analysis* is the Pcons-local method.[10, 39] This approach is almost identical to the confidence assignment step in 3D-SHOTGUN.[42] The idea is simple: to estimate the quality of a residue in a protein model, the whole model is compared to all other models for that protein by superimposing all models and calculate the S-score for each residue. The average S-score for each residue then reflects how well conserved the position of a particular residue is in the whole set of models. It is quite likely that correct positions are well conserved between all models and that incorrect positions are less conserved. Obviously, it is only possible to perform the consensus analysis if there exist a number of models for the same target sequence.

6.5 Performance in CASP7

In CASP7 the importance to be able to predict both the global and local quality was acknowledged by the CASP organizers, and the Quality Assessment category was divided into two parts. The goal in the first section was to predict the global quality of protein models, whereas the goal in the second category was to predict the quality of individual residues.

6.5.1 Evaluation of Global Quality

Here, GDT_TS[43] is used as the evaluation measure of model correctness, as this nowadays is a CASP standard. There are some discrepancies in performance depending on which evaluation measures that are used, in particular in terms of correlation (see Discussion and Wallner, 2007[1]), but for the calculation of the overall quality of the models the difference is marginal.

The analysis is performed using the full-length PDB chains, i.e. not using the domain definitions provided by the CASP organizers. The reason for this is that the predicted quality was the quality for the full-length PDB chain and not the quality of the individual domains.

For assessing the overall performance, two measures were used: (1) average Pearson's correlation coefficient per target and (2) the sum of GDT_TS for the highest ranked model (GDT1). During the evaluation other measure such as the best of the top five ranked models and Receiver Operating Characteristic curves (ROC) were used. However, the overall ranking using any of these measures is consistent with the two measures used.

6.5.2 Evaluation of Local Quality

To assess the methods predicting local quality, a similar analysis protocol as in Wallner *et al.*, 2006 was utilized.[39] This protocol simplifies the comparison of different methods, by restricting the analysis to the 10% highest and the 10% lowest scoring residues from each method and measure the performance by averaging the true local quality measures in the high and low scoring sets, respectively. Here, two local quality measures are used to assess performance: average local deviation between equivalent CA atoms in the model and the native structure and the fraction of wrongly predicted residues defined by residues with local deviation $\geq 3\text{Å}$ or $\leq 3\text{Å}$ for the 10% highest and lowest ranked residues, respectively.

The average local deviation, d^{CA}, was scaled using the following formula when calculating the average:

$$T = \frac{1}{1 + d^{CA}} \tag{6.1}$$

and then transformed back to a distance using

$$\langle d^{CA}_{scaled} \rangle = \frac{1}{\langle T \rangle} - 1 \tag{6.2}$$

In addition, the correlation coefficient between the predicted CA-CA deviation between the native and the actual deviation was calculated for each model. However, it should be noted that the correlation does not take accuracy into account, but it was found that the accuracy in general is acceptable when the correlation coefficient is higher than 0.6.

6.5.3   Results

We participated with three MQAPs in the global and four MQAPs in the local quality assessment category of CASP7, for an overview see Table 6.1. Below we first compare the performance of our global quality predictors (Pcons, ProQ and ProQprof) with each other as well as with the best performing groups. Thereafter, we conduct a similar analysis for the performance of our local quality predictors (Pcons, ProQres, ProQprof and ProQlocal). In the local category only seven groups participated, thus we did not restrict the analysis to the best performing groups. In the official CASP statistics nine groups are listed as having submitted predictions in this category, but two of these have so few predictions that they were impossible to assess.

6.5.4   Global Quality Assessment

The performance for predicting global quality was assessed by comparing all groups using correlation coefficient between predicted quality and GDT_TS (R_GDT) and the sum of GDT_TS for the first ranked models (GDT1). The result for our three MQAPs and the best performing groups are summarized in Table 6.2.

Pcons shows the highest correlation to GDT_TS of all MQAPs (Table 6.2). This high correlation (R = 0.89) to GDT_TS is impressive considering it was trained to predict another quality measure (LGscore) that does not correlate perfectly with GDT_TS (R = 0.90). In this light the correlation of Pcons is probably as close to optimal as it can get, since it cannot be expected to perform better than the quality measure it is trained to predict. The superior Pcons correlation is maintained both for easy and hard targets, but the average correlation is clearly lower for the hard targets, 0.77 vs. 0.96.

Surprisingly, two MQAPs, ABIpro-h and Circle-QA selects models that are equal to or even slightly better than the Pcons model although their correlations are significantly worse. Particularly impressive is ABIpro-h that selects better models than Zhang-Server for hard targets, which no other MQAP is able to do.

ProQ is ranked fourth overall based on correlation and clearly below the top groups in terms of selecting the best model. It performs similar to QA-ModFOLD which is a consensus MQAP that uses ProQ as one of its inputs. As expected, ProQ is not able to perform as good as Pcons (Table 6.2).

The ProQprof method is not a designated global quality predictor; instead the global prediction is derived from the prediction of local quality scores. Clearly, ProQprof is not as successful as Pcons or ProQ. In fact it performs rather poorly, especially for the hard targets where the selected models are no better than randomly selected ones (Table 6.2). By analyzing which models are actually selected by ProQprof it is evident that the routine, for generating the target-template alignment from a model and a template structure using structural alignments, is not optimal. This is most problematic for the hard targets where the most successful predictors use fragments or multiple templates.

Table 6.1 Overview of different MQAPs. The information different MQAPs utilize and the global and local quality measure that is predicted

MQAP	Utilized information	Predicted global quality measure	Predicted local quality measure	Availability
		Our MQAPs		
Pcons	Consensus	LGscore	S-score	http://pcons.net[10]
ProQ	Structure, ANN	LGscore	S-score	http://pcons.net[44]
ProQprof	Sequence similarity	Average S-score	S-score	http://pcons.net[39]
ProQlocal	Structure+Sequence	–	S-score	http://pcons.net[39]
		Other global MQAPs		
LEE	Similarity to own models	GDT_TS	–	N/A
QA-ModFOLD	Combination of other MQAPs	TM-score	–	http://www.biocentre.rdg.ac.uk/bioinformatics/ModFOLD/[30]
ABIpro-H	Structure, ANN	–	–	N/A[45]
Circle-QA	3D-1D structural profiles	–	–	N/A[45]
		Other local MQAPs		
METAMQAP	Combination of primary MQAPs	–	CA-CA distance	https://genesilico.pl/toolkit/
CASPIta-FRST	Structure, dihedrals	–	CA-CA distance	http://protein.cribi.unipd.it/frst/[23]
UCB-SHI	Structure	–	CA-CA distance	N/A[45]
		Standard MQAPs		
Verify3D	3D-1D structural profiles	Σ Profile score	Profile score	http://nihserver.mbi.ucla.edu/Verify_3D/[8]
Prosall	Knowledge-based potential	Z-score	Z-score	https://prosa.services.came.sbg.ac.at/prosa.php[9]

Table 6.2   *Global Quality Assessment. R_GDT, Pearson's correlation coefficient between predicted quality and GDT_TS per target. GDT1, sum of GDT_TS for the highest ranked model for each target. In bold, all numbers that are within one standard deviation from the best MQAP*

MQAP	ALL			EASY		HARD	
	R_GDT	GDT std = 131	GDT1 *unbiased* std = 132	R_GDT	GDT1 std = 117	R_GDT	GDT1 std = 61
Pcons	**0.89**	**59.0**	**58.4**	**0.96**	**73.1**	**0.77**	**37.1**
LEE	0.82	57.6	57.2	0.94	71.8	0.61	35.5
Circle-QA	0.79	58.4	57.4	0.88	72.2	0.65	36.9
ProQ	0.73	56.0	54.7	0.81	70.0	0.59	34.1
Verify3D	0.72	56.9	54.9	0.81	71.0	0.57	35.0
Prosa	0.70	56.7	53.3	0.79	70.5	0.57	35.2
ABIpro-h	0.67	59.3	55.5	0.76	72.2	0.53	39.1
QA-ModFOLD	0.65	56.2	54.8	0.77	70.4	0.46	34.0
ProQprof	0.46	49.3	49.3	0.66	64.8	0.16	25.1
TMscore	0.98	64.0	63.0	0.99	76.4	0.96	44.8
MaxSub	0.97	63.9	62.6	0.99	76.4	0.94	44.3
S-score	0.95	63.5	62.3	0.98	76.2	0.91	43.7
LGscore	0.90	62.2	61.0	0.97	76.0	0.79	40.7
Zhang-Server	–	59.3	–	–	73.5	–	37.2
Perfect (GDT)	1.00	64.5	63.4	1.00	76.8	1.00	45.4
Random	0.00	46.7	45.8	0.00	59.6	0.00	24.3

6.5.5   Local Quality Assessment

Only seven MQAPs participated in the local quality assessment category of CASP7, four of these were ours: Pcons, ProQ, ProQprof and ProQlocal, see Table 6.1 for a summary of all the MQAPs. As in our earlier study the performance for the 10% highest and 10% lowest scoring residues were evaluated in terms of average local CA-CA deviations, $\langle d^{CA}_{scaled} \rangle$ per residue, and fraction of wrongly predicted residues for the high and low scoring residues (Table 6.3). In addition, the correlation coefficient between the predicted CA-CA deviation from native to the actual deviation was calculated for each model (Figure 6.1).

From Table 6.3 it is obvious that Pcons is better than any other MQAP. The average CA-CA deviation for the 10% highest ranked residues by Pcons is 0.62 Å and only 2.4% deviates more than 3 Å. Similarly, for the 10% lowest scoring residues, the average CA-CA deviation, ≥ 30 Å, and almost no (0.2%) residues are wrongly predicted. As a reference, the second best MQAP makes over 15 times as many mistakes as compared to Pcons. Further, more than half of all predictions made by Pcons correlate better than 0.6, while no other MQAP show this correlation for more than 25% of the models (Figure 6.1).

It is also notable that ProQlocal, which combines structural information from ProQ with the evolutionary sequence information from ProQprof, is better than any of these MQAPs. This shows that there is an advantage to use a combined measure over using just one of them. It is particularly interesting that despite the poor performance as a global measure ProQprof is still quite able to predict local quality accurately. In agreement with earlier results, ProQprof seems better at detecting good regions compared to bad regions.[39]

Table 6.3 *Local Quality Assessment. R, correlation coefficient between predicted and actual CA-CA distances calculated per model and averaged over all targets. Measures calculated for the the 10% highest and lowest scoring residues: d^{CA}_{median}, the median CA-CA. FP, false positive rate (fraction of $d^{CA} > Å$). FN, false negative rate (fraction of $d^{CA} < Å$). d^{CA}, CA-CA distance between equivalent residues in the model and the native structure*

	ALL	10% highest		10% lowest	
MQAP	R	d^{CA}_{median}	FP	d^{CA}_{median}	FN
Pcons	0.63	0.61	2.7%	≥15	0.2%
ProQ	0.38	0.42	16.7%	≥15	5.4%
ProQprof	0.38	0.58	12.1%	≥15	10.4%
ProQlocal	0.44	1.7	8.4%	≥15	3.5%
Prosall	0.14	3.2	34.7%	≥15	24.4%
Verify3D	0.26	1.8	21.2%	3.6	9.3%
MetaMQAP	0.38	0.81	9.8%	≥10	7.6%
Casplta-FRST	0.19	0.85	23.1%	≥15	13.2%
UCB-SHI	0.20	1.0	26.4%	5.5	22.0%
Perfect	1.00	0.40	0.0%	≥15	0.0%
Random	0.00	2.8	46.1%	2.8	53.9%

A likely explanation for this is that the profile-profile information that ProQprof uses is optimized to maximize similarity and not dissimilarity .

Besides our MQAPs, MetaMQAP also performed quite well. For the set of MQAPs only using structural information from single models MetaMQAP is actually the best, slightly better than ProQ, i.e. the added structural evaluation information from using a meta

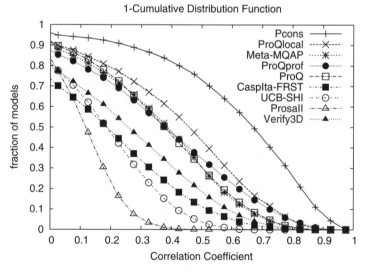

Figure 6.1 *The 1-Culumative Distribution Function for correlation coefficients between predicted and actual CA-CA deviation, i.e. the fraction of models that correlate better than a given correlation coefficient*

approach provides a small improvement over ProQ alone. However, it does not perform better than ProQlocal. In Figure 6.1 it can be seen that ProQlocal and ProQprof make significantly more (but still very few) predictions with a high ($\geq$0.5) correlation coefficient than the structure-based predictions. Again, highlighting the fact that combining structural and sequence information is fruitful. In fact, it is likely that a simple sequence conservation dependent component in MetaMQAP would increase its performance further.

6.6   Discussion

6.6.1   Which is the Best Global MQAP?

Although the Pcons correlation is superior to other MQAPs the highest ranked models are not significantly better. In fact, two MQAPs ABIpro-h and Circle-QA selects models that are equal to or slightly better than the Pcons model. This was analyzed a bit further by investigating from which server method the highest ranked MQAP models originated. It turned out that almost all models from ABIpro-h (80%) and Circle-QA (70%) originated either from three methods (Zhang-Server, ROBETTA or Pmodeller6, which might origin from ROBETTA). As a comparison, Pcons only selects 30% of its models from these methods. Besides the common feature that these three servers are among the best servers (with Zhang-Server clearly being the best), certain properties of the MQAP might be similar to the energy functions used by these three methods. A strong bias towards these particular methods could result in a large drop in performance if these methods are not present. In particular any method that always choose the Zhang-Server model would perform on top of the MQAPs and a method that select Zhang-Server models for easy targets and ROBETTA models for hard targets would perform even better.

To estimate the effect of a selection bias all models from the most frequently selected methods, i.e. Zhang-Server, ROBETTA, Pmodeller6 and Pcons6 were removed, and a new GDT1 *unbiased* score was calculated (Table 6.2). In this test, the decrease in GDT1 for most MQAPs is less than 1.0 and can be explained by the fact that many of the best models are now removed. However, ABIpro-h shows a strong bias towards the removed server methods with a GDT1 decrease of 3.3. In contrast, Circle-QA and Pcons are among the least affected by the removal with minor drops in GDT1 ($\leq$0.4). The conclusion is that the seemingly good performance of ABIpro-h can largely be attributed to the selection of Zhang-Server and ROBETTA models. Circle-QA, on the other hand, seems not to be biased and manages to maintain a good performance even when the best server methods are removed.

Like the MQAPs discussed above ProQ also have a bias towards Zhang-Server and Pmodeller6, close to 40% of the models selected by ProQ are from these two methods. However, the bias does not seem to be crucial for its performance, since the performance is quite well maintained in the GDT1 *unbiased* test. Pmodeller6 uses ProQ in its quality assessment step, so it is not a surprise that models from Pmodeller6 is highly ranked by ProQ. However, ProQ is not trained on Zhang-Server models, indicating that ProQ is able to recognize the high quality protein-like features of the excellent Zhang-Server predictions.

Interestingly, there are two structural based MQAPs, ABIpro-h and Circle-QA, which clearly select better models than ProQ. These methods have not been published before, but from the CASP7 abstract ABIpro-h is claimed to combine many more structural features than ProQ does. In particular, it uses both predicted structural features and a physical and a statistical potential. Circle-QA uses a difficulty dependent 3D-1D-profile, combined with predicted secondary structure. The success of these two methods shows that structural based MQAPs can reach almost the same accuracy as consensus methods. The results for Circle-QA and ABIpro-h on the hard targets are encouraging (Table 6.2).

6.6.2 Future Improvements of MQAPs

One possible improvement might be to make a Pcons version optimized for GDT_TS. This might make Pcons select better models in terms of GDT_TS. However, it is not obvious that this is desirable. This will neither solve the issue described above, when the best models being outliers in the GDT_TS distribution. Therefore, a consensus approach combined with a structural evaluation could be a step towards future improvement of MQAPs. In fact, for three Pcons failures all of the structural based MQAPs (ProQ, ABIpro-h or Circle-QA) select significantly better models than Pcons.[1] The idea of combining consensus and structural evaluation is not new and was actually one of main the reasons for developing ProQ[22] and Pmodeller.[40] In Pmodeller, ProQ is used to re-rank the top hits from Pcons purely based on structural features.[33] However, it seems as if this re-ranking should be restricted to cases when the consensus is weak.

Also, we were quite impressed by two novel structure-based MQAPs, ABIpro-h and Circle-QA, that actually performed on par with Pcons in the ability to select the best models, although their correlation with quality was slightly lower. The result by Circle-QA is especially impressive as its performance is maintained even if the best methods are removed, while ABIpro-h relay heavily on the predictions from Zhang-Server and ROBETTA.

Circle-QA uses relatively simple 3D-1D-profiles similar to the one used by Verify3D,[8] but whereas Verify3D has one single 3D-1D-profile to evaluate all structures, Circle-QA has two, one for easy and one for hard targets. This might be an advantage, since different features might be important to find the best model among mediocre models compared to high quality models. The score from the 3D-1D-profiles is also combined with a secondary structure term that is given a higher weight if the target is difficult, i.e. for difficult targets most of the score will depend on the agreement between predicted and actual secondary structure.

We also noted that the sequence based MQAP, ProQprof did not perform very well as a global measure of quality. In particular, on the 'hard' targets the performance is not better than random. There might be two reason for this (a) to obtain an alignment for the targets we had to perform a structural alignment and that might not represent the alignment used when generating the models and (b) for hard targets the sequence similarity does not contain any information. This is in contrast to the local quality prediction, where the sequence information is quite useful. Here ProQ, ProQprof and MetaMQAP perform quite similar, and the combination of ProQ and ProQprof into ProQlocal shows that structure and evolutionary information can be combined to improve identification of correct and incorrect regions. However, it should be remembered that the performance of any of these methods lags far behind Pcons, i.e. for local quality prediction consensus based methods are still superior.

Acknowledgments

This work was supported by grants from the Swedish Natural Sciences Research Council and the Research School in Functional Genomics and Bioinformatics and the EU 6th Framework Program is gratefully acknowledged for support to the GeneFun project, contract LSHG-CT-2004-503567 and the EMBRACE project, contract LSHG-CT-2004-512092.

References

1. Wallner, B. and E. Elofsson, A. Prediction of global and local quality model in CASP7 using Pcons and ProQ. *Proteins* **69** (S8):184–193, 2007.
2. Chapman, M., Suh, S., Curmi, P., Cascio, D., Smith, W., and Eisenberg, D. Tertiary structure of plant RuBisCO: domains and their contacts. *Science* **241** (4861):71–74, July, 1988.
3. Laskowski, R., MacArthur, M., Moss, D., and Thornton, J. PROCHECK: a program to check the stereochemical quality of protein structures. *J Appl Cryst* **26** (2):283–291, 1993.
4. Hooft, R., Vriend, G., Sander, C., and Abola, E. Errors in protein structures. *Nature* **381** (6580):272, 1996.
5. Brunger, A., Kuriyan, J., and Karplus, M. Crystallographic R factor refinement by molecular dynamics. *Science* **235** (4787):458–460, Jan., 1987.
6. Bujnicki, J., Rychlewski, L., and Fischer, D. Fold-recognition detects an error in the Protein Data Bank. *Bioinformatics* **18** (10):1391–1395, Oct., 2002.
7. Chang, G. Retraction of 'Structure of MsbA from Vibrio cholera: a multidrug resistance ABC transporter homolog in a closed conformation' [*J. Mol. Biol.* (2003) 330 419–430]. *J Mol Biol* **369** (2):596, June, 2007.
8. Lüthy, R., Bowie, J., and Eisenberg, D. Assessment of protein models with three-dimensional profiles. *Nature* **356** (6364):283–285, 1992.
9. Sippl, M. Recognition of errors in three-dimensional structures of proteins. *Proteins* **17** (4):355–362, 1993.
10. Wallner, B., Larsson, P., and Elofsson, A. Pcons.net: Protein Structure Prediction Meta Server. *Nucleic Acids Res* **35** (Web Server issue):W369–374, 2007.
11. Bujnicki, J., Elofsson, A., Fischer, D., and Rychlewski, L. Structure prediction meta server. *Bioinformatics* **17** (8):750–751, 2001.
12. Kurowski, M. and Bujnicki, J. GeneSilico protein structure prediction meta-server. *Nucleic Acids Res* **31** (13):3305–3307, July, 2003.
13. Bowie, J., Lüthy, R., and Eisenberg, D. A method to identify protein sequence that fold into a known three-dimensional structure. *Science* **253**:164–170, 1991.
14. Jones, D., Taylor, W., and Thornton, J. A new appoach to protein fold recognition. *Nature* **358**:86–89, 1992.
15. Miyazawa, S. and Jernigan, R. Residue-residue potentials with a favorable contact pair term and an unfavorable high packing density term, for simulation and threading. *Journal of Molecular Biology* **256** (3):623–644, 1996.
16. Sippl, M. Helmholtz free energy of peptide hydrogen bonds in proteins. *J Mol Biol* **260** (5):644–648, 1996.
17. Colovos, C. and Yeates, T. Verification of protein structures: patterns of nonbonded atomic interactions. *Protein Sci.* **2** (9):1511–1519, 1993.
18. Melo, F. and Feytmans, E. Assessing protein structures with a non-local atomic interaction energy. *J Mol Biol* **277** (5):1141–1152, 1998.
19. Samudrala, R. and Moult, J. An all-atom distance-dependent conditional probability discriminatory function for protein structure prediction. *Journal of Molecular Biology* **275** (5):895–916, 1998.

20. Lu, H. and Skolnick, J. A distance-dependent atomic knowledge-based potential for improved protein structure selection. *Proteins* **44** (3):223–232, 2001.
21. Hooft, R., Vriend, G., Sander, C., and Abola, E. Errors in protein structures. *Nature* **381**:272, 1996.
22. Wallner, B. and Elofsson, A. Can correct protein models be identified? *Protein Sci* **12** (5):1073–1086, May, 2003.
23. Tosatto, S. The victor/FRST function for model quality estimation. *J Comput Biol* **12** (10):1316–1327, Dec, 2005.
24. Cheng, J. Randall, A., Sweredoski, M., and Baldi, P. 3D Structure Prediction Using FOLDpro, 3Dpro, and ABIpro. *CASP7 Abstracts*, 2007.
25. Takaya, D., Terashi, G., Takeda-Shitaka, M., *et al.* CIRCLE for quality assessment in CASP7. *CASP7 Abstracts*, 2007.
26. Krishnamoorthy, B. and Tropsha, A. Development of a four-body statistical pseudo-potential to discriminate native from non-native protein conformations. *Bioinformatics* **19** (12):1540–1548, Aug., 2003.
27. Pontius, J., Richelle, J., and Wodak, S. J. Deviations from standard atomic volumes as a quality measure for protein crystal structures. *J Mol Biol* **264** (1):121–136, Nov, 1996.
28. Lin, K., May, A. C., and Taylor, W. R. Threading using neural nEtwork (TUNE): the measure of protein sequence-structure compatibility. *Bioinformatics* **18** (10):1350–1357, Oct., 2002.
29. Boniecki, M., Rotkiewicz, P., Skolnick, J., and Kolinski, A. Protein fragment reconstruction using various modeling techniques. *J Comput Aided Mol Des* **17** (11):725–738, Nov, 2003.
30. McGuffin, L. J. Benchmarking consensus model quality assessment for protein fold recognition. *BMC Bioinformatics* **8**:345, 2007.
31. Pettitt, C. S., McGuffin, L. J., and Jones, D. T. Improving sequence-based fold recognition by using 3D model quality assessment. *Bioinformatics* **21** (17):3509–3515, Sep, 2005.
32. Lundström, J., Rychlewski, L., Bujnicki, J., and Elofsson, A. Pcons: a neural-network-based consensus predictor that improves fold recognition. *Protein Sci.* **10** (11):2354–2362, 2001.
33. Wallner, B. and Elofsson, A. Pcons5: combining consensus, structural evaluation and fold recognition scores. *Bioinformatics* **21**, Oct., 2005.
34. Cristobal, S., Zemla, A., Fischer, D., Rychlewski, L., and Elofsson, A. A study of quality measures for protein threading models. *BMC Bioinformatics* **2** (5), 2001.
35. Ginalski, K., Elofsson, A., Fischer, D., and Rychlewski, L. 3D-Jury: a simple approach to improve protein structure predictions. *Bioinformatics* **19** (8):1015–1018, 2003.
36. Eisenberg, D., Lüthy, R., and Bowie, J. VERIFY3D: assessment of protein models with three-dimensional profiles. *Methods Enzymol* **277**:396–404, 1997.
37. Kosinski, J., Cymerman, I. A., Feder, M., Kurowski, M. A., Sasin, J. M., and Bujnicki, J. M. A "FRankenstein's monster" approach to comparative modeling: merging the finest fragments of Fold-Recognition models and iterative model refinement aided by 3D structure evaluation. *Proteins* **53** Suppl 6:369–379, 2003.
38. von Grotthuss, M., Pas, J., Wyrwicz, L., Ginalski, K., and Rychlewski, L. Application of 3D-Jury, GRDB, and Verify3D in fold recognition. *Proteins* **53** Suppl 6:418–423, 2003.
39. Wallner, B. and Elofsson, A. Identification of correct regions in protein models using structural, alignment, and consensus information. *Protein Sci* **15** (4):900–913, Apr., 2006.
40. Wallner, B., Fang, H., and Elofsson, A. Automatic consensus-based fold recognition using Pcons, ProQ, and Pmodeller. *Proteins* **53** Suppl 6:534–541, 2003.
41. Tress, M. L., Jones, D., and Valencia, A. Predicting reliable regions in protein alignments from sequence profiles. *J Mol Biol* **330** (4):705–718, Jul, 2003.
42. Fischer, D. 3D-SHOTGUN: a novel, cooperative, fold-recognition meta-predictor. *Proteins* **51** (3):434–441, May, 2003.
43. Zemla, A., Veclovas, C., Moult, J., and Fidelis, K. Processing and analysis of CASP3 protein structure predictions. *Proteins* Suppl 3:22–29, 1999.
44. Wallner, B. and Elofsson, A. Can correct protein models be identified? *Protein Science* **12** (5):1073–1086, 2003.
45. Cozzetto, D., Kryshtafovych, A., Ceriani, M., and Tramontano, A. Assessment of predictions in the model quality assessment category. *Proteins* **69** (S8):175–183, 2007.

7

Prediction of Molecular Interactions from 3D-structures: From Small Ligands to Large Protein Complexes

Kengo Kinoshita, Hidetoshi Kono and Kei Yura

7.1 Introduction

Proteins are involved in almost all processes in the complex biological systems of living organisms. Each protein has its own function, such as an enzymatic reaction, and at the same time it has some roles as part of interaction networks of proteins. The context-free function, the former role, is called the molecular function, while the context-dependent function, the latter role, is designated as the cellular function or biological function of the proteins. Protein functions are realized through interactions with many other molecules, and thus the molecular interaction is the first step in the function of a protein.

In this chapter, we describe the current knowledge about molecular interactions as a part of the function identification or prediction for proteins. The chapter is divided according to the type of partner molecule, small molecules, DNA, RNA and proteins in this order. The aim of this chapter is not to provide the technical details of the prediction methods, but to show an overview of the state of the art of this field. The readers should consult the cited references for details. The web servers that might be convenient for the readers are listed in Table 7.1.

7.2 Protein–Ligand Interactions

Small molecules are used in various aspects of protein functions, as modulator, cofactors or reaction substrates. The roles of small molecules cannot be distinguished from their

Prediction of Protein Structures, Functions, and Interactions   Edited by Janusz Bujnicki
© 2009 John Wiley & Sons, Ltd

Table 7.1 *Web servers for prediction of functionally important residues and binding sites*

Name	URL	Input	Output	Short Description	Ref.
			Ligand structure-based		
eF-site	http://ef-site.hgc.jp	Keywords	Molecular surface, hydrophobicity and electrostatic potential in original format along with Molscript format	A database of electrostatic potential and molecular surface of protein functional sites	132
eF-surf	http://ef-site.hgc.jp/eF-surf	PDB file	Molecular surface, hydrophobicity and electrostatic potential in original format along with Molscript format	A webserver to calculate electrostatic potentials and molecular surface of proteins	
eF-seek	http://ef-site.hgc.jp/eF-seek	PDB file	PDB file with putative ligand	A webserver to predict the ligand binding site by similarity search	
firestar	http://firedb.bioinfo.cnio.es/Php/FireStar.php	PDB file or Amino Acid Sequence	Functional Residues	A web server to predict the functional residues based on the known ligand complexes by similarity search	133
Q-SiteFinder	http://www.bioinformatics.leeds.ac.uk/qsitefinder/	PDB file	List of ligand binding site residues	A webser to predict the ligand binding sites	134
MODPROPEP	http://www.nii.res.in/modpropep.html	Amino Acid sequence of a peptide	PDB file with bound peptide	A webserver to predict the peptide binding structures with protein kinase and MHC proteins	135
			Sequence based		
FORTE	http://www.cbrc.jp/forte	Amino Acid sequence	List of homologous proteins with alignment	A webserver of sequence comparison using profile-profile method	10
COMPASS	http://prodata.swmed.edu/compass	Amino Acid sequence	List of homologous proteins with alignment	A webserver of sequence comparison using profile-profile method	11
HHsenser	http://toolkit.tuebingen.mpg.de/hhsenser	Amino Acid sequence	List of homologous proteins with alignment	A webserver of sequence comparison using HMM-HMM method	13

Name	URL	Input	Output	Description	Ref
DNA sequence-based					
DP-Bind	http://lcg.rit.albany.edu/dp-bind/	Amino Acid sequence	Indication of DNA-binding residues with a confidence level	A webserver to predict DNA binding residues based on sequence information with three machine learning methods: support vector machine SVM, kernel logistic regression KLR, and penalized logistic regression PLR	46
DISIS	http://cubic.bioc.columbia.edu/services/disis	Amino Acid sequence	Indication of DNA-binding residues with a confidence level	A webserver to predict DNA binding residues from amino acid sequence with SVM	48
DBS-PSSM	http://www.netasa.org/dbs-pssm/	Amino Acid sequence	Indication of DNA-binding residues with a confidence level	A webserver to predict DNA binding residues based on PSSM	45
BindN	http://bioinfo.ggc.org/bindn/	Amino Acid sequence	Indication of DNA/RNA-binding residues with a confidence level	A webserver to predict DNA/RNA binding residues from amino acid sequence with SVM	47
DNA structure-based					
DISPLAR	http://pipe.scs.fsu.edu/displar.html	Protein structure	Indication of DNA-binding residues	A webserver to predict DNA/RNA interface out of protein 3D structures with neural network	51
PreDs	http://pre-s.protein.osaka-u.ac.jp/~preds	Protein structure	Indication of DNA-binding protein, electrostatic potential mapping on protein surface, and DNA-binding residues	A webserver to predict double strand DNA binding site based on structural information	136
Protein-Nucleic Acid Interaction Server	http://www.biochem.ucl.ac.uk/bsm/DNA/server/	Protein-DNA complex structure	Tables describing the nature of the interface	A webserver to predict double strand DNA binding site based on structural information	49
ReadOut	http://gibk26.bse.kyutech.ac.jp/jouhou/readout/	Protein-DNA complex structure	Z-scores of direct and indirect readouts for a given complex structure indicating binding specificity	A webserver to calculate direct and indirect readout energy Z-scores for protein-DNA complexes	75

Table 7.1 (continued)

Name	URL	Input	Output	Short Description	Ref.
			RNA sequence based		
SVMProt	http://jing.cz3.nus.edu.sg/cgi-bin/svmprot.cgi	Amino Acid Sequence	List of possible biological function of the sequence	A webserver to predict RNA binding proteins out of amino acid sequences using SVM. The server has a functionality to predict many types of biological functions from amino acid sequences	101
BindN	http://bioinfo.ggc.org/bindn	Amino Acid Sequence	List of RNA-binding residues	A webserver to predict DNA/RNA binding residues out of amino acid sequences using SVM	102
RNABindR	http://bindr.gdcb.iastate.edu/RNABindR	Amino Acid Sequence	List of RNA-binding residues	A webserver to predict RNA binding residues out of amino acid sequences	103, 104
			RNA structure-based		
DISPLAR	http://pipe.scs.fsu.edu/displar.html	PDB file with/without amino acid sequence alignment	PDB file with RNA-binding residues	A webserver to predict DNA/RNA interface out of protein 3D structures	51
KYG	http://cib.cf.ocha.ac.jp/KYG	PDB file with/without amino acid sequence alignment	PDB file with RNA-binding residues	A webserver to predict RNA binding residues out of protein 3D structure with/without multiple sequence alignment	94
			Protein-Protein		
DOMINE	http://domine.utdallas.edu	Keywords	Protein-Protein Interactions	A database of known and predicted protein domain interactions	137
COXPRESdb	http://coxpresdb.hgc.jp	Keywords	co-expression genes and networks	A database of the coexpressed gene network	123
BioGRID	http://www.thebiogrid.org	Keywords	Protein-Protein Interactions	A database of interacting proteins	114

Name	URL	Input	Output	Description	Ref
IntAct	http://www.ebi.ac.uk/intact	Keywords	Protein-Protein Interactions	A database of interacting proteins	115
HPRD	http://www.hprd.org	Keywords	Protein-Protein Interactions for Human	A database of interacting proteins in Human	117
cons-PPISP	http://pipe.scs.fsu.edu/ppisp.html	PDB file	Protein Interaction Site	A webserver to predict the potential protein binding site	138
3D-partner	http://3d-partner.life.nctu.edu.tw	Amino Acid sequence	Poteintial binding protein and its complex structure	A webserver to predict the potential binding proteins and structures based using the homology search against known complexes rmateus	139
PIC	http://crick.mbu.iisc.ernet.in/~PIC/	PDB file	A list of interactions		140
ProMate	http://bioportal.weizmann.ac.il/promate/	PDB file	Potential protein binding sites	A webserver to predict the putative protein binding sites	141
FastContact	http://structure.pitt.edu/servers/fastcontact/	PDB file	Free energy score	A web server to evaluate the direct electrostatic and desolvation free energy between two proteins	142
Others					
Fralanyzer	http://fralanyzer.cse.buffalo.edu/	PDB file or Amino Acid Sequence	Functionally important residues	A web server to visually inspect functionally important residues in a query sequence/structure	143
P-cats	http://p-cats.hgc.jp	PDB file	Functionally important residues	A web server to discriminate functional important residues from conserved residues	144
siteFiNDER|3D	http://sage.csb.yale.edu/sitefinder3d/	PDB file	Functionally important residues	A webserver to predict the location of functionally important sites	145

interactions with our current knowledge, but prediction methods for small molecule binding sites have been extensively studied. The methods can roughly be classified into three types, a docking approach, a similarity search approach, and a pocket search approach. For simplicity, we will call the small molecules ligands in this chapter, regardless of the role of the small molecules.

The docking approach is done by optimizing the evaluation function of the fitness of a ligand molecule on protein structures. The evaluation function can be a certain physical potential, such as an AMBER potential,[1] or may be a statistical potential derived from known complexes. The optimization should be done by considering the flexibility of the ligand and the protein at the same time, but the usual practice is to neglect the flexibility of the protein. This is an approximation to reduce the calculation cost to search for possible interactions. This treatment may be supported by the observation that the structural change upon ligand binding is not large on average,[2] as compared to the structural diversity of ligands.[3] Many methods have been proposed, and a systematic comparison of the methods was reported.[4]

In the similarity search approach, the most reliable information to infer potential binding sites of small molecules is homology information. If proteins with similar sequences can be found in the form of complex structures, then the binding mode will be similar when the sequence similarity is relatively high.[5] In a situation where the sequence-level similarity could not be detected, a fold-level similarity may provide some clues for predicting the potential ligand-binding site.[6, 7] However, the prediction results should be carefully checked when binding site information from the distantly related protein structures is used to predict the binding site. In distant homologs, the binding site may be located in a similar place, but the role of the binding site and/or the kind of the substrate may differ from those of the related proteins.[5] Readers are recommended to refer to the paper by Novotny *et al.* for the methods to search for similar folds.[8] To detect distant homologs without structural information, **PSI-BLAST**[9] is widely used, which compares its position specific score matrix (or profile) and the amino acid sequence. More recently, profile-profile and **HMM-HMM** comparisons, where HMM stands for hidden Markov model, have proved to be more sensitive to detect distance homologs.[10, 13]

The next level of reliable indicator to infer a functional relation is the similarity of the spatial arrangement of atoms. Even though the global folds of two proteins are different, and thus the evolutionary relationship of the two proteins cannot be assumed, the functional sites or the substrate binding sites can sometimes show surprising similarity. The most famous case is found between subtilisin and chymotrypsin, where the global folds are completely different, but the spatial arrangements of the atoms in the active sites of the proteins, the catalytic triads, are quite similar.[14] Another interesting example is the similarity found between cAMP dependent protein kinase and DD-ligase.[15, 16] In this case, the similar part is limited to the adenine recognition and the phosphate recognition, and the functions of the two proteins are different. There are a few other examples where the functional sites are very similar, in spite of the different global folds, but these are rather rare cases. More generally, the similarity of an atomic configuration beyond the superfamily level is infrequently found.[17, 18] In other words, even though structural comparisons are carried out, the result is often equivalent to that obtained by a sophisticated sequence similarity search, although the calculation cost of the structural comparison is far higher than that of the sequence comparison.

The reason why the atomic configuration similarity corresponds to the superfamily relation is not clear, but one important factor is the flexibility of protein structures. Even the same proteins can have *different* atomic positions, due to the fluctuations of protein structure. If this is the case, then a 'template' type approach may be promising.[19, 20] Conceptually, the template is a set of known binding sites, and similarity searches are performed against the template. Small differences due to structural flexibility can be detected if the fluctuated structures exist in the known complex. Another possible reason is that the similarity score is often based on the number of corresponding atoms. In other words, the pairs of binding sites with higher numbers of corresponding atoms are considered to be more similar, and thus the similar main chain trace in the similar fold will generate a higher similarity score than the correspondence of side-chain atoms in the different fold. However, even proteins with different folds can have higher scores by the accidental similarities of the frequently appearing fragments. This possibility may be supported by the observation that the structural elements shared by different proteins usually consist of main chain fragments.[15–17] Correspondences between side-chain atoms may be very difficult to distinguish from accidental similarities. There are several method to assess the significance of the similarity, but more sophisticated methods may be required.

To overcome the limitation in the similarity of atomic configurations, a similarity search of the molecular surface is an option. Originally, Rosen *et al.*[21] developed a method to compare the shapes of molecular surfaces. Kinoshita *et al.* subsequently extended the method and developed a web-based interface to search for similar binding sites on the molecular surfaces along with the electrostatic potentials of the proteins.[22–24] It works well in some cases, but it is one of the final options, when no other information about the binding site can be obtained. Other options that are not described here are available in the review papers, such as Watson *et al.*[7] and Kinoshita and Nakamura.[25]

One of the alternative approaches is to search for a pocket on the protein structure.[26] In this approach, similarity searches against the known structure are not used, and thus the limitation due to the shortage of complex structures is not a problem. However, this type of approach is to search for a potential binding site, and it does not yield information about the potential ligand. Thus, the combination of a pocket search and a similarity search may be promising in the next step of ligand binding site predictions. For the pocket calculations, see Laskowski *et al.*,[27] and Kawabata and Go[28] for details.

All of the approaches described above suffer to an extent from many false positives. It is true that some of the false positives can be real binding sites, because the number of experiments is too small. However, the reliability of the binding site prediction is still not sufficient, especially when the prediction is done only with the structural information, although by combining other source of information, such as an evolutionary trace analysis,[29] the number of false positives can be reduced.

7.3 Protein–DNA Interactions

To date the genomes of more than 600 species have been sequenced and recorded. But to understand how individual genes within a genome are regulated, it is necessary to understand how, at the molecular level, proteins interact with specific DNA sequences to regulate such cellular processes as DNA replication and recombination, and gene expression. We

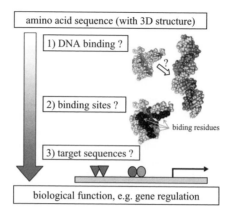

Figure 7.1   *Three ways of predicting protein-DNA interactions. Given an amino acid sequence, sometimes with its 3D-structure, we try to predict (1) whether or not the protein binds DNA and, if so, (2) where the DNA binds and (3) what the target DNA sequences are*

will first discuss what is known about how proteins recognize their target DNA sequences. We will then describe how we compile experimentally obtained data and employ them to (1) select DNA-binding proteins among all proteins, (2) determine DNA binding sites on DNA-binding proteins and (3) determine the DNA sequences targeted by DNA-binding proteins (Figure 7.1).

7.3.1   Thermodynamic and Kinetic Aspects of Protein–DNA Interactions

Although structural studies have provided much information about the important molecular features that contribute to specific protein-DNA recognition and complex formation, it is still difficult to quantify the binding affinity with the structural information. However, this quantity can be obtained through thermodynamic analyses in which protein–DNA interactions are described in terms of free energy, enthalpy and entropy. Some DNA-binding proteins bind to the major groove of the DNA, while others bind to the minor groove. Although both the major and minor groove binders show comparable changes in Gibbs free energy upon binding, their binding mechanisms differ: proteins binding to the major groove do so via an enthalpy driven process, whereas those binding to the minor groove do so via an entropy driven process.[30] This difference in the energetic component is indicative of the distinct features of the hydration of the major and minor grooves. For example, minor groove binding occurs at AT-rich sequences in which water ordering within the groove, the so called hydration spine, is the most prevalent.[31] This suggests that understanding hydration is essential to fully understand the mechanisms by which proteins bind to DNA, especially in the minor groove. Nonetheless, the role of water is not explicitly considered in most of the prediction methods.

To understand the factors important for determining specificity, it is necessary to analyze both specific and nonspecific binding. The structures of both cognate and noncognate protein-DNA complexes provide information about what underlies specificity. We can see some examples in the structures of protein-DNA complexes involving the DNA-binding domain of *lac* repressor, as well as EcoRV, Glucocorticoid repressor and b-ZIP proteins.

Comparison of the structures of these specific and nonspecific complexes suggests that nonspecific complexes are more loosely bound, so that there is room at the protein-DNA interface for water molecules to occupy. This notion was substantiated using the osmotic stress technique, with which one can probe the difference in the number of water molecules within sterically sequestered cavities at the interface of specific and nonspecific protein-DNA complexes.[32] That study showed that nonspecific complexes have more water molecules at the protein-DNA interface than specific ones. Moreover, NMR studies of the dimeric *lac* repressor DNA binding domain showed that the same set of residues can switch from a purely electrostatic interaction with the DNA backbone within a nonspecific complex to a highly specific interaction with bases in the cognate complex.[33] The protein-DNA interface of nonspecific complexes thus appears to be more flexible, with fewer direct protein–DNA interactions, than the cognate complex.

7.3.2 Flexibility and Adaptability within Protein-DNA Complexes

Structural studies of protein-DNA complexes have shown that DNA sequence-specific binding is accompanied by conformational changes in both the protein and DNA component. A good example is CAP, which bends the DNA sharply within the CAP-DNA complex by about 60 to 90 degrees, so that the DNA wraps toward and around the sides of the CAP dimer.[34] As another example, CytR operators bind to their targets having a variable length of the central spacer by changing their conformation in a manner that is dependent on the length of the spacer.[35] One must therefore consider such conformational changes when modeling protein-DNA complexes, which makes it difficult to predict their structures.

To overcome that difficulty, the Varani group developed all-atom statistical potentials to predict protein-DNA interactions from modeled structures and showed an ability to identify 90% of near-native structures within the best-scoring 10% of structures in a decoy set.[36] In addition, the Bujnicki group constructed theoretical models of endonucleases using an approach that entailed merging the finest fragments of fold-recognition models with iterative model refinement.[37, 38] The resultant models were validated experimentally and were shown to identify the residues important for DNA binding. The results demonstrated that modeling the structures of protein-DNA complexes can enable one to infer which residues are located at the DNA interface.

7.3.3 Prediction of DNA-binding Proteins

Structural genomics projects provide the 3D-structures of proteins that have no sequence and structural similarity to those in the current databases. It is challenging to annotate the function of such a protein merely from the determined 3D-structure because the function of newly discovered proteins are generally annotated by searching the databases for proteins that are similar in terms of sequence and/or in structure and have known functions.[39, 40]

Stawiski *et al.* characterized the structural and sequence properties of protein-DNA complexes and found that large, positively charged electrostatic patches coincided with the surfaces of DNA-binding proteins. They further observed that the surface area per residue is the most important factor for the prediction and successfully distinguished between actual DNA-binding proteins and other proteins with large positive electrostatic patches that do not bind nucleic acids.[41] Particularly intriguing was that the distribution of Arg within the positively charged patches was the key to the success of the prediction.

Ahmad and Sarai focused on aspects of the electrostatic properties in more detail.[42] They calculated the electric dipole moment and quadrupole moment tensors as well as the net charge, and showed that those electrostatic properties alone are sufficient to predict DNA-binding proteins.

More recently, the Lu group developed a kernel-based machine learning protocol for predicting DNA-binding proteins and improved the prediction accuracy to 86%.[43] The key factors accounting for the gain in accuracy appear to originate in the ability of the support vector machines (SVMs) to act as a classifier and the consideration of the largest positively charged patch.

7.3.4   Prediction of DNA-binding Residues within DNA-binding Proteins

Sequence-based method. Annotation of protein function is usually carried out on the basis of a comparison of the amino acid sequence. When we find sequences of known function that are similar to a sequence of interest, that function can be assigned to the target sequence. If the sequence comparison yields a sufficient number of similar sequences, we can obtain position-specific scoring matrix (PSSM),[44] which can then be used with several machine learning algorithms to predict DNA-binding residues within DNA-binding proteins (see Table 7.1 for web tools: **DBS-PSSM**,[45] **DB-Bind**,[46] **BindN**[47] and **DISIS**[48]).

Structure-based method. Complementary structure-based methods have also been developed (see Table 7.1 for web tools: Protein-Nucleic Acid Interaction Sever,[49] PreDs[50] and **DISPLAR**[51]). As expected, the DNA-binding interface of proteins appears to take on a positive charge to bind to negatively charged DNA. In addition to the electrostatic score, the size of solvent accessible surface area, amino acid propensity, hydrophobicity and conservation of the interface have also been characterized and used for detecting the binding sites.[49] The electrostatic score is the most essential characteristic when attempting to detect DNA-binding residues, but the accuracy of the prediction can be improved by taking the local and global curvatures of the protein surface into consideration.[50]

7.3.5   Prediction of DNA Sequences Targeted by DNA-binding Proteins

Regulation of gene expression – i.e. their activation and repression – in higher organisms is achieved through the activities of a complex network of transcription factors, and various computational methods have been developed to predict transcription factor binding sites. Those developed so far can be classified as sequence-based and structure-based. Both methods assume that each position within the binding site contributes independently to the binding energy. It is noteworthy, however, that this assumption is not perfect and is only an approximation in most cases.[52]

Sequence-based method. To characterize the binding motif for a given transcription factor, a profile, which is also referred to as PSSM or weight matrix, is generated by aligning a set of binding sites and determining the base frequencies at each position of the alignment. Some 398 matrices have been constructed and complied in **TRANSFAC** release 7.[53] One must be cautious when using the profile because their quality is highly dependent on the sequence alignment. If the aligned sequences are very similar to one another, the

PSSM becomes highly specific and does not allow any variation in the sequence pattern. Conversely, when the aligned sequences are highly divergent, the profile is very sensitive and yields numerous false positives when a genome is scanned for putative binding sites. Several approaches have been devised to reduce the number of false positives. First, when searching for putative binding sites, one can limit the search to the promoter regions. Second, because transcription factors often work with other transcription factors, one can select only those putative-binding sites that have another binding site nearby or within a defined distance[54] or that maximize the joint likelihood of the occurrence of two binding site motifs.[55] Third, one can compare closely related genomes and choose only conserved putative binding sites.[56–58]

It should also be noted that recently developed high-throughput methods, such as cDNA microarray, SAGE and ChIP, enable one to generate profiles by finding similar sequence patterns in the promoter region of coexpressed genes.[59, 60] In this case, the pattern discovery is achieved using the EM algorithm[61] or Gibbs sampling method[62] based on the assumption that coexpressed genes are regulated by the same transcription factors. A disadvantage of the sequence-based method is that a set of patterns for constructing the profile must be somehow obtained for each of transcription factors.

Structure-based method. Structure-based method focuses on the complementarity of physico-chemical properties between protein and DNA interface. Analysis of many protein-DNA complexes showed that there are no one-to-one correspondences between bases and amino acids, but some preferences between them do exist.[63–65] One of the merits against the sequence-based method is that once we convert the preferences into a score or an energy potential, the score or potential is applicable for any transcription factor without constructing a profile for each of transcription factors.

To systematically analyze the geometry of the interactions between amino acids and DNA bases within the structures of large numbers of protein-DNA complexes, certain coordinates have to be defined.[63, 65, 66] Siggers *et al.* devised a score with which to assess the similarity between the interface geometries of protein-DNA complexes. They found that in general the intrafamily interfaces are more similar to one another than interfamily interfaces, and that even the interfaces of noncognate, intrafamily complexes are more similar than interfamily ones. This suggests that each family has a strong driving force to maintain certain contacts.[66] Kono and Sarai as well as Donald *et al.* demonstrated that the potential derived from the distributions of amino acids around bases can be used to find target DNA sites[63, 67] and can potentially be used for novel DNA-binding protein design.[63] Notably, the aforementioned potential can also be derived by calculating the free energy maps of base-amino acid interactions based on molecular mechanics force fields.[68, 69]

Another important factor in specific protein-DNA recognition is the indirect readout, which was demonstrated in experiments showing that mutation of a base not in contact with any amino acid can affect binding affinity (for example, see a review[70] and the references therein). Indirect readout involves water-mediated contacts, specific sequence-dependent conformational features, such as the bending and local geometry of a base pair, and/or binding-induced distortion of DNA. Among these, the contribution made by DNA deformation to the specificity of protein-DNA binding has been assessed.[71–74] Sarai and his colleagues used simplified base-pair step potentials to quantify the specificity of protein-DNA recognition on the basis of direct and indirect readout.[73–75] The results showed that some DNA-binding proteins mainly use direct readout, some mainly use indirect readout and others use both. When they added the two contributions with weighting, they observed that the specificity for the target DNA increased for almost all the DNA-binding proteins tested, suggesting that the direct and indirect readout mechanisms complement one another.[76] Note that for a more detailed description of DNA conformation, sequence-dependent base-step pair parameters were recently derived using molecular dynamics simulations.[77]

7.3.6   Notes for the Application of the Prediction Methods to the Real Issue

We have introduced three prediction methods, each of which has two complementary approaches: those that are sequence-based and those that are structure-based. Because they each have their own advantages and disadvantages, we suggest integrating the two approaches, so as to increase the accuracy of the prediction. In addition, increase in the experimental data on the binding sites and the structures of the protein-DNA complexes will further improve prediction accuracy, and will make it possible to predict DNA sequences targeted by DNA-binding proteins on the basis of the structure of a modeled complex.

7.4   Protein–RNA Interactions

7.4.1   Biological Importance of Protein–RNA Interactions

The central dogma of molecular biology tells that genetic information is transcribed from a gene to a messenger RNA (mRNA) molecule, after which that molecule is translated to a polypeptide on the ribosome. In actual cells, especially eukaryotic cells, however, there are a lot of steps that involve many molecules before the mRNA reaches the ribosome. These post-transcriptional pre-translational processes include, but are not limited to, 3'-polyadenylation, 5'-capping, splicing, editing, repairing, transport, export and degradation.[78–85] During these processes, mRNAs are modified by proteins and/or ribonucleoproteins (RNPs), which are complexes made up of both proteins and RNAs. It is estimated that 3% of the genes in the genome of *Drosophila melanogaster* encode RNA-binding proteins, and the proportion increases to 8% in the genome of Baker's yeast. Application of these ratios to the human genome suggests that between 640 and 2,560 RNA-binding proteins are encoded in the genome.[86] The mechanisms and manner by which these proteins interact with RNA remain unknown, in large part because, despite the importance of protein–RNA interactions in molecular biology, there are only a limited number of methods (see Table 7.1) with which to predict the interactions based on the amino acid sequence and/or the 3D-structure of the proteins.

7.4.2   Prediction of RNA-binding Proteins by Similarity

A recent increase in the number of reported structures of protein-RNA complexes has spurred analyses of the characteristics of RNA-binding proteins. There are several domains commonly used to bind RNA, including the widely used RNA-recognition motif (RRM),[87] K-homology (KH) domain,[88] double stranded RNA binding domain (dsRBD),[89] Piwi Argonaut and Zwille (PAZ) domain[90] and Pumilio homology domain (PUM-HD),[91] among others. In some cases, this structural classification of RNA-binding proteins provides a simple means of predicting RNA-binding proteins and their interfaces: if the protein in question has a 3D-structure or amino acid sequence that is similar to a known RNA-binding protein, it is likely that the protein in question binds RNA in a similar mode. The fact is, however, that there are many cases that defy the logic underlying this prediction method. In addition, there are many other domains that bind RNA. It is therefore clear that structural similarity to a known RNA-binding protein is neither a necessary nor a sufficient condition for a protein to bind RNA.

It would be ideal once the 3D-structure of a protein was known, one could calculate the free energy of protein–RNA complex formation on the basis of a computer simulation and then assess whether the protein could bind RNA. Unfortunately, this method is still not sufficiently reliable to use to make predictions, and the calculation remains beyond the reach of presently available supercomputers. A much easier approach to making predictions is to calculate only the electrostatic potential; indeed, there have been numerous studies in which the electrostatic potential of the surface of a given protein was calculated, and the likelihood that the protein binds DNA/RNA was evaluated. It is logical to assume that the DNA/RNA-binding interface is a patch with a positive potential, as the DNA/RNA molecule carries a negative charge, yet calculation of the electrostatic potential is not sufficiently accurate all the time. This is because the dielectric constants and ionic strength of the solvent are difficult to estimate, and the results of the calculation depend heavily on minute changes in the conformations of the side-chains that may only be known for the crystal structure, which may differ from the structure in solvent.[41, 92]

7.4.3   Extracting Empirical Rules from the Structures of RNA Protein Complexes

An alternative and more practical approach to predicting the interface between RNA-binding proteins and RNA is to make use of empirical rules that can be extracted from data once a sufficient amount has been accumulated. These empirical rules likely average out all the difficult issues that would be encountered by more straightforward methods. Below we will discuss the methods used to extract rules from the structures of protein-RNA complexes and the currently available empirical rules for RNA interfaces on proteins.

Atoms within RNA interfaces on the surfaces of proteins can be defined in one of the following ways. The simpler way is to measure the distance between atoms in the RNA and protein, and if the distance is less than a certain threshold value (e.g. 4 or 5 Å), then the atom in the protein is defined as being at the RNA interface. The other way is to measure the solvent accessible surface area (ASA) of every atom of the protein, with and without the RNA, and if the difference in the ASA (ΔASA) is more than a given threshold (e.g. 0 or 1 Å^2), then the atom is defined as being at the RNA interface. The radius of a water molecule for measuring the ASA is normally set between 1.4 and 1.5 Å. The ASA method tends to put a greater number of atoms at protein-RNA interfaces than the distance method, especially when cavities are involved in the interfaces, but it misses atoms located close to the bound RNA when they are covered by other atoms to prevent access by the solvent. The discrepancy caused by the difference in the methods of defining interface residues does not affect the tendencies discussed below.

Types of preferred amino acid residue in RNA interface have been extensively studied. With such knowledge, one can intuitively judge whether a given patch on the surface of a protein is a likely contributor. A simple count of each type of amino acid at RNA interfaces tells us that positively charged residues such as Arg and Lys frequently appear in the interfaces. However, the numbers of the preferred residue types should be normalized to the numbers on the entire surface, as charged amino acid residues tend to appear at the surface, irrespective of whether or not they are involve in protein–RNA interaction. With that in mind, let us consider two approaches to analyzing amino acid preferences at the RNA interface; one involves a simple count of the amino acid residues,[93, 94] while the other involves weighting the count on the basis of ASA.[95, 96] In each case, the propensity for amino acid type i (P_i) is given by

$$P_i = \frac{\overline{n_i} / \sum_{j=1}^{20} \overline{n_j}}{n_i / \sum_{j=1}^{20} n_j} \text{ or } P_i = \frac{\sum_{j=1}^{\overline{n_i}} \Delta ASA_{i(j)} / \sum_{j=1}^{\overline{n_i}} \sum_{k=1}^{20} \Delta ASA_{k(j)}}{\sum_{j=1}^{n_i} ASA_{i(j)} / \sum_{j=1}^{n_i} \sum_{k=1}^{20} ASA_{k(j)}}, \qquad (7.1)$$

where n_i is the count of amino acid type i on the surface of a protein, $\overline{n_i}$ is the count at the interface, $\Delta ASA_{i(j)}$ is the jth interface area of amino acid type i, and $ASA_{i(j)}$ is the jth surface area of amino acid type i. The simple count method was applied to 86 RNA-protein complexes in Kim *et al.*,[94] while the weighted count method was applied to 32 RNA-protein complexes in Jones *et al.*[95] and to 89 protein in Ellis *et al.*[96] The overall tendencies in propensity P_i were different for some specific amino acid residues, especially ones with relatively large numbers of atoms in their side-chains. A detailed comparison of the tendencies is found in Ellis *et al.*[96] Figure 7.2A shows the application of the simple count method to the latest dataset in the Protein Databank (PDB).[97] Arg residues are the most highly favored at the interface, and they are followed by Lys residues. The overrepresentation of Arg over Lys suggests that more than just electrostatic interactions are involved at the protein-RNA interface, since, electrostatically, Arg and Lys behave similarly. On the other hand, twice as many hydrogen bonds are formed between Arg side-chains and the phosphate oxygens and bases of RNA than are formed by Lys. This tendency to overrepresent Arg is more prominent with RNA than DNA, in part because the RNA-specific 2'-OH group is primarily hydrogen bonded with the carbonyl oxygen atom of the polypeptide backbone and Arg side-chains. In addition, the Arg guanidinium moiety stacks onto the bases of the RNA with a greater preference for uracyl bases than for thymidine bases.[98]

Compared to the propensity of protein–DNA interactions, van der Waals contacts are more prevalent in protein–RNA interactions.[95] RNA prefers to interact with residues having aromatic side-chains. The predominant interactions between RNA and these residues are stacking interactions between a base and an aromatic side-chain. RNA molecules often interact with proteins in single-stranded forms and bases are ready to interact with proteins. As a result, the number of atoms of proteins that interact with RNA bases and RNA backbones are more or less the same. This tendency is quite different from protein–DNA interactions. The double-stranded forms of DNA, in which the bases are stacked one upon another in the interior of the molecule, interact with proteins, and consequently, backbone interactions predominate.[95]

Examination of their locations revealed that amino acid residues at the interface with RNA are often spatially clustered. The obvious implication of this clustering is that if a residue is part of an interface with an RNA molecule, then it is highly likely that its spatial neighbor is also part of the same interface. Because, as outlined above, there is a propensity

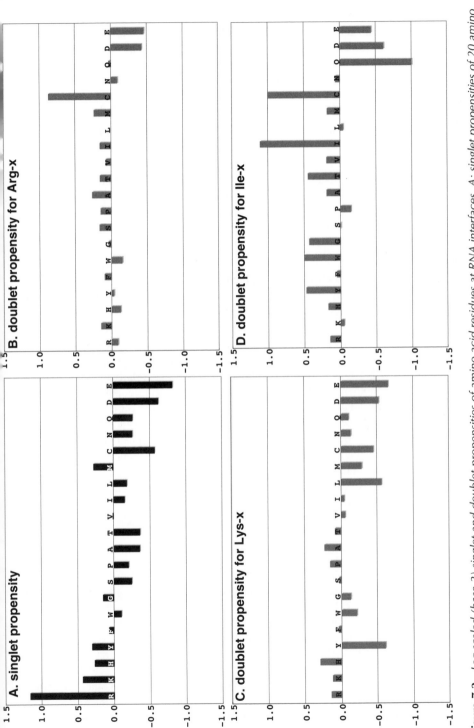

Figure 7.2 Log scaled (base 2) singlet and doublet propensities of amino acid residues at RNA interfaces. A: singlet propensities of 20 amino acid types, B: double propensity for Arg paired with 20 amino acid types, C: double propensity for Lys paired with 20 amino acid types, D: double propensity for Ile paired with 20 amino acid types. The data are derived from 86 chains of RNA-binding proteins stored in the PDB

for certain residues to locate at the interface with RNA, one would also expect there would be preferences for certain neighboring residues at RNA interfaces. Hereafter, the propensity of a single residue to locate at an interface with RNA will be called the singlet propensity and that for a pair of spatial neighbors will be called the doublet propensity. The doublet propensity can be measured using the following equations,[94]

$$Q_{ij} = \frac{\overline{f_{ij}}}{f_{ij}} = \frac{\overline{f_i} \times \overline{f_j} \times C_{ij}}{f_i \times f_j \times D_{ij}} \left(\overline{f_{ij}} = \frac{\overline{n_{ij}}}{\sum_{i=1}^{20} \sum_{j=1}^{20} \overline{n_{ij}}}, \overline{f_i} = \frac{\overline{n_i}}{\sum_{j=1}^{20} \overline{n_j}} \right), \qquad (7.2)$$

where $\overline{f_{ij}}$ is the frequency at which residue types i and j are neighbors at RNA interfaces. Two residues are defined as neighbors when the C_β atoms of the two residues are within 7.0Å. $\overline{f_i}$ is the frequency at which residue type i occurs at RNA interfaces. Hence, C_{ij} is a value reflecting the dependency between the two amino acid types appearing in a pair at RNA interfaces. If $C_{ij} = 1.0$, then the frequency at which amino acid types i and j are neighbors at RNA interfaces is just the product of the frequencies of types i and j, and no preference for pair occurrence is observed. $P_{ij} = \frac{C_{ij}}{D_{ij}}$ is defined as doublet propensity, and plots of part of P_{ij} on a log scale are shown in Figures 7.2B–D. A full description of doublet propensity can be found in Kim *et al.*[94] Note that among the doublet propensities, Cys tends to have high values. This reflects the paucity of data for Cys, and these apparent propensities should not be taken at face value. When interpreting doublet propensity, therefore, it is important to confirm statistical significance, as the number of residue pairs is sometimes very limited at RNA interfaces. Arg and Lys show similar tendencies for pairing with 20 types of amino acids; they both have strong singlet propensities but relatively weak doublet propensities for one another, which means that they do not appear cooperatively at RNA interfaces. An intriguing tendency is found in the doublet propensities for Ile; the Ile-Ile pair has a very high doublet propensity (Figure 7.2D). Although the log singlet propensity of Ile is negative (Figure 7.2A), the log doublet propensity is very highly positive, which means that Ile pairs are more abundant at RNA interfaces than elsewhere on protein surfaces. In that regard, we observed a case in which two Ile side-chains formed a planar structure to accept a uracyl base in a spliceosomal protein.[99] The physicochemical background for this interaction remains unknown.

7.4.4 Prediction of RNA-binding Proteins from Amino Acid Sequence

After the residues at RNA interfaces have been defined and the amino acid preferences determined, one can build a method for predicting RNA-binding proteins based on these preferences. There are three questions to be asked when predicting protein–RNA interactions: (1) Among amino acid sequences predicted from genome nucleotide sequences, which ones are RNA-binding proteins? (2) Where do RNA molecules bind on RNA-binding proteins? (3) How do RNA and protein molecules interact? The first question can be categorized as a general problem of inferring function from amino acid sequences. Information on the general question of predicting function can be found in other textbooks (e.g. von Mering[100]). Han *et al.*[101] has developed a prediction method specific to RNA and implemented it into **SVMProt**, a web server used to predict a large variety of functions on the basis of amino acid sequences. It is expected that the functional characteristics of any protein will be reflected to some extent in its amino acid sequence. SVMProt uses the following properties of each amino acid residue to extract the characteristics of RNA-binding proteins from their amino acid sequences: type of amino acid, hydrophobicity, van der Waals volume, polarity and polarizability. These characteristics are then converted to three types of descriptors: composition, transition and distribution. The composition descriptor is the percentage of each characteristic in the amino acid sequence, the transition is the count of the changes in sign of each characteristic along the amino acid sequence, and the distribution is the percentage of each characteristic in the first fourth, half, three-fourths and all of the amino acid sequence. An SVM was trained to use these descriptors to distinguish between RNA-binding proteins and those that do not bind RNA. SVMProt was reported to single out the amino acid sequences of 94.1% of all known rRNA/tRNA-binding proteins and 79.3% of all known tRNA-binding proteins. The principle underlying these successful predictions is that the known RNA-binding proteins share the characteristics outlined above, which likely reflects the fact that these proteins are derived from a limited number of common ancestors. Consequently, this method may miss new types of RNA-binding proteins that are derived from a different origin.

7.4.5   Prediction of RNA Interface Residues from Amino Acid Sequence

The second question, where do RNA molecules bind on RNA-binding proteins, can be addressed in two ways, depending upon whether one makes use of the amino acid sequence or the 3D-structure of the protein. Prediction of RNA-binding residues from amino acid sequences has been achieved using two machine learning techniques: SVM and naive Bayes classifier. In the SVM method implemented in **BindN**,[102] the characteristics used to identify RNA-binding residues are the side-chain pKa value, hydrophobicity and molecular mass of each amino acid in the sequence. These three characteristics are thought to be indicative of the side-chain ionization necessary for interaction with RNA phosphate groups, the degree to which the residues necessary for interaction with RNA are buried, and the amino acid types, respectively. Descriptors of the target residue are built with the values of five flanking residues on each side along the amino acid sequence (eleven residues in total), and the SVM is trained to use the descriptors to distinguish between residues that bind RNA and those that do not. With their BindN web-server, Wang and Brown achieved 66% sensitivity and 70% specificity.[102]

With the naive Bayes classifier method implemented in **RNABindR**,[103, 104] a stretch of fifteen amino acid residues with the target residue at the center of the stretch is designated as plus if the target residue interacts with RNA, and minus if not. The probability of being a plus or minus is determined based on the observed frequency of an amino acid type at the RNA interface, with the premise that the probability of the stretch is the product of the probabilities of the individual amino acid residues within the stretch. The prediction is performed by dividing the probability of being a plus by that of being a minus; if the value was greater than a certain threshold, then the residue at the center of the stretch was predicted to be at the RNA interface. Terribilini *et al.* reported that they achieved 91% sensitivity and 88% specificity with their RNABindR.[103, 104] Taking into consideration that RNA binds to proteins in three-dimensions, the accuracy of these predictions using only amino acid sequences is surprisingly good. The basis of this accuracy is that RNA-binding residues cluster not only in space, but also within the amino acid sequence,[96, 103] and the

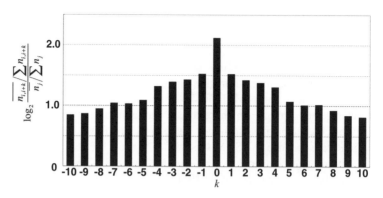

Figure 7.3   *Likelihood that the $i + k$th residue within an amino acid sequences is at the RNA interface when the ith residue is at the RNA interface. The horizontal axis is k, and the vertical axis is the log-odd value. $\overline{n_{i,i+k}}$ is the residue pair count when both the ith and $i + k$th residues within an amino acid sequence are RNA-binding residues, and $\sum n_{i,i+k}$ is the count of all possible residue pairs. In 86 entries of protein-RNA complexes in the PDB, 23% of the residues were identified as RNA-binding, so that the denominator of the equation is 0.23*

premise continues to hold on the latest known 3D-structures of protein-RNA complexes (Figure 7.3). It is natural that neighboring residues within a sequence are close in space, but residues close in space are not necessarily neighbors within a sequence; consequently, the distribution illustrated in Figure 7.3 is not trivial.

7.4.6   Prediction of RNA Interfaces from the 3D-structures of Proteins

The advent of structural genomics has enabled us to use the 3D-structures of RNA-binding proteins as initial input for predicting residues at RNA interfaces. The 3D-structures of protein-RNA complexes have characteristics that cannot be extracted from amino acid sequences, including secondary structures, solvent accessibility, spatial neighbors, and concave and convex structures at the level of the atomic resolution of the RNA interface. These characteristics are expected to provide additional information that will improve the performance of the prediction. **KYG** was the first to include the log-odd score for singlet and doublet propensities together with the amino acid sequence profile to judge whether a residue on the surface of a protein likely interacts with RNA, assuming that the observed frequencies of singlet and doublet within the structure of a protein-RNA complex are equal to the probabilities of singlets and doublets at the RNA interface.[94] Leave-one-out cross-validation experiments showed that the best specificity was around 80%. Without the doublet propensity, the corresponding specificity was around 60%, and the margin of 20% mainly stemmed from the introduction of doublet propensity. KYG can be run without multiple sequence alignment, but the best specificity declines to around 70%. In **DISPLAR**, Tjong and Zhou[51] introduced a neural network technique for predicting DNA/RNA interface residues based on the 3D-structures of proteins. Two types of information about each residue were extracted from the 3D-structure: solvent accessibility and the fourteen spatially neighboring residues. In addition, evolutionary conservation of the residue in question was extracted from multiple sequence alignment in the form of a profile, after which these three characteristics were used to train their neural network. The best specificity achieved with DISPLAR was 57.1%. One needs to have accurate multiple sequence alignment to obtain good performance with either prediction method. The alignment should contain widely varied amino acid sequences, but should not contain sequences with homologues in proteins that do not bind RNA.

7.4.7   Predicting the Structures of Protein-RNA Complexes

The third question, how do RNA and protein molecules interact, relates to problems of docking an RNA onto a protein. Docking methods used by macromolecules are explained in Chapter 9 of this book, and readers should refer to that chapter. One specific issue relating to the structures of protein-RNA complexes involves predicting the conformation of the RNA molecule. For tRNA and rRNA, which have extensive stem structures, rigid body treatment of the RNA is sufficient. Three-dimensional structures of some tRNAs and rRNAs are known, and those structures can serve as a templates for homology modeling of other homologous molecules. For other RNAs, such as mRNAs, small nuclear RNAs (snRNAs) and non-protein-coding RNAs (ncRNAs), however, the 3D-structures have not yet been determined, and they are expected to be quite flexible. We anticipate that efforts currently on the way to predicting the 3D-structures of these RNAs (105-107) will advance our understanding of protein-RNA docking in the near future.

7.5   Protein–Protein Interactions

Each molecular function is realized by a single protein or a stable protein complex, but cellular function is achieved through dynamic interactions among many proteins. There-fore, understanding the protein–protein interactions is the first step to grasp insights into biological functions at the molecular level of proteins. For this purpose, two points should be addressed, that is, (1) to identify which proteins are involved in each biological function, and (2) to determine how they interact with each other.

7.5.1   Which Proteins Interact?

To know which proteins interact with each other is one of the most fundamental problems in the field of molecular biology, and at the same time, it is one of the most difficult problems. There are several methods to judge if a given pair of proteins can interact or not, but it is not straightforward to determine the interacting protein pairs on a genomic scale. For these problems, some high-throughput methods, such as the yeast two-hybrid method,[108] a mass spectrometry-based approach[109] or the tandem-affinity purification (TAP) method[110] can detect potentially interacting proteins on a genomic scale, but the reliability of these methods is still limited.[111] To overcome this difficulty, filtering techniques to find truly interacting pairs using high-throughput data have been proposed.[112, 113] In the filtering, *homologous interactions*, localization in the cell using GO annotation and/or structural information are considered. When a pair of proteins is known to interact in a certain species, and homologs to the protein pairs exist in other species, then the interaction of the pairs in the latter species is considered to be likely and the interactions are called *homologous interactions*.

In spite of the difficulties of large-scale experiments, they are valuable sources to elucidate potentially interacting proteins. Thus, many efforts have been made to collect the known interactions based on the published reports, and several well-designed databases are now available. For example, **BioGRID**[114] and **IntAct**[115] integrate as many interactions as possible, regardless of the experimental scales. The most important point of these databases is that the data are manually curated, using the publications. Therefore, the numbers of entries in the two databases differ somewhat, according to the policy for handling large-scale experiments. There is another type of database that concentrates on human data, aiming at medical applications. Among them, **UniHI**[116] and **HPRD**[117] are updated regularly and are well organized. A comparison of the human protein–protein interaction databases can be found in Mathivivanan *et al.*[118]

Another source of large-scale information can be obtained by DNA microarray analyses. Although these analyses do not yield direct information about protein-protein interactions, the data are strongly correlated with the protein interactions.[119] In addition, the data uniformity and the DNA microarray databases are now improving, and the quality of the data is now critically evaluated by MAQC (Micro Array Quality Control) projects.[120] Therefore, the reliability of microarray data has greatly improved. Actually, in the *Arabidopsis* field,[121, 122] microarray data have been used extensively to identify the potentially interacting proteins. As compared with the *Arabidopsis* field, the utilization of array data is rather limited in the mammalian field. Thus, we have constructed **COXPRESdb**,[123] which provides coexpressed gene networks for human and mouse, in order to facilitate the use of DNA microarray data in the public databases such as **NCBI GEO**.[124]

When each interaction is integrated, a large interaction network will emerge. What can we learn from the large interaction network? One important observation is that the interaction network is like a scale-free network,[125] where only a small number of proteins can interact with many different partners, and others only interact with a few partners. The proteins with multiple partners are called hub proteins, and they are considered to play a critical role in maintaining the interaction network, because disrupting the hub protein can dramatically change the topology of the network.[125] Han *et al.*[126] further classified the hub proteins into *date hubs*, whose interactions with partners are simultaneous, and *party hubs*, which interact with other partners a different times or locations, by using the coexpression profiles obtained by DNA microarray analyses.[126] The date hubs are identified as the proteins with a relatively low average correlation with their partners, while the party hubs are considered to have relatively higher correlated expression patterns with their partners. The work done by Han *et al.* is very interesting from the viewpoint of *network dynamics*. The protein–protein interactions are more or less dynamic; that is, the interaction partners will change according to the state of the cell, but the current large scale experimental analyses lack the dynamic viewpoint, and thus a computational approach will be necessary to elucidate the real dynamic view of the protein–protein interaction network. For this purpose, coexpression networks will be valuable sources of information, and they are extensively used in the plant fields.[127, 128]

7.5.2   How Do They Interact?

Once we have the pair of interacting proteins and their three-dimensional structures, the next problem is to determine their complex structures. This type of problem is known as

(a)   (b)

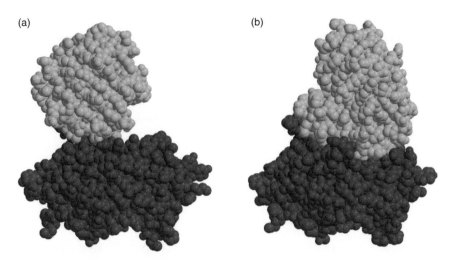

Figure 7.4   *An example of (A) the native complex structure with (B) a decoy complex, de-picted by space-filling model. The black and grey colored subunit are xylanase inhibitor protein 1 and xylanase, respectively. The backbone structure of xylanase inhibitor was superimposed and the complex structure was shown in the same orientation*

the docking problem. In this textbook, the docking approach is described in Chapter 9, and thus here we will briefly describe the concept and the problems from our perspectives.

The docking problem consists of two steps, sampling and selection. The sampling is to search for the possible binding modes in all possible relative rotation and translation space, and the selection is to choose the answer from the candidates that were enumerated in the sampling step. In many cases, these steps are carried out at the same time, but we think there are different types of problems in the two steps.

Docking methods have been critically assessed through the CAPRI community wide experiment on the comparative evaluation of protein-protein docking, or the Critical As-sessment of PRediction of Interactions.[129] According to the CAPRI evaluation, for the rigid body docking, where the flexibility of proteins is neglected, the sampling step is not a serious problem with the current sophisticated algorithms, such as **ZDOCK**,[130] but the selection step still has many problems. In short, we may be able to create lists of possible candidates in many cases, and usually the answer is included in the list, but it cannot be selected before the answer is obtained. Protein flexibility is one of the problems, because proteins may change their structures upon complex formation. However, the main problem is that our knowledge about complex structures is rather limited. The number of heterocomplex in the PDB is quite small. For example, only 15% of all entries in the PDB are heterocomplexes as of May 2007, and the number of non-redundant heterodimer is only around 130. Therefore, a statistical approach and/or a similarity search approach to select an answer from the many candidates has not worked well so far, in contrast to the ligand-binding site of proteins. In addition, when we observe the differences between a native complex with decoy complexes obtained in the sampling step, the differences are surprisingly small on average (Figure 7.4), and in some cases, we could find *better interac-tions* in the decoy complex as compared with the native structure. For example, in the case of the complex between xylanase inhibitor protein I and xylanase (PDB: 1tex1), we found a decoy with a contact area somewhat larger (817 Å^2) than that of the native complex (376 Å^2), and with a completely different binding mode (*r.m.s.d.* value of the backbone structure is 15.5 Å). This is one of the extreme cases, but when we generate possible decoy structures using our own docking method,[131] we can find many decoys with larger contact area than the native ones. In short, because our knowledge about the interface of proteins is short, the native heterocomplexes are evaluated as a bad model compared with the decoy structures.

References

1. D. A. Case, T. E. Cheatham, 3rd, T. Darden, *et al*. The Amber biomolecular simulation programs, *J Comput Chem*, **26**, 1668–1688 (2005).
2. A. Gutteridge and J. Thornton, Conformational changes observed in enzyme crystal structures upon substrate binding, *J Mol Biol*, **346**, 21–28 (2005).
3. G. R. Stockwell and J. M. Thornton, Conformational diversity of ligands bound to proteins, *J Mol Biol*, **356**, 928–944 (2006).
4. G. L. Warren, C. W. Andrews, A. M. Capelli, *et al*. A critical assessment of docking programs and scoring functions, *J Med Chem*, **49**, 5912–5931 (2006).
5. A. E. Todd, C. A. Orengo and J. M. Thornton, Evolution of function in protein superfamilies, from a structural perspective, *J Mol Biol*, **307**, 1113–1143 (2001).
6. J. M. Thornton, A. E. Todd, D. Milburn, N. Borkakoti and C. A. Orengo, From structure to function: approaches and limitations, *Nat Struct Biol*, **7 Suppl**, 991–994 (2000).
7. J. D. Watson, R. A. Laskowski and J. M. Thornton, Predicting protein function from sequence and structural data, *Curr Opin Struct Biol*, **15**, 275–284 (2005).
8. M. Novotny, D. Madsen and G. J. Kleywegt, Evaluation of protein fold comparison servers, *Proteins*, **54**, 260–270 (2004).
9. S. F. Altschul, T. L. Madden, A. A. Schaffer, *et al*. Gapped BLAST and PSI-BLAST: a new generation of protein database search programs, *Nucleic Acids Res*, **25**, 3389–3402 (1997).
10. K. Tomii and Y. Akiyama, FORTE: a profile-profile comparison tool for protein fold recognition, *Bioinformatics*, **20**, 594–595 (2004).
11. R. I. Sadreyev, D. Baker and N. V. Grishin, Profile-profile comparisons by COMPASS predict intricate homologies between protein families, *Protein Sci*, **12**, 2262–2272 (2003).
12. A. J. Reid, C. Yeats and C. A. Orengo, Methods of remote homology detection can be combined to increase coverage by 10% in the midnight zone, *Bioinformatics*, **23**, 2353–2360 (2007).
13. J. Soding, Protein homology detection by HMM-HMM comparison, *Bioinformatics*, **21**, 951–960 (2005).
14. A. Barth, K. Frost, M. Wahab, H. D. Schadler and R. Franke, Classification of serine proteases derived from steric comparisons of their active sites, part II: 'Ser, His, Asp arrangements in proteolytic and nonproteolytic proteins', *Drug Des Discov*, **12**, 89–111 (1994).
15. N. Kobayashi and N. Go, ATP binding proteins with different folds share a common ATP-binding structural motif, *Nat Struct Biol*, **4**, 6–7 (1997).
16. K. A. Denessiouk and M. S. Johnson, When fold is not important: a common structural framework for adenine and AMP binding in 12 unrelated protein families, *Proteins*, **38**, 310–326 (2000).
17. K. Kinoshita, K. Sadanami, A. Kidera and N. Go, Structural motif of phosphate-binding site common to various protein superfamilies: all-against-all structural comparison of protein-mononucleotide complexes, *Protein Eng*, **12**, 11–14 (1999).
18. N. D. Gold and R. M. Jackson, Fold independent structural comparisons of protein-ligand binding sites for exploring functional relationships, *J Mol Biol*, **355**, 1112–1124 (2006).

19. A. C. Wallace, N. Borkakoti and J. M. Thornton, TESS: a geometric hashing algorithm for deriving 3D coordinate templates for searching structural databases. Application to enzyme active sites, *Protein Sci*, **6**, 2308–2323 (1997).
20. R. A. Laskowski, J. D. Watson and J. M. Thornton, Protein function prediction using local 3D templates, *J Mol Biol*, **351**, 614–626 (2005).
21. M. Rosen, S. L. Lin, H. Wolfson and R. Nussinov, Molecular shape comparisons in searches for active sites and functional similarity, *Protein Eng*, **11**, 263–277 (1998).
22. N. Handa, T. Terada, Y. Kamewari, *et al*. Crystal structure of the conserved protein TT1542 from Thermus thermophilus HB8, *Protein Sci*, **12**, 1621–1632 (2003).
23. K. Kinoshita and H. Nakamura, Identification of protein biochemical functions by similarity search using the molecular surface database eF-site, *Protein Sci*, **12**, 1589–1595 (2003).
24. K. Kinoshita, Y. Murakami and H. Nakamura, eF-seek: prediction of the functional sites of proteins by searching for similar electrostatic potential and molecular surface shape, *Nucleic Acids Res*, **35**, W398–402 (2007).
25. K. Kinoshita and H. Nakamura, Protein informatics towards function identification, *Curr Opin Struct Biol*, **13**, 396–400 (2003).
26. A. T. Laurie and R. M. Jackson, Methods for the prediction of protein-ligand binding sites for structure-based drug design and virtual ligand screening, *Curr Protein Pept Sci*, **7**, 395–406 (2006).
27. R. A. Laskowski, N. M. Luscombe, M. B. Swindells and J. M. Thornton, Protein clefts in molecular recognition and function, *Protein Sci*, **5**, 2438–2452 (1996).
28. T. Kawabata and N. Go, Detection of pockets on protein surfaces using small and large probe spheres to find putative ligand binding sites, *Proteins*, **68**, 516–529 (2007).
29. I. Mihalek, I. Res and O. Lichtarge, A family of evolution-entropy hybrid methods for ranking protein residues by importance, *J Mol Biol*, **336**, 1265–1282 (2004).
30. P. L. Privalov, A. I. Dragan, C. Crane-Robinson, K. J. Breslauer, D. P. Remeta and C. A. Minetti, What drives proteins into the major or minor grooves of DNA?, *J Mol Biol*, **365**, 1–9 (2007).
31. M. L. Kopka, A. V. Fratini, H. R. Drew and R. E. Dickerson, Ordered water structure around a B-DNA dodecamer. A quantitative study, *J Mol Biol*, **163**, 129–146 (1983).
32. N. Y. Sidorova, S. Muradymov and D. C. Rau, Differences in hydration coupled to specific and nonspecific competitive binding and to specific DNA Binding of the restriction endonuclease BamHI, *J Biol Chem*, **281**, 35656–35666 (2006).
33. C. G. Kalodimos, N. Biris, A. M. Bonvin, M. M. Levandoski, M. Guennuegues, R. Boelens and R. Kaptein, Structure and flexibility adaptation in nonspecific and specific protein-DNA complexes, *Science*, **305**, 386–389 (2004).
34. C. L. Lawson, D. Swigon, K. S. Murakami, S. A. Darst, H. M. Berman and R. H. Ebright, Catabolite activator protein: DNA binding and transcription activation, *Curr Opin Struct Biol*, **14**, 10–20 (2004).
35. V. Tretyachenko-Ladokhina, M. J. Cocco and D. F. Senear, Flexibility and adaptability in binding of E. coli cytidine repressor to different operators suggests a role in differential gene regulation, *J Mol Biol*, **362**, 271–286 (2006).
36. T. A. Robertson and G. Varani, An all-atom, distance-dependent scoring function for the prediction of protein-DNA interactions from structure, *Proteins*, **66**, 359–374 (2007).
37. K. J. Skowronek, J. Kosinski and J. M. Bujnicki, Theoretical model of restriction endonuclease HpaI in complex with DNA, predicted by fold recognition and validated by site-directed mutagenesis, *Proteins*, **63**, 1059–1068 (2006).
38. A. A. Chmiel, M. Radlinska, S. D. Pawlak, D. Krowarsch, J. M. Bujnicki and K. J. Skowronek, A theoretical model of restriction endonuclease NlaIV in complex with DNA, predicted by fold recognition and validated by site-directed mutagenesis and circular dichroism spectroscopy, *Protein Eng Des Sel*, **18**, 181–189 (2005).
39. P. Bork and E. V. Koonin, Predicting functions from protein sequences–where are the bottlenecks?, *Nat Genet*, **18**, 313–318 (1998).
40. C. Chothia and A. M. Lesk, The relation between the divergence of sequence and structure in proteins, *Embo J*, **5**, 823–826 (1986).

41. E. W. Stawiski, L. M. Gregoret and Y. Mandel-Gutfreund, Annotating nucleic acid-binding function based on protein structure, *J Mol Biol*, **326**, 1065–1079 (2003).
42. S. Ahmad and A. Sarai, Moment-based prediction of DNA-binding proteins, *J Mol Biol*, **341**, 65–71 (2004).
43. N. Bhardwaj, R. E. Langlois, G. Zhao and H. Lu, Kernel-based machine learning protocol for predicting DNA-binding proteins, *Nucleic Acids Res*, **33**, 6486–6493 (2005).
44. M. Gribskov, A. D. McLachlan and D. Eisenberg, Profile analysis: detection of distantly related proteins, *Proc Natl Acad Sci USA*, **84**, 4355–4358 (1987).
45. S. Ahmad and A. Sarai, PSSM-based prediction of DNA binding sites in proteins, *BMC Bioinformatics*, **6**, 33 (2005).
46. I. B. Kuznetsov, Z. Gou, R. Li and S. Hwang, Using evolutionary and structural information to predict DNA-binding sites on DNA-binding proteins, *Proteins*, **64**, 19–27 (2006).
47. L. Wang and S. J. Brown, BindN: a web-based tool for efficient prediction of DNA and RNA binding sites in amino acid sequences, *Nucleic Acids Res*, **34**, W243–248 (2006).
48. Y. Ofran, V. Mysore and B. Rost, Prediction of DNA-binding residues from sequence, *Bioinformatics*, **23**, i347–353 (2007).
49. S. Jones, H. P. Shanahan, H. M. Berman and J. M. Thornton, Using electrostatic potentials to predict DNA-binding sites on DNA-binding proteins, *Nucleic Acids Res*, **31**, 7189–7198 (2003).
50. Y. Tsuchiya, K. Kinoshita and H. Nakamura, Structure-based prediction of DNA-binding sites on proteins using the empirical preference of electrostatic potential and the shape of molecular surfaces, *Proteins*, **55**, 885–894 (2004).
51. H. Tjong and H. X. Zhou, DISPLAR: an accurate method for predicting DNA-binding sites on protein surfaces, *Nucleic Acids Res*, **35**, 1465–1477 (2007).
52. P. V. Benos, M. L. Bulyk and G. D. Stormo, Additivity in protein-DNA interactions: how good an approximation is it?, *Nucleic Acids Res*, **30**, 4442–4451 (2002).
53. V. Matys, O. V. Kel-Margoulis, E. Fricke, *et al.* TRANSFAC and its module TRANSCompel: transcriptional gene regulation in eukaryotes, *Nucleic Acids Res*, **34**, D108–110 (2006).
54. K. Frech, G. Herrmann and T. Werner, Computer-assisted prediction, classification, and delimitation of protein binding sites in nucleic acids, *Nucleic Acids Res*, **21**, 1655–1664 (1993).
55. D. GuhaThakurta and G. D. Stormo, Identifying target sites for cooperatively binding factors, *Bioinformatics*, **17**, 608–621 (2001)
56. S. Levy, S. Hannenhalli and C. Workman, Enrichment of regulatory signals in conserved non-coding genomic sequence, *Bioinformatics*, **17**, 871–877 (2001) .
57. Y. Pilpel, P. Sudarsanam and G. M. Church, Identifying regulatory networks by combinatorial analysis of promoter elements, *Nat Genet*, **29**, 153–159 (2001).
58. Z. Zhu, Y. Pilpel and G. M. Church, Computational identification of transcription factor binding sites via a transcription-factor-centric clustering (TFCC) algorithm, *J Mol Biol*, **318**, 71–81 (2002).
59. H. J. Bussemaker, H. Li and E. D. Siggia, Regulatory element detection using correlation with expression, *Nat Genet*, **27**, 167–171 (2001).
60. W. W. Wasserman, M. Palumbo, W. Thompson, J. W. Fickett and C. E. Lawrence, Human-mouse genome comparisons to locate regulatory sites, *Nat Genet*, **26**, 225–228 (2000).
61. C. E. Lawrence and A. A. Reilly, An expectation maximization (EM) algorithm for the identification and characterization of common sites in unaligned biopolymer sequences, *Proteins*, **7**, 41–51 (1990).
62. C. E. Lawrence, S. F. Altschul, M. S. Boguski, J. S. Liu, A. F. Neuwald and J. C. Wootton, Detecting subtle sequence signals: a Gibbs sampling strategy for multiple alignment, *Science*, **262**, 208–214 (1993).
63. H. Kono and A. Sarai, Structure-based prediction of DNA target sites by regulatory proteins, *Proteins*, **35**, 114–131 (1999).
64. N. M. Luscombe, R. A. Laskowski and J. M. Thornton, Amino acid-base interactions: a three-dimensional analysis of protein-DNA interactions at an atomic level, *Nucleic Acids Res*, **29**, 2860–2874 (2001).

65. C. O. Pabo and L. Nekludova, Geometric analysis and comparison of protein-DNA interfaces: why is there no simple code for recognition?, *J Mol Biol*, **301**, 597–624 (2000).
66. T. W. Siggers, A. Silkov and B. Honig, Structural alignment of protein–DNA interfaces: insights into the determinants of binding specificity, *J Mol Biol*, **345**, 1027–1045 (2005).
67. J. E. Donald, W. W. Chen and E. I. Shakhnovich, Energetics of protein-DNA interactions, *Nucleic Acids Res*, **35**, 1039–1047 (2007).
68. F. Pichierri, M. Aida, M. M. Gromiha and A. Sarai, Free-energy maps of base-amino acid interactions for DNA-protein recognition, *Journal of the American Chemical Society*, **121**, 6152 (1999).
69. K. Sayano, H. Kono, M. M. Gromiha and A. Sarai, Multicanonical monte carlo calculation of free-energy map for base-amino acid interaction, *J Comp Chem*, **21**, 954–962. (2000).
70. R. S. Hegde, The papillomavirus E2 proteins: structure, function, and biology, *Annu Rev Biophys Biomol Struct*, **31**, 343–360 (2002).
71. N. R. Steffen, S. D. Murphy, L. Tolleri, G. W. Hatfield and R. H. Lathrop, DNA sequence and structure: direct and indirect recognition in protein-DNA binding, *Bioinformatics*, **18 Suppl 1**, S22–30 (2002).
72. G. Paillard and R. Lavery, Analyzing protein-DNA recognition mechanisms, *Structure*, **12**, 113–122 (2004).
73. M. M. Gromiha, J. G. Siebers, S. Selvaraj, H. Kono and A. Sarai, Intermolecular and Intramolecular Readout Mechanisms in Protein-DNA Recognition, *J Mol Biol*, **337**, 285–294 (2004).
74. A. Sarai and H. Kono, Protein-DNA Recognition Patterns and Predictions, *Annu Rev Biophys Biomol Struct*, **34**, 379–398 (2005).
75. S. Ahmad, H. Kono, M. J. Arauzo-Bravo and A. Sarai, ReadOut: Structure-based Calculation of Direct and Indirect Readout Energies and Specificities for Protein-DNA Recognition, *Nucleic Acids Res*, **34**, w124–127 (2006).
76. A. Sarai, J. Siebers, S. Selvaraj, M. M. Gromiha and H. Kono, Integration of bioinformatics and computational biology to understand protein-DNA recognition mechanism, *J Bioinform Comput Biol*, **3**, 1–15 (2005).
77. S. Fujii, H. Kono, S. Takenaka, N. Go and A. Sarai, Sequence-dependent DNA deformability studied using molecular dynamics simulations, *Nucleic Acids Res*, **35**, 6063–6074 (2007).
78. H. Hieronymus and P. A. Silver, A systems view of mRNP biology, *Genes Dev*, **18**, 2845–2860 (2004).
79. M. S. Jurica and M. J. Moore, Pre-mRNA splicing: awash in a sea of proteins, *Mol Cell*, **12**, 5–14 (2003).
80. R. Bock, Sense from nonsense: how the genetic information of chloroplasts is altered by RNA editing, *Biochimie*, **82**, 549–557 (2000).
81. B. L. Bass, *RNA editing*, Oxford University Press (2001).
82. T. J. Begley and L. D. Samson, Molecular biology: A fix for RNA, *Nature*, **421**, 795–796 (2003).
83. G. Dreyfuss, V. N. Kim and N. Kataoka, Messenger-RNA-binding proteins and the messages they carry, *Nat Rev Mol Cell Biol*, **3**, 195–205 (2002).
84. S. Kindler, H. Wang, D. Richter and H. Tiedge, RNA transport and local control of translation, *Annu Rev Cell Dev Biol*, **21**, 223–245 (2005).
85. A. Kohler and E. Hurt, Exporting RNA from the nucleus to the cytoplasm, *Nat Rev Mol Cell Biol*, **8**, 761–773 (2007).
86. J. D. Keene, Ribonucleoprotein infrastructure regulating the flow of genetic information between the genome and the proteome, *Proc Natl Acad Sci USA*, **98**, 7018–7024 (2001).
87. G. M. Rubin, M. D. Yandell, J. R. Wortman, *et al.* Comparative Genomics of the Eukaryotes, *Science*, **287**, 2204–2215 (2000).
88. H. A. Lewis, K. Musunuru, K. B. Jensen, *et al.* Sequence-specific RNA binding by a Nova KH domain: implications for paraneoplastic disease and the Fragile X syndrome, *Cell*, **100**, 323 (2000).
89. A. Ramos, S. Grunert, J. Adams, *et al.*, RNA recognition by a Staufen double-stranded RNA-binding domain, *Embo J*, **19**, 997–1009 (2000).

90. J.-B. Ma, K. Ye and D. J. Patel, Structural basis for overhang-specific small interfering RNA recognition by the PAZ domain, *Nature*, **429**, 318 (2004).

91. X. Wang, J. McLachlan, P. D. Zamore and T. M. Hall, Modular recognition of RNA by a human pumilio-homology domain, *Cell*, **110**, 501–512 (2002).

92. F. B. Sheinerman, R. Norel and B. Honig, Electrostatic aspects of protein-protein interactions, *Curr Opin Struct Biol*, **10**, 153–159 (2000).

93. H. Kim, E. Jeong, S. W. Lee and K. Han, Computational analysis of hydrogen bonds in protein-RNA complexes for interaction patterns, *FEBS Lett*, **552**, 231–239 (2003).

94. O. T. Kim, K. Yura and N. Go, Amino acid residue doublet propensity in the protein-RNA interface and its application to RNA interface prediction, *Nucleic Acids Res*, **34**, 6450–6460 (2006).

95. S. Jones, D. T. Daley, N. M. Luscombe, H. M. Berman and J. M. Thornton, Protein-RNA interactions: a structural analysis, *Nucleic Acids Res*, **29**, 943–954 (2001).

96. J. J. Ellis, M. Broom and S. Jones, Protein–RNA interactions: structural analysis and functional classes, *Proteins*, **66**, 903–911 (2007).

97. H. M. Berman, J. Westbrook, Z. Feng, *et al.* The Protein Data Bank, *Nucleic Acids Res*, **28**, 235–242 (2000).

98. J. Allers and Y. Shamoo, Structure-based analysis of protein-RNA interactions using the program ENTANGLE, *J Mol Biol*, **311**, 75–86 (2001).

99. I. Vidovic, S. Nottrott, K. Hartmuth, R. Luhrmann and R. Ficner, Crystal structure of the spliceosomal 15.5kD protein bound to a U4 snRNA fragment, *Mol Cell*, **6**, 1331–1342 (2000).

100. C. von Mering, *Inferring Protein Function from Genomic Context*, Wiley-VCH Verlag GmbH & Co KGaA, Weinheim, 2007.

101. L. Y. Han, C. Z. Cai, S. L. Lo, M. C. Chung and Y. Z. Chen, Prediction of RNA-binding proteins from primary sequence by a support vector machine approach, *Rna*, **10**, 355–368 (2004).

102. L. Wang and S. J. Brown, BindN: a web-based tool for efficient prediction of DNA and RNA binding sites in amino acid sequences, *Nucl Acids Res*, **34**, W243–248 (2006).

103. M. Terribilini, J. H. Lee, C. Yan, R. L. Jernigan, V. Honavar and D. Dobbs, Prediction of RNA binding sites in proteins from amino acid sequence, *RNA*, **12**, 1450–1462 (2006).

104. M. Terribilini, J. D. Sander, J.-H. Lee, *et al.* RNABindR: a server for analyzing and predicting RNA-binding sites in proteins, *Nucl Acids Res*, gkm294 (2007).

105. F. Jossinet, T. E. Ludwig and E. Westhof, RNA structure: bioinformatic analysis, *Curr Opin Microbiol*, **10**, 279–285 (2007).

106. P. S. Klosterman, D. K. Hendrix, M. Tamura, S. R. Holbrook and S. E. Brenner, Three-dimensional motifs from the SCOR, structural classification of RNA database: extruded strands, base triples, tetraloops and U-turns, *Nucl Acids Res*, **32**, 2342–2352 (2004).

107. B. A. Shapiro, Y. G. Yingling, W. Kasprzak and E. Bindewald, Bridging the gap in RNA structure prediction, *Curr Opin Struct Biol*, **17**, 157–165 (2007).

108. T. Strachan and A. P. Read, *Human Molecular Genetics*, Garland Science, Taylor & Francis Group, 2003.

109. R. M. Ewing, P. Chu, F. Elisma, *et al.* Large-scale mapping of human protein-protein interactions by mass spectrometry, *Mol Syst Biol*, **3**, 89 (2007).

110. T. A. Brown, *Genomes Three*, Garland Science, New York, 2006.

111. C. von Mering, R. Krause, B. Snel, M. Cornell, S. G. Oliver, S. Fields and P. Bork, Comparative assessment of large-scale data sets of protein-protein interactions, *Nature*, **417**, 399–403 (2002).

112. C. M. Deane, L. Salwinski, I. Xenarios and D. Eisenberg, Protein interactions: two methods for assessment of the reliability of high throughput observations, *Mol Cell Proteomics*, **1**, 349–356 (2002).

113. A. Patil and H. Nakamura, Filtering high-throughput protein-protein interaction data using a combination of genomic features, *BMC Bioinformatics*, **6**, 100 (2005).

114. C. Stark, B. J. Breitkreutz, T. Reguly, L. Boucher, A. Breitkreutz and M. Tyers, BioGRID: a general repository for interaction datasets, *Nucleic Acids Res*, **34**, D535–539 (2006).

115. S. Kerrien, Y. Alam-Faruque, B. Aranda, *et al.* IntAct–open source resource for molecular interaction data, *Nucleic Acids Res*, **35**, D561–565 (2007).

116. G. Chaurasia, Y. Iqbal, C. Hanig, H. Herzel, E. E. Wanker and M. E. Futschik, UniHI: an entry gate to the human protein interactome, *Nucleic Acids Res*, **35**, D590–594 (2007).
117. G. R. Mishra, M. Suresh, K. Kumaran, *et al*. Human protein reference database–2006 update, *Nucleic Acids Res*, **34**, D411–414 (2006).
118. S. Mathivanan, B. Periaswamy, T. K. Gandhi, *et al*. An evaluation of human protein-protein interaction data in the public domain, *BMC Bioinformatics*, **7 Suppl 5**, S19 (2006).
119. R. Jansen, D. Greenbaum and M. Gerstein, Relating whole-genome expression data with protein-protein interactions, *Genome Res*, **12**, 37–46 (2002).
120. L. Shi, L. H. Reid, W. D. Jones, *et al*. The MicroArray Quality Control (MAQC) project shows inter- and intraplatform reproducibility of gene expression measurements, *Nat Biotechnol*, **24**, 1151–1161 (2006).
121. T. Obayashi, K. Kinoshita, K. Nakai, *et al*. ATTED-II: a database of co-expressed genes and cis elements for identifying co-regulated gene groups in Arabidopsis, *Nucleic Acids Res*, **35**, D863–869 (2007).
122. S. Y. Rhee, W. Beavis, T. Z. Berardini, *et al*. The Arabidopsis Information Resource (TAIR): a model organism database providing a centralized, curated gateway to Arabidopsis biology, research materials and community, *Nucleic Acids Res*, **31**, 224–228 (2003).
123. T. Obayashi, S. Hayashi, M. Shibaoka, M. Saeki, H. Ohta and K. Kinoshita, COXPRESdb: a database of coexpressed gene networks in mammals, *Nuc Acid Res* (2008) Jan. 36 (Database issue):D77-82. [Epub 2007, Oct 11].
124. R. Edgar and T. Barrett, NCBI GEO standards and services for microarray data, *Nat Biotechnol*, **24**, 1471–1472 (2006).
125. H. Jeong, S. P. Mason, A. L. Barabasi and Z. N. Oltvai, Lethality and centrality in protein networks, *Nature*, **411**, 41–42 (2001).
126. J. D. Han, N. Bertin, T. Hao, *et al*. Evidence for dynamically organized modularity in the yeast protein-protein interaction network, *Nature*, **430**, 88–93 (2004).
127. K. Aoki, Y. Ogata and D. Shibata, Approaches for extracting practical information from gene co-expression networks in plant biology, *Plant Cell Physiol*, **48**, 381–390 (2007).
128. M. Altaf-Ul-Amin, Y. Shinbo, K. Mihara, K. Kurokawa and S. Kanaya, Development and implementation of an algorithm for detection of protein complexes in large interaction networks, *BMC Bioinformatics*, **7**, 207 (2006).
129. J. Janin, Assessing predictions of protein-protein interaction: the CAPRI experiment, *Protein Sci*, **14**, 278–283 (2005).
130. C. Zhang, S. Liu and Y. Zhou, Docking prediction using biological information, ZDOCK sampling technique, and clustering guided by the DFIRE statistical energy function, *Proteins*, **60**, 314–318 (2005).
131. E. Kanamori, Y. Murakami, Y. Tsuchiya, D. M. Standley, H. Nakamura and K. Kinoshita, Docking of protein molecular surfaces with evolutionary trace analysis, *Proteins*, **69**, 832–838 (2007).
132. K. Kinoshita and H. Nakamura, eF-site and PDBjViewer: database and viewer for protein functional sites, *Bioinformatics*, **20**, 1329–1330 (2004).
133. G. Lopez, A. Valencia and M. L. Tress, firestar–prediction of functionally important residues using structural templates and alignment reliability, *Nucleic Acids Res*, **35**, W573–577 (2007).
134. A. T. Laurie and R. M. Jackson, Q-SiteFinder: an energy-based method for the prediction of protein-ligand binding sites, *Bioinformatics*, **21**, 1908–1916 (2005).
135. N. Kumar and D. Mohanty, MODPROPEP: a program for knowledge-based modeling of protein-peptide complexes, *Nucleic Acids Res*, **35**, W549–555 (2007).
136. Y. Tsuchiya, K. Kinoshita and H. Nakamura, PreDs: a server for predicting dsDNA-binding site on protein molecular surfaces, *Bioinformatics*, **21**, 1721–1723 (2005).
137. B. Raghavachari, A. Tasneem, T. M. Przytycka and R. Jothi, DOMINE: a database of protein domain interactions, *Nucleic Acids Res* (2007).
138. H. Tjong, S. Qin and H. X. Zhou, PI2PE: protein interface/interior prediction engine, *Nucleic Acids Res*, **35**, W357–362 (2007).
139. Y. C. Chen, Y. S. Lo, W. C. Hsu and J. M. Yang, 3D-partner: a web server to infer interacting partners and binding models, *Nucleic Acids Res*, **35**, W561–567 (2007).

140. K. G. Tina, R. Bhadra and N. Srinivasan, PIC: Protein Interactions Calculator, *Nucleic Acids Res*, **35**, W473–476 (2007).
141. H. Neuvirth, R. Raz and G. Schreiber, ProMate: a structure based prediction program to identify the location of protein-protein binding sites, *J Mol Biol*, **338**, 181–199 (2004).
142. P. C. Champ and C. J. Camacho, FastContact: a free energy scoring tool for protein-protein complex structures, *Nucleic Acids Res*, **35**, W556–560 (2007).
143. H. K. Saini and D. Fischer, FRalanyzer: a tool for functional analysis of fold-recognition sequence-structure alignments, *Nucleic Acids Res*, **35**, W499–502 (2007).
144. K. Kinoshita and M. Ota, P-cats: prediction of catalytic residues in proteins from their tertiary structures, *Bioinformatics*, **21**, 3570–3571 (2005).
145. C. A. Innis, siteFiNDER|3D: a web-based tool for predicting the location of functional sites in proteins, *Nucleic Acids Res*, **35**, W489–494 (2007).

8

Structure-based Prediction of Enzymes and Their Active Sites

James W. Torrance and Janet M. Thornton

8.1 Introduction

Structures can be available for proteins whose function is either not fully understood, or wholly unknown. The number of such structures has been greatly increased by structural genomics projects. For this reason, it is useful to be able to predict the function of a protein based on its structure. This chapter considers those methods for structure-based function prediction that are applicable to proteins in general, and then focuses on those methods which are more specifically suited to identifying enzymes. The chapter concludes by discussing the best strategy for using these methods together to predict function.

There are several levels at which the function of an enzyme can be described. The different function prediction methods described below vary in the level at which they predict function. Enzyme function can be described in terms of the Enzyme Commission (EC) classification.[1] This is a numerical, hierarchical classification with four levels. However, many methods for function prediction aim to predict the function of non-enzymes as well, and for this reason they frequently make use of the Gene Ontology (GO), which is a vocabulary of terms for describing the biochemical functions of gene products, and the biochemical processes and cellular components with which they are associated.[2] Some methods predict the general location of the active site of the enzyme. Other methods attempt to identify individual residues with key roles in catalysis, and may attempt to discriminate between those residues which are involved in binding to the substrate, and those which play a chemical role in catalysis.

Prediction of Protein Structures, Functions, and Interactions   Edited by Janusz Bujnicki
© 2009 John Wiley & Sons, Ltd

Protein function can be predicted by identifying a homologue of similar function. Because protein structure is more conserved than protein sequence, it can be possible to detect homologues by a comparison of overall protein structure which could not be detected by sequence comparison methods. These methods depending on overall structure comparison are relevant to both enzymes and non-enzymes. Because the function of enzymes is dependent on the location of small numbers of residues which are grouped around their active site, it is possible to identify enzymes and predict their function by searching protein structures for groups of residues resembling the active sites of enzymes of known function, in the hope of identifying homologues and cases of convergent evolution. Finally, it is possible to attempt to identify the location of the catalytic site and the positions of individual catalytic residues on the basis of their structural properties without deriving this information from homologues, although this is a difficult task. Function prediction meta-servers exist which integrate several of these approaches.

8.2 Identifying Homologues Using Overall Protein Structure

It is possible to predict the function of a protein by identifying homologues whose function is known. Methods for detecting homologues by sequence searching have been discussed in Chapters 1 and 4. Although far fewer protein structures have been determined than protein sequences, it is possible to discover more remote homologues by comparing the overall fold of a protein. Webservers for searching databases of protein structures for those with a similar fold to a query structure are described in Table 8.1. (Various similar methods exist for obtaining a structural alignment between a pair of proteins, but only those which carry out database searching are listed here. Structure comparison methods have been reviewed in detail by Koehl.[3]) Most of these fold searching methods do not explicitly provide functional predictions, although AnnoLite[4] provides GO annotation and EC classification predictions, along with probability values scoring their degree of certainty.

An assessment by Novotny *et al.*[12] of the ability of structure alignment methods to detect homologues concluded that there was no one overall 'best' server, although they found that Dali,[5] VAST,[6] CE[7] and Matras[8] all performed well. They recommend using several servers to confirm the results. A study by Sierk and Pearson[13] analysed the ability of structure alignment methods to discriminate homologues from non-homologous proteins with the same topology; this found that Dali was the most discriminating.

How accurate a prediction of enzyme function can be made on the basis of homology? As the level of sequence identity between a pair of enzymes declines, so the probability that they have the same EC function declines. Different analyses have reached significantly different conclusions concerning how rapidly the probability of functional similarity declines as sequence similarity decreases,[14–16] but all agree that below 40% sequence identity the likelihood of two enzymes sharing the same function as described by the full EC code declines swiftly.

For some enzyme homologous superfamilies, function is well conserved, and in these cases identification of a protein as a member of the superfamily on the basis of structure can imply a shared function. Furthermore, even in functionally diverse superfamilies, some elements of catalytic function are often conserved. Todd *et al.* analysed evolution within 31 enzyme superfamilies, each of which included enzymes with a range of functions.[14]

Table 8.1 *Methods for using overall protein structure to identify homologues*

Method	URL	Description
Dali[5]	http://www.ebi.ac.uk/dali	Describes protein structures as a residue-residue distance matrix. This is broken down into submatrices describing distances between hexapeptide fragments; similar submatrices are aligned, and an overall alignment is built up from these fragment alignments
VAST[6]	http://www.ncbi.nlm.nih.gov/Structure/VAST/vast.shtml	Proteins which have secondary structural elements with a similar topology and orientation are detected using a graph approach
CE[7]	http://cl.sdsc.edu	Initially aligns fragments of the protein using characteristics of the local geometry, defined by vectors between Cα positions. The resulting aligned fragment pairs are joined to form an overall alignment, with gaps if required
Matras[8]	http://biunit.naist.jp/matras	Obtains an initial alignment of proteins by comparing secondary structural elements. Alignments are scored by using a Markov transition model of evolution to calculate the probability of one structure evolving into another
CATHEDRAL[9]	http://cathwww.biochem.ucl.ac.uk/cgi-bin/cath/CathedralServer.pl	Initially uses a fast graph theory algorithm to align secondary structural elements. Subsequently obtains a more accurate alignment using a residue-based method
SSM[10]	http://www.ebi.ac.uk/msd-srv/ssm	Describes a protein structure as a graph of secondary structural elements, and uses a rapid graph-matching algorithm to match protein structures
FATCAT[11]	http://fatcat.burnham.org/fatcat	Combines aligned fragment pairs in a manner that accounts for the flexibility of proteins
AnnoLite[4]	http://www.salilab.org/DBAli	Calculates a similarity measure between the Cα atom positions of hexapeptide pairs in the structures under comparison. Dynamic programming is used on the resulting similarity matrix to find an alignment

Details of catalysis were available for 27 of these superfamilies. The analysis found that catalytic mechanism was conserved in four of these 27 superfamilies, and that mechanism is 'semi-conserved' (meaning that a common chemical strategy is used in the context of different overall transformations) in a further 18. Nevertheless, if a protein has homologues of known function identified using structure comparison that cannot be recognised using sequence-based methods, generally only a very weak speculation can be made concerning its function. However, analyses of protein structure that focus on details around the putative active site can provide stronger evidence concerning protein function.

8.3   Recognising Distant Homologues and Cases of Convergent Evolution Using Template Matching Methods

The function of an enzyme depends upon the geometry of the catalytic residues and substrate binding residues. These key residues can have well-conserved geometries in distantly-related proteins of similar function.[17] If residues with a similar arrangement are found in a distant homologue, this suggests that it may have a similar function; conversely, if key catalytic residues are missing or have a different geometry, this suggests a change in function. There are also cases where unrelated enzymes have evolved similar arrangements of residues which carry out similar functions in catalysis (convergent evolution). Thus, if a group of residues in a protein of unknown function is found to resemble a group of catalytic residues in a non-homologous enzyme, this might indicate a similar function- although it is difficult to assess how practical this is as an approach to function prediction, as discussed below.

A number of methods exist which search protein structures for groups of residues that have a particular spatial arrangement. In this chapter, the term 'structural template' is used to describe a predefined spatial pattern of residues which these methods can search for within a protein. (This should not be confused with the unrelated idea of a template structure in homology modelling.) These structural templates can correspond to the catalytic residues of an enzyme (although they can also correspond to any component of a protein structure).

Template matching methods can be used with individual templates created by the user. However, it can be useful to have a library of structural templates corresponding to known functional sites. A number of efforts have been made to create such libraries in a systematic or automated manner; these will be discussed later in this chapter.

The template matching process can be made clearer with an example. There are a number of unrelated hydrolase enzymes which make use of a combination of Ser, His and Asp residues playing equivalent roles in catalysis.[18] These residues can occur at widely separated locations in the protein sequence, and can occur in a different order in the sequence in different, nonhomologous enzymes. However, they generally occur in very similar positions in space relative to one another. A structural template can represent the relative spatial positions of the three residues in one of these enzymes. This structural template can be used to search other protein structures for occurrences of these three residues in similar arrangements, which may imply the existence of a catalytic site performing a similar function.[19] We will begin by discussing methods that have the potential to detect cases of convergent evolution as well as distant relatives; that is, those methods which match similar patterns of atoms independently of residue order in the protein sequence, and independently of larger structural features such as protein surface clefts. First we discuss the extent to which these methods can be useful for predicting function, then we describe the details of how individual methods function.

8.3.1 Usefulness of Templates for Predicting Function

Template searching is a useful complement to methods of function prediction that are based on recognising homologues using sequence or overall structure, for the following reasons.

1. There are instances when proteins have independently evolved the same configuration of catalytic residues for carrying out similar reactions. In these cases of convergent evolution it may be possible to predict the common function on the basis of common catalytic residue conformation.
2. Catalytic residue conformation in homologous enzymes of similar function may remain conserved when the rest of the protein structure has diverged to the extent that it cannot be used to predict function.
3. Even when distant homologues can be identified using sequence methods, their correct sequence alignment may be ambiguous; a structural comparison of catalytic sites may resolve the ambiguity and thus suggest which residues are most likely to be involved in catalysis.
4. Even for homologues identifiable by sequence methods, identifying similar catalytic sites that are spread over multiple protein chains may be simpler using structural similarity of catalytic sites than by using sequence comparison.
5. Even if two enzymes are clearly recognisable as homologues on the basis of their sequence or overall structure, they may still have different functions[20]. If they do have different functions, they will often not retain the same catalytic residue conformation– so a consideration of catalytic residue conformation will permit the hypothesis of similar function to be rejected.

8.3.2 Usefulness of Templates for Identifying Cases of Convergent Evolution

The usefulness of structural templates for detecting cases of convergent evolution of catalytic residues depends upon how frequently this type of convergent evolution occurs, and how preciely the convergently evolved sites resemble one another. Unfortunately, these questions are currently difficult to answer.

There are many cases of convergent evolution of overall enzyme function. In principle, it might be the case that the active sites of these enzymes resemble one another and that these similarities might form a basis for predicting function by comparing non-homologous structures. A comparison of the SCOP structural classification and the EC classification by Galperin and Koonin found 34 EC numbers which occurred in more than one SCOP superfamily, implying that convergent evolution had occurred.[21] However, because the EC classification only describes the substrates and products of a reaction, these enzymes may have entirely different mechanisms and active sites.

There are cases where unrelated enzymes catalyse similar reactions using residues with similar geometries in similar dispositions relative to one another. The best studied example is the use of combinations of Ser, His, and Asp residues to catalyse hydrolysis reactions by a nucleophilic substitution mechanism. This has independently evolved on at least six occasions, and a number of chemically similar residue groups exist, such as Cys-His-Asp triads.[18] Template matching methods can be used to detect such similarities, and in fact several of the template matching methods described below have been shown to be capable of detecting the similar Ser-His-Asp groups in unrelated hydrolase enzymes. Wallace *et al.* analysed Ser-His-Asp catalytic triads from several convergently evolved groups of hydrolases.[22] They found that in the majority of these triads, the distance between the functional oxygens of the Ser and Asp residues was within 1.4 Å of the consensus distance over all triads. They also found that few non-catalytic Ser-His-Asp associations had this conformation.

However, at the time of writing no general study exists that assesses how practical the detection of such cases of convergent evolution is for function prediction purposes. For this reason, a template match should be interpreted with great caution if there is no other evidence of homology between the structure used to generate the template and the structure which was matched.

8.3.3   Usefulness of Templates for Identifying Distant Homologues

Active sites are usually highly conserved in sequence and structure in homologous enzymes of similar function. Nevertheless, the ability of structural templates to identify homologues depends on the degree of structural variability of these active sites in homologous enzymes. The more structurally variable these sites, the more difficult it will be to discriminate between matches to homologous active sites and matches which are meaningless. An assessment has been carried out of the degree of structural variability occurring in homologous active sites, and the ability of structural templates to recognise equivalent catalytic sites in homologues, through an analysis[17] of a library of 147 structural templates representing catalytic sites in the Catalytic Site Atlas (CSA)[23] (this library is further described below, and is available at http://www.ebi.ac.uk/thornton-srv/databases/CSS).

The CSA is a database of catalytic residues in proteins of known structure; the annotation in this database is manually derived from the scientific literature. For each of the 147 literature-derived catalytic sites in the CSA that were considered in this study, relatives were identified using the sequence searching program PSI-BLAST.[24] Relatives were only considered if they had identical residue types to the catalytic residues occurring in the literature entry (according to the sequence entry alignment). Most of these catalytic residues in homologues were found to differ by less than 1 Å atom coordinate root mean square deviation (RMSD), even in very distant relatives.

This analysis further examined how well these structural templates could discriminate between matches to the catalytic sites of homologues of the structure which was the basis for the template and random matches to non-relatives. The analysis found that these structural templates could discriminate sites in homologues from random matches with over 85% sensitivity and predictive accuracy. This high performance is a consequence of the high degree of structural conservation of catalytic sites described in the previous paragraph. However, these homologues were identified using PSI-BLAST to identify homologues recognisable from their sequence, so this study did not address the question of how well these structural templates can identify homologues which could not also be identified using sequence searching.

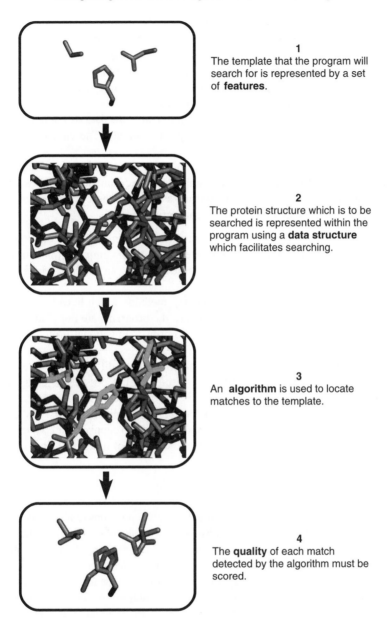

1
The template that the program will search for is represented by a set of **features**.

2
The protein structure which is to be searched is represented within the program using a **data structure** which facilitates searching.

3
An **algorithm** is used to locate matches to the template.

4
The **quality** of each match detected by the algorithm must be scored.

Figure 8.1 *Components of a template matching method. This figure was prepared using Pymol (www.pymol.org)* (See insert for color representation of figure)

8.3.4   Components of a Template Matching Method

A method for template matching can be divided into four components. These are shown in Figure 8.1, and listed here.

1. The method must specify the features that are to be used to represent the template. Methods which abstract away some of the details of template structure will be less specific; this may result in more matches that are not biologically meaningful, but may also allow the detection of functionally similar sites with minor variations in structure.
2. Most methods represent the geometry of the protein structure which is to be searched using a data structure that aids the search process.
3. The method must have an algorithm for searching within the data structure representing the protein for groups of residues resembling the template. The choice of algorithm determines the speed of the method.
4. The quality of the match between the template and the protein structure must then be scored. The choice of scoring measure is important, since smaller templates will often have large numbers of matches, and the scoring assigned to these is crucial for deciding which of these matches are biologically meaningful.

Relevant template matching methods are summarised in terms of these four components in Table 8.2, which also provides literature references. Each component is considered in more detail below. Webservers through which these methods can be accessed are described in Table 8.3.

There are a number of other template matching methods which have been developed, but which are not described in detail here because the programs are not readily available for use. These include TESS,[19] Fuzzy Functional Forms,[25] Functional-group 3D motifs,[26] Triads,[27] and a method developed by Singh and Saha.[28]

There have been no studies comparing the effectiveness of different template matching methods for predicting function, and for this reason it is not possible to suggest which methods are the most useful, and it is not possible to comment on the best choices for each of the four components described above, except in broad qualitative terms. The choice of methods made by a user will depend on whether they wish to supply their own template or make use of a template library, and if the latter, which types of functional site they wish to search for.

8.3.5   Features of the Template

A template could be represented by describing the relative positions of all atoms in all relevant residues. However, it can be useful to abstract away some of the details, as this reduces the amount of storage and memory space required for representing the protein structures to be searched. It also simplifies and thus speeds up the search process. Abstracting away sidechains avoids problems due to ambiguities in experimental identification of atoms (such as sidechain oxygen and nitrogen atoms in asparagine and glutamine residues) and atom nomenclature (such as the equivalent oxygen atoms in aspartate and glutamate residues). Abstraction also makes it easier for the template to match structures which have minor differences in sidechain conformation. The disadvantage of abstraction is the possibility of lowering template specificity.

Some template matching methods treat the structures as sets of atoms, independent of residues, and allow the user to decide which atoms to employ in their template definition (NeedleHaystack,[37] Jess[38]).

Most methods, however, treat structures as sets of residues and define how each residue is to be represented. PDBSiteScan[39] employs backbone atoms; this can be justified by the

Table 8.2 Methods for template matching. Methods are listed in order of the first associated publication

Method	Features used	Data structure and search method	Scoring results
ASSAM[29, 30]	Each residue is represented by two pseudo-atoms, one near the start of the sidechain and one near its end	The pseudoatoms form the nodes in a graph; the edges are the interatomic distances. Subgraph matching is carried out using Ullmann's subgraph isomorphism algorithm	Maximum difference in distances between an atom pair in the template and an equivalent atom pair in the protein
PINTS[31–33]	Each residue is represented by a pseudo-atom at the average position of a set of functional atoms	Recursive search through all possible residue combinations	E-value calculated using general a priori formula
SPASM/RIGOR[34, 35]	Each residue is represented by its Cα atom and a pseudo-atom at the average position of all sidechain atoms	Recursive search through all possible residue combinations	RMSD
SITEMINE[36]	Each residue is represented by a single pseudo-atom at the average position of all atoms for that residue	Difference-distance matrix between pseudo-atoms	RMSD
NeedleHaystack[37]	A template is a set of atoms	A set of three widely separated atoms from the template forms an 'anchor'; every possible match to that anchor in the protein is tried. Template and proteins are superposed using the anchor, and template atoms are paired with their nearest equivalents	'Skip-penalised' RMSD, which increases if all template atoms are not matched

(continued overleaf)

Table 8.2 (continued)

Method	Features used	Data structure and search method	Scoring results
Jess[38]	A template is a set of constraints over a number of atoms	The structure is represented using a KD-tree. Potential matches in the protein are identified by building on partial solutions	E-value based on fitting RMSD data from a given template to a normal or bimodal distribution
PDBSiteScan[39, 40]	Each residue is represented by its protein backbone atoms	Triplets of residues are compared with template residues, and potential matches are expanded until they match the total number of residues in the template	Largest single distance between any pair of equivalent atoms in superposed templates
Query3d[41]	Each residue is represented by the $C\alpha$ atom and a pseudo-atom at the average position of all sidechain atoms	All possible single-residue matches are considered; each match is extended to discover all possible two-residue matches, and so on until all template residues are matched	RMSD

Table 8.3 *Webservers for template matching*

Method	Type of input	URL
ASSAM[29, 30]	Template	http://grafss.imfr.net/assam/index.html
PINTS[31–33]	Template or structure	http://www.russell.embl-heidelberg.de/pints
SPASM[34, 35]	Template	http://portray.bmc.uu.se/cgi-bin/spasm/scripts/spasm.pl
RIGOR (as part of ProKnow)[42]	Structure	http://proknow.mbi.ucla.edu/index.php
SITEMINE[36]	Template or structure	http://www.ebi.ac.uk/msd-srv/MSDtemplate
NeedleHaystack[37]	Template and structure	http://bioinf.charite.de/haystack
Jess (as part of CSS)[17, 38]	Structure	http://www.ebi.ac.uk/thornton-srv/databases/CSS
Jess (as part of ProFunc)[38, 43]	Structure	http://www.ebi.ac.uk/thornton-srv/databases/ProFunc
PDBSiteScan[41]	Structure	http://wwwmgs.bionet.nsc.ru/mgs/gnw/pdbsitescan
Query3d (as part of PDBfun)[41, 44]	Template	http://pdbfun.uniroma2.it

Note: The 'Type of input' column indicates what the user supplies: Do they supply a template which is used to search a library of structures, or do they supply a structure which is searched using a library of templates? Where either of these possibilities is an option, the 'Type of input' column contains the text 'Template or structure'; where the user has to submit a single structure and a single template in order to search the one using the other, the 'Type of input' column contains the text 'Template and structure'.

argument that backbone conformation is better conserved and more accurately experimentally determined than sidechain positions. Alternatively, a residue may be represented by one or more pseudo-atoms. These are points which do not fall at the position of any one atom, but at the average position of all atoms in the residue (SITEMINE[36]), the average position of a set of functional atoms (PINTS,[31–33] Query3d[41, 44]) or at some other position derived from the residue atom coordinates (ASSAM,[29, 30] SPASM[34, 35]).

Some methods only permit residues to match with residues with the same amino acid type. However, many methods allow similar residues to match one another (for example, Asp may match Glu).

The only analysis which has compared the function prediction effectiveness of templates that represent residue positions in different ways is the study of 147 structural templates derived from the CSA which was described in the section above on using structural templates to identify distant homologues. This compared the effectiveness of templates which represented residues using atoms likely to be directly involved in catalysis with the effectiveness of templates using Cα and Cβ atoms to represent residues.[17] Templates using Cα and Cβ atoms were found to be slightly more effective for discriminating relatives from random matches.

8.3.6 Data Structures and Algorithms

Template searching methods must use some internal representation of the structure of the protein(s) being searched (a 'data structure'). The choice of data structure is closely tied to the nature of the algorithm used for template searching.

Some methods employ a relatively direct approach of searching through all possible combinations of matching residues, without using any elaborate data structure (PINTS, SPASM, SITEMINE, Query3d). The relatively thorough approach taken by these methods is possible partly because they represent each residue using only one or two atoms. Additionally, PINTS only uses conserved hydrophilic residues. All these methods search through the possible combinations of matching residues, discarding any potential matches as soon as they are found to involve a single distance pair differing by more than a given cutoff. Both PINTS and SPASM use a depth-first recursive search pattern to go through the possible matches.

NeedleHaystack uses a set of three atoms in the template to carry out an initial search. The three atoms in the template which are most widely separated from one another are selected for this purpose. All sets of equivalent atoms with similar distances in the structure being searched are selected. For each of these initial matches, the template and structure are superposed, and the remaining atoms in the template are assigned the nearest equivalent atoms in the structure being searched. The superposition of the template and the match is then optimised using all atoms.

ASSAM makes use of graph theory. There are a number of well-studied algorithms in graph theory for characterising and comparing subgraphs. If protein structures can be represented as graphs, then the template matching problem can be reduced to an easily tractable problem in graph theory. ASSAM's graph representation of a structure treats the pseudo-atoms representing residues as vertices in the graph, with edges representing distances between pairs of atoms. ASSAM uses Ullmann's subgraph isomorphism algorithm to carry out template matching.

Jess treats the template as a set of constraints that have to be met by a number of atoms, and apply these constraints stepwise to potential matches. Jess builds up solutions atom by atom, finding all those atom combinations in the search proteins that satisfy all constraints on a subset of atoms from the template, then adding one more template atom to the subset. In order to increase the efficiency of the search, Jess stores atom data in a data structure known as a KD-tree. This is suited to handling the geometric aspect of the problem.

8.3.7   Scoring Results

Template matching methods typically return substantial numbers of results. A measure of the accuracy of the results must be used to distinguish close from distant matches. The choice of measure is not generally related to the data structures and algorithms used by the method. Many methods use the coordinate RMSD between equivalent atoms in the optimally superposed structures of the template and the match. These include TESS, SPASM, SITEMINE, and Query3d. NeedleHaystack uses a slight variation on RMSD. ASSAM uses the maximum difference in distances between a pseudoatom pair in the template and an equivalent pseudoatom pair in the protein. PDBSiteScan uses the largest single distance between any pair of equivalent atoms in the superposed structures of the template and the match.

Whilst RMSD is the standard measure of similarity between two molecular structures, it suffers from not being comparable between different template matches. The other similarity measures detailed above have the same problem. Template matches with a different number of atoms are not comparable (since matches with a larger number of atoms will tend to have a higher RMSD). Differences in the frequency of residues mean that the residue composition also affects the significance of a given RMSD.

PINTS attempts to assign an E-value to a template match (an E-value is the number of matches of the same quality which one would expect to occur at random in the database being searched). PINTS converts RMSD values into E-values using a general method that can cope with templates of any size and residue type. The formula used is based on the geometry of a match with a given RMSD level. Several parameters are derived empirically from data.

The two webservers that currently employ Jess use different methods. The CSS web-server uses the same method for providing E-values as PINTS. The ProFunc webserver ranks matches by comparing the protein environment of the match with that of the original template. In order to carry out this environment comparison, residues in equivalent positions in a 10 Å sphere around the template match are paired up. These paired residues are filtered to include only those where the residues are in the same relative sequence order. These remaining pairs are scored in a manner that takes into account the number of paired residues and the number of insertions that would be required in either protein sequence to bring these residues into alignment. This method of scoring is better suited to detecting distant homologues than detecting cases of convergent evolution.

A study of 147 structural templates derived from the CSA (described in the section above on using structural templates to identify distant homologues) compared the effectiveness of RMSD and E-values calculated using the PINTS method for discriminating between matches to homologues and random matches.[17] This study found that the two scoring methods were similar in their effectiveness. However, the set of templates used in this study was largely composed of templates made up of three or four residues. Since one of the problems of RMSD is that it is not comparable between templates of different sizes, it is possible that E-values would be more useful when comparing matches between templates that differ considerably in size.

8.3.8 Template Libraries

When a researcher wishes to search a protein structure of unknown function for the presence of one specific functional site (perhaps derived from a distant homologue) they will be able to supply their own template. However, if they wish to search a protein structure of unknown function for any functional sites it may possess (perhaps when a homologue of known function cannot be identified), they will need a library of templates. Several of the template matching methods described above are associated with such a template library. Attempts to identify enzymes require a library which relates either to catalytic residues, or to ligand binding.

The program SPASM described above searches a set of protein structures for matches to a user-defined template. RIGOR is a second program by the same author which performs the opposite operation, searching a query structure for matches to a library of templates.[42] RIGOR uses a database of templates that meet one of three criteria: they consist of a sequential run of residues of the same type (such as four arginines), they are a spatial cluster of residues that share a property (all hydrophobic or all hydrophilic), or they all contact a ligand.[34] This last category is the most relevant to enzyme identification.

PDBSiteScan templates are based on the SITE records of PDB files. These SITE records are an optional section of the PDB file where its depositors can record any notable features of the structure, and as such they can correspond to many different types of functional site. The PINTS server offers three template databases: one where templates are derived from the SITE records of PDB files, one that focuses on surface residues, and one containing templates based on residues that are within 3 Å of bound ligands. The Query3d method is part of the pdbFun resource;[44] this permits users to construct templates from a range of functional sites in proteins, including clefts, ligand binding sites, and catalytic residues derived from the CATRES[45] resource.

A data mining approach was used by the author of SITEMINE to identify common geometric combinations of residue atoms. This takes into account the symmetry of certain residues (such as aspartate and phenylalanine) and the fact that groups in some residues are equivalent (such as the amide groups in asparagine and glutamine). The matches from this data mining are significant in a geometric sense, but may not be biologically interesting. The data mining output includes information on ligand interactions in order to aid the identification of biologically meaningful matches. Each data mining match is converted into a template representing the average positions of the equivalent atoms from the data mining match.

As described above, the CSA is a database of catalytic residues in proteins of known structure. A subset of the catalytic sites described in the CSA have been used as the basis for a library of structural templates. It is possible to search protein structures for instances of these templates using the Jess template matching method via the webserver Catalytic Site Search (CSS).

The ProFunc webserver contains the same set of structural templates derived from the CSA, and also includes templates representing ligand binding sites and DNA binding sites. Furthermore, each structure used as a query by ProFunc is used to create a set of 'reverse templates', which are used to search a representative subset of the structures in the PDB. Like the CSS webserver, ProFunc carries out template matching using the Jess program, although as noted above the two webservers differ significantly in how they score template matches.

There have been no formal comparisons of the effectiveness of different template libraries for predicting protein function. However, the templates derived from the CSA are the most likely to be relevant to searches for groups of catalytic residues, whereas the template libraries associated with RIGOR and PINTS which consist of residues involved in contacting ligands are the most likely to be relevant to searches for ligand binding sites.

8.4   Other Methods for Matching Parts of Structures

There are several methods for matching portions of protein structures which differ significantly from the type of template matching methods discussed above. Webservers for these other methods are summarised in Table 8.4.

There are methods which require residues to have the same sequence order in the structures being compared. These methods have been discussed by Eidhammer *et al.* in their general review of structure comparison methods.[46] Conklin reviewed some early approaches to discovering this type of structure-sequence motif, and developed his own

Table 8.4 *Webservers for matching portions of proteins by methods other than template matching*

Method	Type of input	URL	Description
SiteEngine[51]	Site and structure	http://bioinfo3d.cs.tau.ac.il/SiteEngine	Identifies surface patches with similar physicochemical properties
SURFACE[52]	Structure	http://cbm.bio.uniroma2.it/surface	Compares surface patches to a library of residues lining major clefts in proteins by comparing coordinates of residues lining clefts
pvSOAR[55]	Structure	http://pvsoar.bioengr.uic.edu	Compares surface patches to a library of residues lining major clefts in proteins by comparing sequences of residues lining the cleft
eF-seek[57]	Structure	http://eF-site.hgc.jp/eF-seek	Predicts ligand binding sites in protein structures by comparing the shape and electrostatic potential of the surface of the query protein with a database of ligand binding clefts
webFEATURE[59]	Structure	http://feature.stanford.edu/webfeature/index.php	Represents functional sites in terms of the average physicochemical properties of a set of concentric spheres surrounding the site

Note: The 'type of input' column indicates what the user supplies: do they supply a structure that is searched against a predefined library of functional sites ('Structure'), or do they have to supply a structure and also a functional site to be used for searching ('Site and structure')?

machine learning approach.[47] The methods SPratt2[48] and TRILOGY[49] discover common motifs of this type in groups of protein structures.

Methods exist that focus on protein surface patch or cleft properties; these have the potential to identify similar active sites. These methods have been reviewed by Via *et al.*[50] SiteEngine[51] is a method that identifies surface patches with similar physicochemical properties. SURFACE[52] is a webserver that matches surface patches in a manner resembling the template matching methods described above. Residues in a template patch are represented using their C_α atom and a pseudo-atom at the geometric centre of the sidechain atoms. The program searches through all possible residue pairs in the query structure and the template patch, looking for matches with similar inter-residue distances. SURFACE has a surface patch template library constructed using the program SURFNET[53] to identify residues lining major clefts in proteins; these clefts are annotated with functional information from the PROSITE[54] sequence motif database, and with information concerning ligands bound in the cleft. pvSOAR[55] is a webserver that represents a given surface cleft by concatenating the sequences of all the portions of the protein lining the cleft.[56] This concatenated sequence is then compared with a database of such cleft-lining sequences using a dynamic programming sequence comparison method. eF-seek is a method that predicts ligand binding sites in protein structures by comparing the shape and electrostatic potential of the surface of the query protein with a database of ligand binding clefts.[57] FEATURE[58] is a template matching program which represents functional sites in terms of the average physicochemical properties of a set of concentric spheres surrounding the site. It is made available through the webserver webFEATURE.[59]

8.5   Function Prediction without Inferring Function from Homologues

It is also possible to attempt to predict a function for an enzyme, or to attempt to identify its catalytic residues, without inferring these functional details from homologues of known function. This is a highly difficult task. Note that some of the methods discussed in this section do involve the identification of homologues, but these homologues are used purely for identifying conserved regions; they are not used as sources of functional annotation. Methods for function prediction which do not infer this function from homologues and are publicly available are summarised in Table 8.5.

It is possible to make an informed guess at the general location of the catalytic site on a protein structure without identifying homologues of known function. A study[63] using the cleft analysis program SURFNET[53] found that in single chain enzymes, the ligand is bound in the largest cleft in the protein in over 83% of cases. PatchFinder[60] is a method that identifies conserved patches of surface residues; the authors found that in 63% of cases, at least half the residues in PDB file SITE records are in the main patch identified by PatchFinder. These patches are relatively large, averaging 29 residues in size. Evolutionary trace[64, 65] is a method that uses a phylogenetic analysis to identify residues which are either conserved across all relatives of a protein or else conserved within clusters of relatives; these residues are likely to be under evolutionary pressure and are therefore likely to be of functional significance. Functional sites are predicted where clusters of these trace residues occur in proximity to one another in the protein structure. The largest cluster identified by evolutionary trace has been found to have a statistically significant overlap with the enzyme

Table 8.5 Methods for predicting enzyme function which do not infer function from homologues

Method	Availability	URL	Description
SURFNET[53]	Software	http://www.biochem.ucl.ac.uk/~roman/surfnet/surfnet.html	Locates clefts in protein surfaces
PatchFinder[60]	Software	http://bioinfo.tau.ac.il/~nimrodg/patchfinder/patchfinder.html	Identifies conserved patches of surface residues
Evolutionary Trace Report Maker[61]	Webserver	http://mammoth.bcm.tmc.edu/server.html	Identifies residues which are either conserved across all relatives of a protein or else conserved within clusters of relatives
SARIG[62]	Webserver	http://bioinfo2.weizmann.ac.il/~pietro/SARIG/V3/index.html	Predicts active site residues using a graph representation of the protein structure.

active site in 97% of enzymes analysed.[65] These trace residue clusters can be quite large, so often only a general region of the protein is identified. Thus both cleft and conservation methods allow a rough identification of the general region of the active site; however, this does not provide any information about whether a protein is an enzyme, what the enzyme function is, or the precise catalytic residues.

There have been a number of attempts to identify individual catalytic residues on the basis of their structural properties and residue conservation alone. Gutteridge *et al.*[66] attempted to predict catalytic residues on the basis of their residue type, conservation, depth within the protein, solvent accessibility, secondary structure type and the size of the cleft which they occur in. A neural network was trained with these parameters using a training set of known catalytic residues. This neural network can successfully predict 56% of catalytic residues (sensitivity); however, only one in seven predicted residues is catalytic (specificity). The non-catalytic predicted residues are, however, often close to the catalytic site. This suggests that residues around the active site can be distinguished on this basis, but that these parameters do not permit one to distinguish specific catalytic residues. Another study using machine learning to predict catalytic residues using a similar set of parameters (this time using a support vector machine approach) had a similar level of success.[67] SARIG[62] takes a different approach to active site residue prediction. SARIG represents proteins as graphs, where residues are the nodes and edges represent interactions between them. Analysis of these graphs has shown that the 'closeness' of a residue – the mean distance of its graph node to all other nodes – is significantly higher for catalytic and ligand-binding residues. Using an optimal closeness threshold, it was possible to predict catalytic residues with 46.5% sensitivity – but only 9.4% specificity.

8.6 Meta-servers for Function Prediction

As described above, function can vary considerably within homologous superfamilies, so establishing homology between a query protein and a protein of known function on the basis of structure alone using the overall structural comparison methods described at the start of this chapter can be a poor guide to function. Furthermore, since convergent evolution is a relatively poorly studied phenomenon, it is difficult to make functional predictions on the basis of local similarities of residue groups or clefts in the absence of evidence of homology. For this reason, it is useful to combine methods for establishing homology between proteins with methods that carry out local comparisons.

There are a number of meta-servers for protein function prediction which combine the results of different methods in this manner. These meta-servers run a number of individual function prediction methods separately on a single protein, and assemble the results in a format that is convenient for users to inspect. The result is the same as if the user had run the methods separately, but using the meta-server can save considerable time and effort. There are two meta-servers which incorporate information from protein structures: ProKnow[42] (http://proknow.mbi.ucla.edu/index.php) and ProFunc[43] (http://www.ebi.ac.uk/thornton-srv/databases/ProFunc).

ProKnow incorporates fold comparison using DALI, and template matching using RIGOR, along with a range of sequence-based methods. ProKnow associates GO annotations with the results of each of these search methods (for example, each fold that

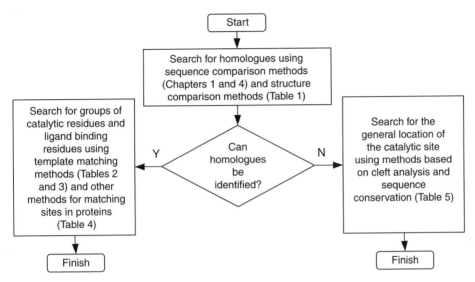

Figure 8.2   *Flowchart describing a strategy for function prediction*

could be matched by DALI is associated with a set of GO annotations, as is each template that could be matched by RIGOR). When a structure is searched using ProKnow, probabilities are assigned to the GO terms returned by the various search methods which it incorporates. These probabilities are calculated using a Bayesian approach based on the strength of each match to a fold/template etc. annotated with a given GO term, the number of different matches per method which are annotated with that GO term, and the number of different methods which return that GO term.

ProFunc makes use of the template matching program Jess to search query structures against structural templates representing a range of functional sites, as described above. ProFunc also employs a range of other structure-based methods for predicting function, including fold comparison using SSM, surface cleft analysis using SURFNET, and searching for potentially DNA-binding helix-turn-helix motifs. Furthermore, it includes various sequence-based methods, including searching for similar sequences in the PDB and the UniProt sequence database[68] using BLAST, and searching several motif and family databases. The results from these methods are laid out in a convenient format, but they are not integrated in any way to produce a general function prediction.

8.6.1   A Strategy for Function Prediction

How should this range of function prediction methods be applied in practice? Each protein structure presents a unique function prediction challenge, but it is possible to provide broad guidelines, which are summarised by Figure 8.2. These guidelines refer only to types of method; as described above, it is difficult to recommend individual methods, since there are few studies comparing the effectiveness of these structure-based methods for function prediction, and where such studies do exist (as for the methods for identifying homologues through overall structure comparison) they are unable offer a conclusive recommendation.

The most important task is to identify homologues of known function. This can be approached using the fold comparison methods described in Table 8.1, together with the sequence comparison methods discussed in Chapters 1 and 4.

If it is possible to identify homologous proteins whose function is known, then predicting function involves examining the literature concerning these homologues, and examining local similarities and differences in sequence and structure between the query protein and these homologues. As part of this process, it may be useful to employ local structural comparison methods using structural templates (Tables 8.2 and 8.3) or other methods (Table 8.4) in order to ascertain whether key functional sites in the homologues are conserved in the query protein. A structural template could be derived from these homologues of known function, or a template library could be employed if it contains templates associated with the homologues of known function.

If homologues of known function cannot be identified, then it is still possible that searching against a template library using the local structural comparison methods described in Tables 8.2–8.4 might identify very distant relatives or cases of convergent evolution of functional sites. However, any function prediction made on this basis would be highly speculative. Generally, in the absence of homologues of known function the most that can be obtained is an assessment of the rough location of the active site using the cleft and sequence conservation methods described in Table 8.5.

All of these different methods of function prediction are complementary, and should be considered together. The meta-servers described in the previous section provide a convenient way to obtain information from many function prediction methods simultaneously.

References

1. E.B. Webb. Enzyme nomenclature, in *Recommendations of the Nomenclature Committee of the International Union of Biochemistry and Molecular Biology*. Academic Press, San Diego (1992).
2. M. Ashburner, C.A. Ball, J.A. Blake, *et al.*, Gene ontology: tool for the unification of biology. The Gene Ontology Consortium. *Nat Genet*, **25**, 25–9 (2000).
3. P. Koehl, Protein structure similarities. *Curr Opin Struct Biol*, **11**, 348–53 (2001).
4. M.A. Marti-Renom, A. Rossi, F. Al-Shahrour, *et al.*, The AnnoLite and AnnoLyze programs for comparative annotation of protein structures. *BMC Bioinformatics*, **8 Suppl 4**, S4 (2007).
5. L. Holm and C. Sander, Protein structure comparison by alignment of distance matrices. *J Mol Biol*, **233**, 123–38 (1993).
6. T. Madej, J.F. Gibrat, and S.H. Bryant, Threading a database of protein cores. *Proteins*, **23**, 356–69 (1995).
7. I.N. Shindyalov and P.E. Bourne, Protein structure alignment by incremental combinatorial extension (CE) of the optimal path. *Protein Eng*, **11**, 739–47 (1998).
8. T. Kawabata, MATRAS: A program for protein 3D structure comparison. *Nucleic Acids Res*, **31**, 3367–9 (2003).
9. F.M. Pearl, C.F. Bennett, J.E. Bray, *et al.*, The CATH database: an extended protein family resource for structural and functional genomics. *Nucleic Acids Res*, **31**, 452–5 (2003).
10. E. Krissinel and K. Henrick, Secondary-structure matching (SSM), a new tool for fast protein structure alignment in three dimensions. *Acta Crystallogr D Biol Crystallogr*, **60**, 2256–68 (2004).
11. Y. Ye and A. Godzik, FATCAT: a web server for flexible structure comparison and structure similarity searching. *Nucleic Acids Res*, **32**, W582–5 (2004).
12. M. Novotny, D. Madsen, and G.J. Kleywegt, Evaluation of protein fold comparison servers. *Proteins*, **54**, 260–70 (2004).

13. M.L. Sierk, and W.R. Pearson, Sensitivity and selectivity in protein structure comparison. *Protein Sci*, **13**, 773–85 (2004).
14. A.E. Todd, C.A. Orengo, and J.M. Thornton, Evolution of function in protein superfamilies, from a structural perspective. *J Mol Biol*, **307**, 1113–43 (2001).
15. B. Rost, Enzyme function less conserved than anticipated. *J Mol Biol*, **318**, 595–608 (2002).
16. W. Tian and J. Skolnick, How well is enzyme function conserved as a function of pairwise sequence identity? *J Mol Biol*, **333**, 863–082 (2003).
17. J.W. Torrance, G.J. Bartlett, C.T. Porter, and J.M. Thornton, Using a Library of Structural Templates to Recognise Catalytic Sites and Explore their Evolution in Homologous Families. *J Mol Biol*, **347**, 565–81 (2005).
18. G. Dodson and A. Wlodawer, Catalytic triads and their relatives. *Trends Biochem Sci*, **23**, 347–52 (1998).
19. A.C. Wallace, N. Borkakoti, and J.M. Thornton, TESS: a geometric hashing algorithm for deriving 3D coordinate templates for searching structural databases. Application to enzyme active sites. *Protein Sci*, **6**, 2308–23 (1997).
20. J.A. Gerlt and P.C. Babbitt, Divergent evolution of enzymatic function: mechanistically diverse superfamilies and functionally distinct suprafamilies. *Annu Rev Biochem*, **70**, 209–46 (2001).
21. M.Y. Galperin, D.R. Walker, and E.V. Koonin, Analogous enzymes: independent inventions in enzyme evolution. *Genome Res*, **8**, 779–90 (1998).
22. A.C. Wallace, R.A. Laskowski, and J.M. Thornton, Derivation of 3D coordinate templates for searching structural databases: application to Ser-His-Asp catalytic triads in the serine proteinases and lipases. *Protein Sci*, **5**, 1001–13 (1996).
23. C.T. Porter, G.J. Bartlett, and J.M. Thornton, The Catalytic Site Atlas: a resource of catalytic sites and residues identified in enzymes using structural data. *Nucleic Acids Res*, **32 Database issue**, D129–33 (2004).
24. S.F. Altschul, T.L. Madden, A.A. Schaffer, *et al.* Gapped BLAST and PSI-BLAST: a new generation of protein database search programs. *Nucleic Acids Res*, **25**, 3389–402 (1997).
25. J.S. Fetrow and J. Skolnick, Method for prediction of protein function from sequence using the sequence-to-structure-to-function paradigm with application to glutaredoxins/thioredoxins and T1 ribonucleases. *J Mol Biol*, **281**, 949–68 (1998).
26. Ye. Y., T. Xie, and Ding D., Protein functional-group 3D motif and its applications. *Chinese Sci Bull*, **45**, 2044–2052 (2000).
27. T. Hamelryck, Efficient identification of side-chain patterns using a multidimensional index tree. *Proteins*, **51**, 96–108 (2003).
28. R. Singh and M. Saha, Identifying structural motifs in proteins. *Pac Symp Biocomput*, 228–39 (2003).
29. P.J. Artymiuk, A.R. Poirrette, H.M. Grindley, D.W. Rice, and P. Willett, A graph-theoretic approach to the identification of three-dimensional patterns of amino acid side-chains in protein structures. *J Mol Biol*, **243**, 327–44 (1994).
30. R.V. Spriggs, P.J. Artymiuk, and P. Willett, Searching for patterns of amino acids in 3D protein structures. *J Chem Inf Comput Sci*, **43**, 412–21 (2003).
31. R.B. Russell, Detection of protein three-dimensional side-chain patterns: new examples of convergent evolution. *J Mol Biol*, **279**, 1211–27 (1998).
32. A. Stark, S. Sunyaev, and R.B. Russell, A model for statistical significance of local similarities in structure. *J Mol Biol*, **326**, 1307–16 (2003).
33. A. Stark and R.B. Russell, Annotation in three dimensions. PINTS: Patterns in Non-homologous Tertiary Structures. *Nucleic Acids Res*, **31**, 3341–4 (2003).
34. G.J. Kleywegt, Recognition of spatial motifs in protein structures. *J Mol Biol*, **285**, 1887–97 (1999).
35. D. Madsen and G.J. Kleywegt, Interactive motif and fold recognition in protein structures. *J Appl Cryst*, **35**, 137–139 (2002).
36. T.J. Oldfield, Data mining the protein data bank: residue interactions. *Proteins*, **49**, 510–28 (2002).
37. A. Hoppe and C. Frommel, NeedleHaystack: a program for the rapid recognition of local structures in large sets of atomic coordinates. *J Appl Cryst*, **36**, 1090–1097 (2003).

38. J.A. Barker and J.M. Thornton, An algorithm for constraint-based structural template matching: application to 3D templates with statistical analysis. *Bioinformatics*, **19**, 1644–9 (2003).
39. V.A. Ivanisenko, S.S. Pintus, D.A. Grigorovich, and N.A. Kolchanov, PDBSiteScan: a program for searching for active, binding and posttranslational modification sites in the 3D structures of proteins. *Nucleic Acids Res*, **32**, W549–54 (2004).
40. V.A. Ivanisenko, S.S. Pintus, D.A. Grigorovich, and N.A. Kolchanov, PDBSite: a database of the 3D structure of protein functional sites. *Nucleic Acids Res*, **33**, D183–7 (2005).
41. G. Ausiello, A. Via, and M. Helmer-Citterich, Query3d: a new method for high-throughput analysis of functional residues in protein structures. *BMC Bioinformatics*, **6 Suppl 4**, S5 (2005).
42. D. Pal and D. Eisenberg, Inference of protein function from protein structure. *Structure*, **13**, 121–30 (2005).
43. R.A. Laskowski, J.D. Watson, and J.M. Thornton, ProFunc: a server for predicting protein function from 3D structure. *Nucleic Acids Res*, **33**, W89–93 (2005).
44. G. Ausiello, A. Zanzoni, D. Peluso, A. Via, and M. Helmer-Citterich, pdbFun: mass selection and fast comparison of annotated PDB residues. *Nucleic Acids Res*, **33**, W133–7 (2005).
45. G.J. Bartlett, C.T. Porter, N. Borkakoti, and J.M. Thornton, Analysis of catalytic residues in enzyme active sites. *J Mol Biol*, **324**, 105–21 (2002).
46. I. Eidhammer, I. Jonassen, and W.R. Taylor, Structure comparison and structure patterns. *J Comput Biol*, **7**, 685–716 (2000).
47. D. Conklin, Machine discovery of protein motifs. *Machine Learning*, **21**, 125–150 (1995).
48. I. Jonassen, I. Eidhammer, D. Conklin, and W.R. Taylor, Structure motif discovery and mining the PDB. *Bioinformatics*, **18**, 362–7 (2002).
49. P. Bradley, P.S. Kim, and B. Berger, TRILOGY: Discovery of sequence-structure patterns across diverse proteins. *Proc Natl Acad Sci U S A*, **99**, 8500–5 (2002).
50. A. Via, F. Ferre, B. Brannetti, and M. Helmer-Citterich, Protein surface similarities: a survey of methods to describe and compare protein surfaces. *Cell Mol Life Sci*, **57**, 1970–7 (2000).
51. A. Shulman-Peleg, R. Nussinov, and H.J. Wolfson, Recognition of functional sites in protein structures. *J Mol Biol*, **339**, 607–33 (2004).
52. F. Ferre, G. Ausiello, A. Zanzoni, and M. Helmer-Citterich, SURFACE: a database of protein surface regions for functional annotation. *Nucleic Acids Res*, **32**, D240–4 (2004).
53. R.A. Laskowski, SURFNET: a program for visualizing molecular surfaces, cavities, and intermolecular interactions. *J Mol Graph*, **13**, 323–30, 307–8 (1995).
54. N. Hulo, C.J. Sigrist, V. Le Saux, *et al.*, Recent improvements to the PROSITE database. *Nucleic Acids Res*, **32**, D134–7 (2004).
55. T.A. Binkowski, P. Freeman, and J. Liang, pvSOAR: detecting similar surface patterns of pocket and void surfaces of amino acid residues on proteins. *Nucleic Acids Res*, **32**, W555–8 (2004).
56. T.A. Binkowski, L. Adamian, and J. Liang, Inferring functional relationships of proteins from local sequence and spatial surface patterns. *J Mol Biol*, **332**, 505–26 (2003).
57. K. Kinoshita, Y. Murakami, and H. Nakamura, eF-seek: prediction of the functional sites of proteins by searching for similar electrostatic potential and molecular surface shape. *Nucleic Acids Res*, **35**, W398–402 (2007).
58. L. Wei and R.B. Altman, Recognizing complex, asymmetric functional sites in protein structures using a Bayesian scoring function. *J Bioinform Comput Biol*, **1**, 119–38 (2003).
59. M.P. Liang, D.R. Banatao, T.E. Klein, D.L. Brutlag, and R.B. Altman, WebFEATURE: An interactive web tool for identifying and visualizing functional sites on macromolecular structures. *Nucleic Acids Res*, **31**, 3324–7 (2003).
60. G. Nimrod, F. Glaser, D. Steinberg, N. Ben-Tal, and T. Pupko, In silico identification of functional regions in proteins. *Bioinformatics*, **21 Suppl 1**, i328–37 (2005).
61. I. Mihalek, I. Res, and O. Lichtarge, Evolutionary trace report_maker: a new type of service for comparative analysis of proteins. *Bioinformatics*, **22**, 1656–7 (2006).
62. G. Amitai, A. Shemesh, E. Sitbon, M. Shklar, D. Netanely, I. Venger, and S. Pietrokovski, Network analysis of protein structures identifies functional residues. *J Mol Biol*, **344**, 1135–46 (2004).
63. R.A. Laskowski, N.M. Luscombe, M.B. Swindells, and J.M. Thornton, Protein clefts in molecular recognition and function. *Protein Sci*, **5**, 2438–52 (1996).

64. O. Lichtarge, H.R. Bourne, and F.E. Cohen, An evolutionary trace method defines binding surfaces common to protein families. *J Mol Biol*, **257**, 342–58 (1996).

65. H. Yao, D.M. Kristensen, I. Mihalek, M.E. Sowa, C. Shaw, M. Kimmel, L. Kavraki, and O. Lichtarge, An accurate, sensitive, and scalable method to identify functional sites in protein structures. *J Mol Biol*, **326**, 255–61 (2003).

66. A. Gutteridge, G.J. Bartlett, and J.M. Thornton, Using a neural network and spatial clustering to predict the location of active sites in enzymes. *J Mol Biol*, **330**, 719–34 (2003).

67. N.V. Petrova and C.H. Wu, Prediction of catalytic residues using Support Vector Machine with selected protein sequence and structural properties. *BMC Bioinformatics*, **7**, 312 (2006).

68. The UniProt Consortium, The Universal Protein Resource (UniProt). *Nucleic Acids Res*, **35**, D193-7 (2007).

9

The Prediction of Macromolecular Complexes by Docking

Sjoerd J. de Vries, Marc van Dijk and Alexandre M.J.J. Bonvin

9.1   Introduction

Macromolecular complexes are the molecular machines of the cell. In order to fully understand how the various units work together to fulfill their tasks, structural knowledge at the atomic level is required. Only in this way, functional mechanisms such as binding specificity, signal transduction through conformational change, and molecular scaffolding, can be fully understood. The number of expected macromolecular complexes will, however, exceed the number of proteins in a proteome by at least one order of magnitude; a significant fraction of these will be extremely difficult to study using classical structural methods such as NMR and X-ray crystallography.

Therefore, the importance of computational approaches such as docking, the process of predicting the three-dimensional (3D) structure of a complex based on its known constituents, is evident. In recent years, docking has emerged as an important method, complementary to experimental structural methods. To monitor the performance of current protein-protein docking methods, CAPRI (Critical Assessment of Predicted Interactions), a community-wide blind docking experiment, has been established (http://capri.ebi.ac.uk). In this experiment, participants are asked to predict by docking a recently solved protein-protein complex a few weeks prior to its publication. Ten models may be submitted, and successful predictions are awarded one to three stars depending on their accuracy. Figure 9.1 shows examples of successful CAPRI predictions of various accuracies. The accuracy of a prediction is defined based on the fraction of native intermolecular contacts correctly predicted (Fnat) and on the positional root mean square deviation (RMSD)

Prediction of Protein Structures, Functions, and Interactions   Edited by Janusz Bujnicki
© 2009 John Wiley & Sons, Ltd

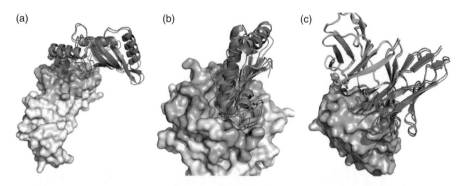

Figure 9.1 *Example predictions from the CAPRI experiment overlaid onto the experimental crystal structure. **a:** a one-star prediction (HADDOCK model #1 for target 27,[92] PDB code 2O25); **b:** a two-star prediction (HADDOCK model #1 for target 26,[93] PDB code 2HQS); **c:** a three-star prediction (HADDOCK model #1 for target 13,[94] PDB code 1YNT).*

calculated on all residues within 5Å of the partner molecule termed interface RMSD (i-RMSD): one-star (acceptable): Fnat $\geq$ 10% and i-RMSD $\leq$ 4.0Å; two-star (good): Fnat $\geq$ 30% and i-RMSD $\leq$ 2.0Å and three-star (high quality): Fnat $\geq$ 50% and i-RMSD $\leq$ 1.0Å.

The previous rounds of CAPRI challenge have shown considerable progress.[1, 2] In the earlier rounds, most targets were unbound-bound complexes, meaning that for only one of the proteins, the free form structure was available, while for the other protein, the structure within the complex was supplied by the organizers. Although this is a useful computational exercise, it is not a truly blind prediction. In contrast, all but one targets of the last CAPRI meeting had to be predicted using only unbound structures or even homology models. Also, recent targets have a higher representation of biologically interesting signal transduction complexes, which are difficult to dock. Facing these challenges, successful predictions were made for several targets that were considered beyond the limits of docking methodology a few years ago.

Despite this progress, there still are major problems to be addressed.[3–6] In the context of recent CAPRI rounds, these issues have become more pronounced. Here we will discuss two major bottlenecks: first, molecular flexibility and conformational changes, and second, the inclusion of experimental or bioinformatic data on the nature of the interaction. In addition, we will also address protein–DNA docking: compared to protein–protein docking, development in this particular type of biomolecular docking has lagged behind, despite the important role of these complexes in recognition and gene expression. A discussion of the third major bottleneck, the discrimination of correct docking solutions from incorrect ones (the scoring problem) lies beyond the scope of this chapter, but it will be occasionally mentioned in relation to other issues.

Finally, we will discuss several practical aspects that must be considered when docking is performed. In discussing these aspects, we will mainly focus on the docking program HADDOCK, which was developed in our group; however, they are equally valid for other docking programs.

9.2 Flexibility

Initially, most protein–protein docking approaches have been developed based on rigid-body docking algorithms, thus ignoring any conformational change that might occur upon binding. Typically, a large number of possible solutions are considered (searching/sampling), which are subsequently ranked according to quantitative criteria (scoring). The most widely used sampling approach is the Fast Fourier Transform (FFT) algorithm,[7] where the protein is represented as one or more values at every point on a discrete grid. Typically, one protein is rotated, while the other is kept fixed. Docking is then performed by overlaying the grids of the two proteins using FFT, calculating all possible overlays in a single step. The values of the grid are chosen such that overlap (interpenetration) between surface points is favored, while overlap between the protein interiors is penalized.

Other methods to perform a rigid-body docking search are rigid-body energy minimization[8] and rigid-body Monte Carlo minimization.[9, 10] Unlike FFT, representing the proteins as grids is optional, and both rotations and translations can be searched simultaneously, although neither of them in a single step.

The rigid-body approximation does not work well for every complex. Although for targets that show only small backbone conformational changes, excellent predictions can be obtained, targets for which conformational changes take place upon binding are extremely challenging (even for backbone RMSD changes as small as 2 Å!).

Therefore, the realization of the importance of flexibility in docking is leading to new developments. Flexibility can be introduced at several levels: implicitly, by smoothing the protein surfaces or allowing some degree of interpenetration (soft docking) or by performing multiple docking runs from various conformations (cross or ensemble docking); or explicitly, by allowing side-chain and/or backbone flexibility, either during docking or in a refinement step. Finally, it is possible to perform multi-body or incremental rigid-body docking, cutting the protein at flexible hinge regions, which must be determined *a priori*.

9.2.1 Implicit Flexibility

In grid representations, side-chain (and small backbone) rearrangements can be modeled by thickening the surface layer for which interpenetration is scored favorably,[11] or by trimming long side-chains.[12] Snapshots of a Molecular Dynamics (MD) simulation have been used to build grids for docking in which only grid points consistently occupied by all conformations are considered, thus excluding mobile regions from the construction of the grid.[13] One of the drawbacks of this kind of flexibility treatment is that severe steric clashes are frequently introduced when returning to full-atom representations, so that further optimization is required.

Alternatively, implicit treatment of flexibility can be achieved by performing rigid-body docking of ensembles of conformations. In such cases, the docking process is repeated from various combinations of starting structures (ensemble or cross docking). The ensembles can be taken from NMR structures, or generated by MD simulations or any other conformational sampling method. These ensembles can span various degrees of flexibility, from small, mainly side-chain rearrangements to large-scale global backbone motions.

Even conformational sampling limited to side-chain rearrangements can be beneficial for improving docking predictions. It has been shown that residues important for molecular recognition will usually sample bound conformations during MD simulations of the unbound protein.[14, 15] Similarly, exposed side-chains in an ensemble of NMR structures usually sample various conformations; their use in data-driven docking was shown to increase both accuracy and hit rate.[16] Two rather systematic studies have been published that investigate the use of MD structures in ensemble docking.[17, 18] Both studies indicate that ensemble docking improves the performance in terms of an increased number of (near) native solutions, but makes it more difficult to distinguish correct from wrong solutions.

9.2.2   Explicit Flexibility

While rigid body docking has only six degrees of freedom (three rotations and three translation), this number increases explosively if flexibility is explicitly considered. Therefore, the amount of flexibility is necessarily restricted. In most docking approaches, flexibility (often in the form of a short energy minimization) aims not at improving the structure towards the bound form, but rather at removing clashes and improving the scoring.[19] Flexibility can be explicitly introduced in the docking process. For example, ATTRACT[20, 21] uses a reduced protein representation together with multiple side-chain copies: switching between rotamers is performed at various stages during docking and the best conformation is selected for the next few subsequent EM steps. HADDOCK[8] allows explicit flexibility of both side-chains and backbone during the MD simulated annealing refinement stage. For the successful predictions obtained with HADDOCK, explicit inclusion of backbone and side-chain flexibility both increased the quality of the solutions and improved the scoring of the resulting complexes.[6, 22] However, the quality increase was mostly caused by improvements in the fraction of native contacts and not that much by improved backbone conformations: on average, the backbone got closer to the bound form, but in some cases it moved away.

In addition, several promising approaches have been reported that allow the sampling of large conformational changes. Normal mode analysis methods use a graph representation of proteins to predict large-scale motions, potentially including motions that lead to the transition from the unbound and to the bound conformation ([23] and references therein). Motions along principal components are treated as additional degrees of freedom in ATTRACT, allowing the structures to deform along soft harmonic modes to facilitate the binding process.[21, 24] Normal modes have also been applied to the optimization of complexes against electron density maps[25, 26] and the refinement of protein–DNA models.[27] Such approaches should lead to improved docking results, provided that the identified modes are relevant to the binding process; this is again related to our ability to predict motions.

9.2.3   Multiple Docking

Large loop rearrangements are also difficult to model and can thus hinder docking, even when some degree of backbone flexibility is introduced. A mean-field approach has been proposed to deal with this problem where multiple loop conformations are considered simultaneously.[28, 29] Even more problematic is the case of protein sub-domains separated by a flexible hinge or linker. In this case, very large conformational changes can occur.

FlexDock[30] solves this problem by dissecting the protein into sub-domains: once hinge regions have been identified, a fast two-body rigid docking is performed with the various combinations of sub-domains. The resulting fragments are then assembled using a graph theory algorithm. This approach successfully modeled the very large conformational change that occurs upon the binding of calmodulin to a target peptide. However, this approach requires the *a priori* knowledge of hinge regions whose prediction is far from trivial. Recently, the same group was involved in the development of HingeProt, an algorithm to predict such regions.[23]

9.2.4 Flexibility in Protein-DNA Docking

Flexibility is an inherent property of DNA. DNA often exhibits large conformational changes upon binding to a protein, which can significantly alter the interaction interface compared to the initial encounter complex. This dynamic behavior is most pronounced when describing DNA global flexibility in terms of bending and twisting. The global conformational changes arises, however, from a series of local changes at the level of individual base pairs and of the sugar phosphate backbone. It is a major challenge in protein-DNA docking to account for these local and global conformational changes during the docking while maintaining a relevant (B-)DNA conformation.

Various methods have attempted to incorporate DNA flexibility into docking in different ways. Implicit flexibility has been used in various forms, ranging from the use of pre-bent and twisted DNA starting models in rigid body docking to interface overlap in soft-body docking or a combination of both.[31] The use of different starting models in a rigid-body docking step allows to quickly sample large fractions of the conformational space but is unable to predict the specific base pair conformations that give rise to the global deformation. In contrast, the use of soft-body DNA models is able to provide insight into local conformational changes that improve complex formation. But, as a drawback, this results in severe steric clashes.

A few approaches incorporate flexibility during the docking using molecular dynamics. Tzou *et al.*[32] modelled the CAP-DNA and Rep-DNA systems from the repressors in their bound conformation and from canonical B-DNA in a series of molecular mechanics and dynamics simulations using distance restraints derived from a statistical analysis of homologous protein-DNA complexes. Knegtel *et al.* developed MONTY,[33] which uses a Monte Carlo search allowing for flexibility in both protein and DNA, and can account for experimentally determined contacts to guide the docking. The initial position of the protein in the predicted complex should, however, not deviate too much from that of the actual complex; even small deviations can result in DNA curling around the protein. HADDOCK uses an information driven approach to circumvent the search through sequence space and drive the docking; when starting from canonical B-DNA the method is able to predict the global conformation that the DNA tends to adopt in the complex. This information is subsequently used to model pre-bent and twisted DNA starting structures used as input for a second refinement docking round. As such the method is successful in modelling large conformational changes without the risk of loosing helical structure (see Figure 9.2). In addition, the second refinement docking round allows the sampling of local base pair conformations, thereby improving the convergence of the results.

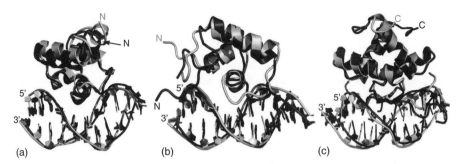

Figure 9.2   *Best solutions of unbound flexible docking with HADDOCK starting from a library of pre-bent and twisted DNA structures. The HADDOCK model is shown in light gray, superimposed onto the experimentally solved structure in dark gray: Cro-O1R (a), Lac-O1 (b) and Arc-operator (c). Adapted from van Dijk et al.[34] with permission from Oxford University Press, Oxford, United Kingdom.*

All of the docking procedures discussed above are able to make predictions that were representative of the published complexes in terms of spatial disposition. Only a few methods allow for flexibility of the DNA and protein side-chains during the docking. They, however, require extensive knowledge to position the two components relative to each other[32] and problems were encountered in the absence of such information.[33] The combination of information-driven docking, molecular dynamics and DNA libraries in HADDOCK[34] has solved some of these problems but does require convergence of the global DNA conformation in order to generate custom DNA libraries for the second docking step.

9.3   The Use of Data in Docking

Ab initio docking programs generate docking models based on the coordinates of the (free) proteins, disregarding any further knowledge of the system under study. However, inclusion of such knowledge can be very beneficial for the docking progress by eliminating a large number of possible outcomes. This has been an important factor for success in the CAPRI experiment.[35]

A large number of experimental techniques can give useful information to assist the docking process (for a detailed review, see van Dijk *et al.*[5]) such as, for example, mutagenesis data, NMR chemical shift perturbation (CSP) data, NMR residual polar couplings (RDC) and NMR or mass spectrometry-derived hydrogen/deuterium (H/D) exchange data. Many docking methods have been used in combination with experimental data to filter the generated solutions. For example, the FFT programs Hex and GRAMM have been used with mutagenesis data,[36–39] DOT has been used with H/D exchange data,[40, 41] and FTDOCK has been used with NMR CSP and RDC data.[42, 43] In addition, in recent years a large number of methods have been developed that aim at predicting protein-protein interface regions, exploiting for example sequence conservation (for a review, see Zhou and Qin,[44] De Vries and Bonvin,[45] and Chapter 7 of this book by Kinoshita *et al.*). Several of these prediction methods have also been used in combination with docking, for example WHISCY,[46] ProMate[47] and cons-PPISP.[48]

Next to filtering, experimental and/or predicted data can also be used directly in the sampling stage of docking. Compared to simple filtering of solutions, the advantage is then that 'correct' or 'near-correct' configurations should be enriched, provided of course that the information is correct. This becomes especially important when the number of configurations is too large to be adequately sampled, as is often the case when flexibility is introduced.

In several FFT methods, information can be used *a priori*, for example by limiting the rotational search to certain angles, or by up-weighting[49-51] or blocking[52] given residues in the generation of the grid. However, there are very few methods in which data are used to directly drive the docking. Among the methods that have participated in CAPRI there is only one, HADDOCK.[8, 22] Most experimental and/or predicted data are highly ambiguous and only provide information about putative interface residues, but not about the specific contacts made. To reflect this, such data are incorporated into HADDOCK in the form of ambiguous interaction restraints (AIRs). Prior to docking, the user must supply for every molecule a list of active residues (residues that are known to make contact within the complex) and passive residues (residues that potentially make contact). For every active residue, a single AIR restraint is defined between that residue and all active and passive residues on the partner. An explicit AIR energy term is introduced into the calculation through a harmonic potential (becoming linear after a given cutoff distance) that depends on an *effective distance*. The latter is calculated through the following formula:

$$d_{iAB}^{eff} = \left(\sum_{m_{iA}=1}^{Natoms} \sum_{k=1}^{NresB} \sum_{n_{kB}=1}^{Natoms} \frac{1}{d_{m_{iA}n_{kB}}^6} \right)^{-\frac{1}{6}} \tag{9.1}$$

where N_{atoms} indicates all atoms of a given residue and N_{res} the sum of active and passive residues for a given protein. An upper limit to the effective distance (typically 2 Å) is enforced by HADDOCK. If this limit is exceeded, the AIR energy becomes positive and the active residue experiences an attractive force towards the active and passive residues of the partner molecule. If not, the restraint is satisfied and the AIR energy and attractive force are zero for that restraint. Since many atom-atom distances inversely contribute to the effective distance, an AIR restraint is typically satisfied if a residue comes within 4–5 Å of any active or passive residue of the partner molecule.

In this way (putative) interface residues are enforced to make contact with (a surface region on) the partner protein, but not with any specific partner residue. These ambiguous restraints drive the docking in the same way that nuclear Overhauser effect (NOE) distance restraints drive the calculation of an NMR structure. In fact, if NOEs have been measured they can be directly included. HADDOCK can deal with a large variety of experimental data, including among others mutagenesis, NMR chemical shift perturbation data, residual dipolar coupling, H/D exchange data, cross-linking and NMR relaxation data, but also interface prediction data.

9.3.1   The Use of Data in Protein-DNA Docking

Proteins can form highly specific complexes with DNA, often based on subtle properties emerging from the base sequence. This specificity can, however, not be described in terms of a simple recognition code.[53] DNA recognition in an initial encounter complex is not only specific in terms of residue-to-base interactions but also affected by surrounding base pairs that can facilitate conformational changes leading to a strengthening of the interaction. This sequence-to-shape interplay is poorly reflected in the geometrical and physico-chemical properties along the native, highly charged DNA structure.[54] This makes it particularly difficult for an *ab initio* docking method to resolve a unique conformation, as both sequence and conformational space need to be sampled at once. The amount of possible combinations in this scenario quickly explodes.

Most of the current protein-DNA docking methods however are not *ab initio*. They either focus on the sequence aspect of recognition or use information driven approaches to bias the search through sequence and/or conformation space. Methods that focus on sequence often follow a threading approach in which the base sequence of a fixed canonical DNA structure is changed.[55] Information driven approaches use biochemical and biophysical information such as conservation data and data from DNA footprinting and mutagenesis experiments to bias or completely circumvent the search through sequence space.[56–58] Sequences or DNA conformations that favor interactions in both approaches are resolved based on energy scoring functions, knowledge based filters or amino acid to base pairing propensities. Most of these methods still rely on a rigid body docking approach to facilitate complex formation and, as such, are often only successful in predicting complexes in which the DNA does not change conformation much. Flexibility can however not be ignored since it is an inherent property of DNA. Most efficient protein-DNA docking methods therefore incorporate some sort of flexibility in the docking process.

9.4   State-of-the-art Methodology in Macromolecular Docking

Table 9.1 shows an overview of state-of-the-art docking programs and their characteristics. Table 9.2 shows the top-scoring docking predictors for recent CAPRI targets. While there is a wide diversity in underlying methodology, they all have several common aspects: all methods use an initial rigid-body stage, followed by a scoring step. ZDOCK,[19, 59] MolFit,[7] DOT[41] and PIPER[60, 61] are all FFT algorithms, while in HADDOCK[8, 22] the rigid-body search is performed by data-driven energy minimization. RosettaDock[10] and ICM-DISCO[9] use a Monte Carlo minimization in their initial stage. Also, most methods account for flexibility at some stage. In ICM-DISCO, docking against a soft grid is followed by Monte Carlo (MC) optimization of the ligand side-chains.[9] This procedure is quite successful in reproducing induced changes in surface side-chains as long as no large backbone rearrangements take place. In RosettaDock,[10] after the initial low-resolution search, the side-chains are repacked and further optimized in an MC search that includes rigid-body displacements. Recent developments include the sampling of off-rotamer side-chain conformations[62] and a new solvated rotamer library. Rather impressive results on the accuracy of side-chain positioning were obtained for previous CAPRI targets that exhibit only minor backbone conformational changes upon binding.[63, 64] As shown in Figure 9.3, HADDOCK takes into account backbone as well as side-chain flexibility. This is achieved through the use of Molecular Dynamics (MD) simulations: first, a semi-flexible simulated annealing of the interface, followed by a refinement in explicit solvent. In CAPRI, MD simulations were used as well by the teams of Smith[65] and Zhou.[44]

Table 9.1 Overview of current docking methods

Docking method	
ATTRACT[20, 21]	Reduced-representation EM algorithm. Allows large-scale motions during docking. Considers multiple conformations simultaneously
ClusPro[82, 83]	Clustering and scoring of docking solutions. Web server available, http://nrc.bu.edu/cluster
DOT[41]	FFT algorithm. Makes use of atomic contact potentials. Software is free (open source), http://www.sdsc.edu/CCMS/DOT
GRAMM[84]	FFT algorithm. Web server (GRAMM-X) available, software is free, http://vakser.bioinformatics.ku.edu/resources/gramm
HADDOCK[8, 22]	Full-atom EM/MD algorithm. Data-driven approach, suitable for a wide range of biomolecular complexes and experimental data. Implements both side-chain and backbone flexibility during the protocol. Software is free (source code) for academic use, http://www.nmr.chem.uu.nl/haddock
ICM-DISCO[9]	Full-atom MC/MD algorithm with side-chain optimization. Part of the ICM program for protein modeling, commercially available, http://www.molsoft.com
MolFit[7]	FFT algorithm. Allows the use of data to limit the search space. Allows incremental docking of subunits or domains. Software is free (binaries), http://www.weizmann.ac.il/Chemical_Research_Support/molfit
PatchDock/ FlexDock[30, 85]	Geometric algorithm, very fast. Allows incremental docking of subunits or domains. Web server available, software is free (binaries) for academic use, http://bioinfo3d.cs.tau.ac.il/PatchDock
PIPER[60, 61]	FFT algorithm. Makes use of atomic contact potentials
pyDock[86, 87]	Scoring of docking solutions. Makes use of atomic contact potentials. Has performed well in the separate CAPRI scoring experiment
RosettaDock[10, 88]	Full-atom MC/MD algorithm with side-chain optimization. Regions of low energy are searched extensively. Part of the Rosetta software package for protein prediction, free (source code) for academic use, http://www.rosettacommons.org. Computationally intensive
ZDOCK[19, 59]	FFT algorithm. Allows the use of data to limit the search space. Software is free for academic use, http://zlab.bu.edu/zdock

Table 9.2   *Docking methods that scored at least four stars among the most recent CAPRI targets (round 6–13). For the criteria for one-, two- and three-star predictions, see the main text. 0: zero-star (incorrect) prediction. – : predictor did not participate for this target*

Predictor	T20	T21	T24	T25[a]	T26	T27	T28	T29[a]	Total
ZDOCK[19, 59]	0	*	*	**	*	**	0	*	6/2/0
HADDOCK[8, 22]	0	**	0	*	**	*	0	**	5/3/0
Smith[65b]	0	0	0	**	**	*	0	*	4/2/0
MolFit[7]	0	0	0	***	*	*	0	*	4/1/1
DOT[41]	–	**	0	0	0	*	0	**	3/2/0
PIPER[60, 61]	0	*	0	**	**	0	0	0	3/2/0
RosettaDock[10, 89c]	*	0	0	0	**	*	0	0	3/1/0
ICM-DISCO[9]	–	–	*	**	*	–	–	–	3/1/0
Zhou[44d]	–	–	0	0	–	**	0	**	2/2/0
pyDock[86, 87e]	–	–	0	**	0	0	0	**	2/2/0
RosettaDock[10, 88f]	0	**	0	0	**	0	0	0	2/2/0

[a]targets 25 and 29 are bound-unbound docking, not a blind prediction.
[b]combination of methods: MolFit and FTDOCK[43] for sampling, steered MD and RosettaDock for refinement.
[c]Baker team.
[d]combination of methods: ZDOCK, ClusPro[82, 83] (ZDOCK+DOT) and HADDOCK for sampling, MD for refinement.
[e]combination of methods: ZDOCK and FTDOCK for sampling, followed by a scoring step.
[f] Gray team

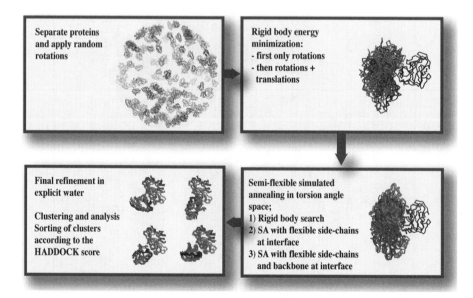

Figure 9.3   *Overview of the HADDOCK docking protocol. Typically, in the order of 10,000 solutions are sampled at the rigid body docking stage with only 1000 saved to disk. The top 20% is then subjected to the semi-flexible refinement and final refinement in explicit solvent.*

For CAPRI targets with large conformational changes, an incremental, multi-body, multistage docking strategy was successfully applied using MolFit.[66] In target 20, it was clear that a loop must undergo significant conformational changes. The only acceptable solution was obtained using RosettaDock, by simply ignoring the loop during docking and rebuilding it in the context of the complex in a loop modeling step. Success was also achieved (after the submission) using HADDOCK by docking the loop, the rest of the protein and its partner as three separate bodies.[3, 22]

Experimental data are also widely used to assist in docking. ZDOCK[44] offers the possibility to block certain regions, whereas in HADDOCK experimental data directly drives the docking. Both options were used by the team of Zhou,[44] using a combination of experimental data and interface predictions. In addition, MolFit offers the possibility to up weight putative interface residues in the FFT grid. pyDOCK and Smith's team also use experimental data along with physicochemical properties at the scoring stage. Since CAPRI is not fully automated and human intervention is allowed in selecting the ten submitted models, experimental data play an important role in the visual inspection and selection of the solutions, as many groups have stated.

In terms of performance, ZDOCK and HADDOCK are the most successful methods, with ZDOCK achieving the largest number of successful predictions (6/8 targets) and HADDOCK achieving the most two-star predictions (3/8 targets). Good predictions were also made by predictor teams that do not rely on one particular method, but instead select and refine solutions generated by different methods (Smith, Zhou and pyDOCK).

The only three-star prediction was scored by MolFit for target 25, the easiest of the two bound-unbound targets. Three-star docking predictions have not yet been submitted for any unbound CAPRI target. However, RosettaDock's two-star prediction for target 26 was nearly of three-star quality; also, three-star docking solutions were generated for target 27 by HADDOCK in the initial stage. For that target, there is some disagreement between the crystal structure and mutagenesis data from literature, which why the three-star predictions were not scored at the top.[22] The above observations indicate that three-star predictions for unbound docking are getting within reach.

In conclusion, accurate docking solutions are now routinely obtained for 'easy' targets, i.e. proteins that do not undergo large conformational changes, or for which good experimental data are available. Methods that take explicit flexibility into account and use some experimental (or predicted) data tend to outperform methods that do not; highly accurate predictions (three-star according to the CAPRI criteria) are getting within reach for blind, unbound predictions. However, the prediction of large backbone conformational changes remains a formidable challenge. Although methods to deal with such large changes have been and are being developed, success is not yet guaranteed.

9.5 Performance of Docking Servers

Table 9.3 shows the performance of various docking servers in recent rounds of CAPRI. Although the easiest target, target 25 (bound-unbound), was well predicted, the success of the servers has been very limited for other targets. In previous CAPRI rounds, the ClusPro server scored 2 one-star, 1 two-star and even a three-star prediction among the nine targets, however, all of those successes were bound-unbound docking.[2, 67] This suggests that while

Table 9.3 *Docking servers that scored at least one star among the most recent CAPRI targets (round 6–13). For the criteria for one-, two- and three-star predictions, see the main text. 0: zero-star (incorrect) prediction. – : server did not participate for this target*

Server	T20	T21	T24	T25[a]	T26	T27	T28	T29[a]	Total
PatchDock[85]	–	–	0	**	*	–	–	0	2/1/0
ClusPro[82, 83]	0	0	0	*	0	*	0	0	2/0/0
GRAMM-X[84]	–	0	0	**	–	0	0	0	1/1/0
SKE-DOCK[90]	–	0	0	**	0	0	0	0	1/1/0
SmoothDock[91]	0	0	0	**	0	0	0	0	1/1/0

[a] targets 25 and 29 are bound-unbound docking, not a blind prediction.

docking can be fully automated in some easy cases, more improvement is needed before reliable results can be obtained for the average unbound docking target without human supervision.

9.6 Practical Aspects of Docking

There are many important practical aspects that need to be considered in the docking of macromolecular complexes. This begins already at the preparation of a docking run. While almost all docking programs accept the input structures in PDB format, this format is very loose, and various programs may enforce additional rules. For example, in HADDOCK, it is required that every PDB ends with an END statement, and that the segid (column 73–76) is empty. Non-protein atoms such as ions are allowed, but their nomenclature must be compatible to the definitions in the provided topologies; further their electric charge state should be specified.

Also, the input structures should be of sufficient quality. Missing atoms, clashes, improbable conformations and similar irregularities may be tolerated to some extent, but must otherwise be fixed in the unbound starting structure prior to docking (recommended). In case there is no unbound structure of the protein available, the user must instead generate a homology model of sufficiently high quality. Some energies and scores utilized by docking programs, for example shape complementarity and van der Waals energy, are highly dependent on the quality of the structure. Structure validation programs such as ProCheck,[68] WHATIF[69] and MolProbity[70] can assess the quality of structures and homology models.

The nature of the system under study plays an important role as well. Most docking methods specialize in protein-protein complexes, but a few programs offer support for protein-DNA complexes.[34, 38, 41, 43] Only one program, HADDOCK, has been applied to a wide variety of biomolecular complexes: protein-protein,[71, 72] protein-peptide,[73, 74] protein-nucleic acid,[75, 76] protein-oligosaccharide[77] and protein-small ligand[78, 79] complexes (see also http://www.nmr.chem.uu.nl/haddock/publications.html for a list of applications). Each type of complex has characteristic practical aspects that must be considered. In particular, small ligands (and oligosaccharides) are difficult for docking methods that use a full-atom representation. Appropriate force field parameters, such as atom partial charges and allowed bond angle parameters, are well defined for proteins, but they must be approximated for

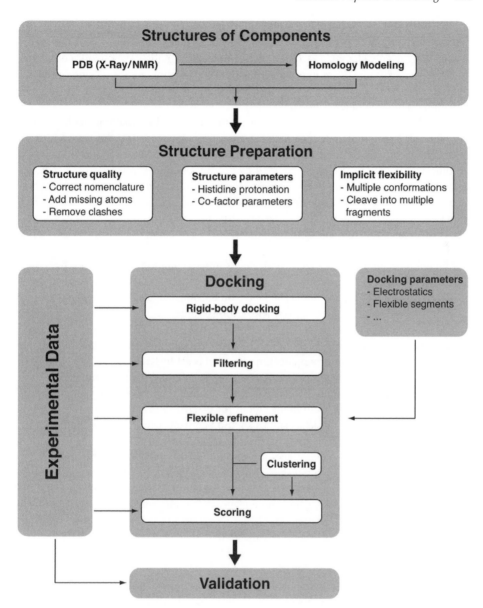

Figure 9.4   *Flowchart of macromolecular docking.*

small ligands. Moreover, small ligands typically enjoy considerable conformational freedom, up to the point that their conformation may be completely unknown. Fortunately, there are several programs and web servers for small ligands that can provide information and parameters for docking such as PRODRG,[80] HIC-Up[81] and CORINA (http://www.molecular-networks.com/online_demos/corina_demo.html). These can also be helpful in case of small molecules that are part of proteins, such as co-factors and post-translationally modified amino acids. These molecules often play a crucial role in the interaction energetics and, therefore, must be modeled as accurately as possible. In many cases (ions, ATP, phosphorylated residues), a net charge is carried by these molecules. In the large majority of docking programs, electrostatics is an important factor in the sampling and/or scoring of docking solutions. Therefore, partial charges must be accurately distributed among the atoms in order to properly compute electrostatic energies.

The problem of proper electrostatics is by no means limited to small molecules. Neutral histidines, for example, can be protonated at either the delta or epsilon nitrogen. A histidine can also be positively charged and thus protonated at both nitrogens. Within HADDOCK, histidines are treated as positively charged by default, but they can be specified as neutral (delta- or epsilon-protonated) by the user. Again, this may have significant effects on the electrostatic energies of docking solutions. The protonation state of histidines can be estimated from the pH at which the starting structure was solved, and programs such as WHATIF[69] can estimate them from the protein hydrogen bonding network of the starting structure.

A final important aspect of electrostatics is its implementation in the chosen force field, in particular how the electrostatic energy declines with the distance between charges and the screening effect of the medium. The latter can be approximated by varying the value of the dielectric constant (or epsilon parameter): for example, for vacuum epsilon $= 1$, in the interior of a protein epsilon $= 2$–3 and for water epsilon $= 70$–80. In some cases, epsilon is made distance-dependent, resulting in a stronger screening of electrostatic interactions. In HADDOCK, for protein-protein docking, epsilon is set to 10 for the vacuum part of the protocol (a compromise between the interior of a protein and the screening effect of the solvent) and to 1 for the final explicit solvent refinement. Other docking programs may use different values, or use a different model for electrostatics altogether. Regardless of the docking program, proper electrostatics should be given special attention in the case of unusual systems, such as for example a transmembrane protein.

As discussed previously, many docking programs take into account flexibility of the macromolecules that are being docked. This may require a considerable amount of input from the user's side. Docking programs may accept multiple starting structures, but they must be generated by the user, using some of the techniques outlined previously. Ideally, some of the starting conformation should be closer to the bound form than the initial unbound structure, but this is notoriously difficult to achieve *a priori* to docking.

In case of explicit flexibility, additional guidance by the user may be required. For example, HADDOCK consists of an initial rigid body minimization followed by a simulated annealing phase and an explicit solvent refinement (Figure 9.3). Flexibility is introduced in a stepwise manner during refinement: first only for side-chains, and later for both side-chains and backbone of the interface residues; finally the entire system becomes flexible in the final cooling phase of the explicit solvent refinement. The user may specify the semi-flexible regions manually, but also choose to let HADDOCK determine them automatically from an analysis of contacting residues. Fully flexible segments can also be defined: these are treated as fully flexible during the entire simulated annealing phase; this might be appropriate in cases where parts of a structure are disordered or unstructured, for flexible loops, or when docking small flexible ligands or peptides onto a protein.

Experimental or predicted data used in docking must be treated carefully as well. Interface prediction data and some experimental data such as H/D exchange data are very fuzzy, in the sense that not every identified residue may actually take part in the interaction. This can be translated into a more liberal filtering scheme if data are used to filter the docking results. In HADDOCK, where data are used to drive the docking, a percentage of the data can be discarded at random for each docking trial.

Data need not to be limited to lists of interface residues. HADDOCK is a data-driven docking method rooted in NMR structure determination. Therefore, it can deal with a wide variety of restraints besides AIRs, such as classical NOEs but also symmetry restraints. Also, flexibility in HADDOCK can be controlled through the use of restraints between atoms *within* one protein. For example, this allows the user to impose secondary structure restraints to prevent the unfolding of a fully flexible segment, or distance restraints to prevent a co-factor to drift away. It also allows the use of HADDOCK for *ab initio* structure calculations of complexes when classical NMR restraints are available to drive the folding. It is further possible in HADDOCK to cut a protein into different pieces at hinge points and to define peptide bond and angle restraints between the segments. The various pieces are then docked simultaneously (multi-body docking), providing a natural way of modeling large conformational changes that can take place upon binding. This is different from the approach followed by FlexDock[30] and MolFit,[7] where the components are docked incrementally and the solutions are filtered rather than restrained.

The result of a docking search is always a large number of possible complexes rather than a single solution. The choice of the correct model is not trivial and is known as the scoring problem. Docking programs may employ many different scores, filters and energies in order to rank the solutions. Some of these, but usually not all of them, may have been used to drive the docking as well. Although a discussion of the various scores is beyond the scope of this chapter, it should be kept in mind that their relative usefulness depends on the system under study. There is thus, at present, no general, universal scoring function. For example, desolvation energy is an excellent score for homodimer complexes, in contrast, in the case of enzyme-inhibitors, electrostatics play a more important role, and desolvation energy is inappropriate in a membrane environment. Also, many docking programs allow the use of experimental data or interface predictions to filter the docking solutions during scoring; their caveats have been discussed above.

Finally, clustering of solutions is often applied to facilitate scoring. The size of a cluster may even be used as a score by itself. In HADDOCK, clustering parameters (RMSD cutoff and minimum cluster size) can be specified by the user, and scoring is done on a per-cluster rather than per-solution basis. In addition, it is recommended to calculate cluster averages on a similar number of solutions to avoid cluster size effects and allow easier comparison.

9.7 The Interpretation of Docking Results

After a docking run, a small number of solutions or clusters is typically selected and presented to the user. How can one know which of these models, if any, is correct? One way is to validate them against additional experimental data that were not used to derive the docking models. This could for example data describing the shape of the complex, such as electron microscopy or SAXS data, or mutagenesis data not use in the docking. Strikingly, docking models can also be used to guide and predict the outcome of experiments. In the original HADDOCK paper,[8] the most favorable cluster contained a specific salt bridge.

Mutating one of the residues to an opposite charge abolished the interaction, but mutating and swapping the identity of both residues restored the interaction. Examples such as this, as well as recent results in CAPRI, indicate that docking is becoming more and more reliable, and can be complementary to experimental approaches in the unraveling of the structural genome.

Acknowledgments

This work was supported by the Netherlands Organization for Scientific Research (VICI grant #700.56.442 to A.B.) and by the European Community (Integrated Project SPINE2-COMPLEX contract no. 032220 and STREP Project Extend-NMR contract no. 18988).

List of Abbreviations

3D = Three-dimensional. AIR = Ambiguous Interaction Restraint. CAPRI = Critical Assessment of Predicted Interactions. EM = Energy Minimization. FFT = Fast Fourier Transform. MC = Monte Carlo. MD = Molecular Dynamics. PDB = Protein Data Bank. RMSD = Root Mean Square Deviation.

References

1. J. Janin, Assessing predictions of protein-protein interaction: the CAPRI experiment, *Protein Sci*, **14**, 278–283 (2005).
2. R. Méndez, R. Leplae, M. F. Lensink and S. J. Wodak, Assessment of CAPRI Predictions in Rounds 3–5 Shows Progress in Docking Procedures, *Proteins*, **60**, 150–169 (2005).
3. A. M. Bonvin, Flexible protein-protein docking, *Curr Opin Struc Biol*, **16**, 194–200 (2006).
4. S. Vajda and C. J. Camacho, Protein-protein docking: is the glass half-full or half-empty?, *Trends Biotechnol*, **22**, 110–116 (2004).
5. A. D. van Dijk, R. Boelens and A. M. Bonvin, Data-driven docking for the study of biomolecular complexes, *Febs J*, **272**, 293–312 (2005).
6. A. D. J. van Dijk, S. J. de Vries, C. Dominguez, H. Chen, H. X. Zhou and A. M. J. J. Bonvin, Data-driven docking: HADDOCK's adventures in CAPRI, *Proteins-Structure Function and Bioinformatics*, **60**, 232–238 (2005).
7. E. Katchalski-Katzir, I. Shariv, M. Eisenstein, A. A. Friesem, C. Aflalo and I. A. Vakser, Molecular-surface recognition – determination of geometric fit between proteins and their ligands by correlation techniques, *P Natl Acad Sci USA*, **89**, 2195–2199 (1992).
8. C. Dominguez, R. Boelens and A. M. J. J. Bonvin, HADDOCK: A protein-protein docking approach based on biochemical or biophysical information, *J Am Chem Soc*, **125**, 1731–1737 (2003).
9. J. Fernandez-Recio, M. Totrov and R. Abagyan, ICM-DISCO docking by global energy optimization with fully flexible side-chains, *Proteins*, **52**, 113–117 (2003).
10. J. J. Gray, S. Moughon, C. Wang, *et al.* Protein-protein docking with simultaneous optimization of rigid-body displacement and side-chain conformations, *J Mol Biol*, **331**, 281–299 (2003).
11. M. Eisenstein and E. Katchalski-Katzir, Geometric recognition as a tool for predicting structures of molecular complexes, *Letters in Peptide Science*, **5**, 365–369 (1998).
12. A. Heifetz and M. Eisenstein, Effect of local shape modifications of molecular surfaces on rigid-body protein-protein docking, *Protein Eng*, **16**, 179–185 (2003).

13. B. Ma, T. Elkayam, H. Wolfson and R. Nussinov, Protein-protein interactions: structurally conserved residues distinguish between binding sites and exposed protein surfaces, *Proc Natl Acad Sci U S A*, **100**, 5772–5777 (2003).
14. X. Li, O. Keskin, B. Ma, R. Nussinov and J. Liang, Protein-protein interactions: hot spots and structurally conserved residues often locate in complemented pockets that pre-organized in the unbound states: implications for docking, *Journal of Molecular Biology*, **344**, 781–795 (2004).
15. D. Rajamani, S. Thiel, S. Vajda and C. J. Camacho, Anchor residues in protein-protein interactions, *Proc Natl Acad Sci U S A*, **101**, 11287–11292 (2004).
16. C. Dominguez, A. M. J. J. Bonvin, G. S. Winkler, F. M. A. van Schaik, H. T. M. Timmers and R. Boelens, Structural model of the UbcH5B/CNOT4 complex revealed by combining NMR, mutagenesis, and docking approaches, *Structure*, **12**, 633–644 (2004).
17. R. Grünberg, J. Leckner and M. Nilges, Complementarity of structure ensembles in protein-protein binding, *Structure*, **12**, 2125–2136 (2004).
18. G. R. Smith, M. J. E. Sternberg and P. A. Bates, The relationship between the flexibility of proteins and their conformational states on forming protein-protein complexes with an application to protein-protein docking, *J Mol Biol*, **347**, 1077–1101 (2005).
19. L. Li, R. Chen and Z. Weng, RDOCK: Refinement of Rigid-body Protein Docking Predictions, *Proteins, Structure Function and Bioinformatics*, **53**, 693–707 (2003).
20. M. Zacharias, Protein-protein docking with a reduced protein model accounting for side-chain flexibility, *Protein Sci*, **12**, 1271–1282 (2003).
21. M. Zacharias, Rapid protein-ligand docking using soft modes from molecular dynamics simulations to account for protein deformability: binding of FK506 to FKBP, *Proteins, Structure Function and Bioinformatics*, **54**, 759–767 (2004).
22. S. J. de Vries, A. D. van Dijk, M. Krzeminski, *et al.* HADDOCK versus HADDOCK: New features and performance of HADDOCK2.0 on the CAPRI targets, *Proteins* (2007).
23. U. Emekli, D. Schneidman-Duhovny, H. J. Wolfson, R. Nussinov and T. Haliloglu, HingeProt: Automated prediction of hinges in protein structures, *Proteins* (2007).
24. A. May and M. Zacharias, Accounting for global protein deformability during protein-protein and protein-ligand docking., *Biochimica and Biophysica Acta* (2005) Dec 30; 1754(1–2):225–31 [Epub 2005 Sep 12].
25. K. Hinsen, N. Reuter, J. Navaza, D. L. Stokes and J.-J. Lacapère, Normal mode-based fitting of atomic structure into electron density maps: application to sarcoplasmic reticulum Ca-ATPase, *Biophys J*, **88**, 818–827 (2005).
26. F. Tama, O. Miyashita and C. L. Brooks III, Flexible mutli-scale fitting of atomic structures into low-resolution electron density mapes with elastic network mode analysis, *Journal of Molecular Biology*, **337**, 985–999 (2004)
27. E. Lindahl and M. Delarue, Refinement of docked protein-ligand and protein-DNA structures using low frequency normal mode amplitude optimization, *Nucleic Acids Res*, **33**, 4496–4506 (2005).
28. K. Bastard, C. Prevost and M. Zacharias, Accounting for loop flexibility during protein-protein docking, *Proteins*, **62**, 956–969 (2006).
29. K. Bastard, A. Thureau, R. Lavery and C. Prévost, Docking macromolecules with flexible segments, *J Comput Chem*, **24**, 1910–1920 (2003).
30. D. Schneidman-Duhovny, Y. Inbar, R. Nussinov and H. J. Wolfson, Geometry-based flexible and symmetric protein docking, *Proteins, Structure Function and Bioinformatics*, **60**, 224–231 (2005).
31. C. Sandmann, F. Cordes and W. Saenger, Structure model of a complex between the factor for inversion stimulation (FIS) and DNA: modeling protein-DNA complexes with dyad symmetry and known protein structures., *Proteins*, **25**, 486–500 (1996)
32. W. S. Tzou and M. J. Hwang, Modeling helix-turn-helix protein-induced DNA bending with knowledge-based distance restraints, *Biophys J*, **77**, 1191–1205 (1999).
33. R. M. Knegtel, R. Boelens and R. Kaptein, Monte Carlo docking of protein-DNA complexes: incorporation of DNA flexibility and experimental data, *Protein Eng*, **7**, 761–767 (1994).

34. M. van Dijk, A. D. van Dijk, V. Hsu, R. Boelens and A. M. Bonvin, Information-driven protein-DNA docking using HADDOCK: it is a matter of flexibility, *Nucleic Acids Res*, **34**, 3317–3325 (2006).

35. R. Mendez, R. Leplae, L. De Maria and S. J. Wodak, Assessment of blind predictions of protein-protein interactions: current status of docking methods, *Proteins*, **52**, 51–67 (2003).

36. Y. Azuma, L. Renault, J. A. Garcia-Ranea, A. Valencia, T. Nishimoto and A. Wittinghofer, Model of the Ran-RCC1 interaction using biochemical and docking experiments, *Journal of Molecular Biology*, **289**, 1119–1130 (1999).

37. C. Gaboriaud, J. Juanhuix, A. Gruez, *et al.* The crystal structure of the globular head of complement protein C1q provides a basis for its versatile recognition properties, *J Biol Chem*, **278**, 46974–46982 (2003).

38. D. W. Ritchie and G. J. L. Kemp, Protein docking using spherical polar Fourier correlations, *Proteins-Structure Function and Genetics*, **39**, 178–194 (2000).

39. I. A. Vakser, Protein docking for low-resolution structures, *Protein Eng*, **8**, 371–377 (1995).

40. G. S. Anand, D. Law, J. G. Mandell, *et al.* Identification of the protein kinase A regulatory R-I alpha-catalytic subunit interface by amide H/H-2 exchange and protein docking, *P Natl Acad Sci USA*, **100**, 13264–13269 (2003).

41. J. G. Mandell, V. A. Roberts, M. E. Pique, *et al.* Protein docking using continuum electrostatics and geometric fit, *Protein Engineering*, **14**, 105–113 (2001).

42. A. Dobrodumov and A. M. Gronenborn, Filtering and selection of structural models: Combining docking and NMR, *Proteins-Structure Function and Genetics*, **53**, 18–32 (2003).

43. H. A. Gabb, R. M. Jackson and M. J. E. Sternberg, Modelling protein docking using shape complementarity, electrostatics and biochemical information, *Journal of Molecular Biology*, **272**, 106–120 (1997).

44. S. Qin and H. X. Zhou, A holistic approach to protein docking, *Proteins* (2007).

45. S. J. de Vries and A. M. Bonvin, How proteins get in touch: Interface prediction of protein-protein complexes. *Curr Protein Pept Sci.* (2008) Aug; **9** (4): 394–406.

46. S. J. de Vries, A. D. van Dijk and A. M. Bonvin, WHISCY: what information does surface conservation yield? Application to data-driven docking, *Proteins*, **63**, 479–489 (2006).

47. H. Neuvirth, R. Raz and G. Schreiber, ProMate: a structure based prediction program to identify the location of protein-protein binding sites, *J Mol Biol*, **338**, 181–199 (2004).

48. H. X. Zhou and Y. Shan, Prediction of protein interaction sites from sequence profile and residue neighbor list, *Proteins*, **44**, 336–343 (2001).

49. E. Ben-Zeev and M. Eisenstein, Weighted geometric docking: Incorporating external information in the rotation-translation scan, *Proteins*, **52**, 24–27 (2003).

50. E. Ben-Zeev, R. Zarivach, M. Shoham, A. Yonath and M. Eisenstein, Prediction of the structure of the complex between the 30S ribosomal subunit and colicin E3 via weighted-geometric docking, *J Biomol Struct Dyn*, **20**, 669–675 (2003).

51. R. Zarivach, E. Ben-Zeev, N. Wu, *et al.* On the interaction of colicin E3 with the ribosome, *Biochimie*, **84**, 447–454. (2002)

52. R. Chen, L. Li and Z. Weng, ZDOCK: an initial-stage protein-docking algorithm, *Proteins*, **52**, 80–87 (2003).

53. C. O. Pabo and L. Nekludova, Geometric analysis and comparison of protein-DNA interfaces: why is there no simple code for recognition?, *J Mol Biol*, **301**, 597–624 (2000).

54. M. J. Sternberg, H. A. Gabb and R. M. Jackson, Predictive docking of protein-protein and protein-DNA complexes, *Curr Opin Struct Biol*, **8**, 250–256 (1998).

55. F. Fanelli and S. Ferrari, Prediction of MEF2A-DNA interface by rigid body docking: a tool for fast estimation of protein mutational effects on DNA binding., *J Struct Biol.*, **153**, 278–283 (2006).

56. V. A. Roberts, Case, D.A. & Tsui, V., Predicting interactions of winged-helix transcription factors with DNA, *Proteins*, **57**, 172–187 (2004).

57. P. Aloy, G. Moont, H. A. Gabb, E. Querol, F. X. Aviles and M. J. Sternberg, Modelling repressor proteins docking to DNA, *Proteins*, **33**, 535–549 (1998).

58. L. De Luca, G. Vistoli, A. Pedretti, M. L. Barreca and A. Chimirri, Molecular dynamics studies of the full-length integrase-DNA complex, *Biochem Biophys Res Commun*, **336**, 1010–1016 (2005).

59. B. Pierce and Z. Weng, ZRANK: reranking protein docking predictions with an optimized energy function, *Proteins*, **67**, 1078–1086 (2007).

60. D. Kozakov, R. Brenke, S. R. Comeau and S. Vajda, PIPER: an FFT-based protein docking program with pairwise potentials, *Proteins*, **65**, 392–406 (2006).

61. Y. Shen, R. Brenke, D. Kozakov, S. R. Comeau, D. Beglov and S. Vajda, Docking with PIPER and refinement with SDU in rounds 6-11 of CAPRI, *Proteins* (2007).

62. C. Wang, O. Schueler-Furman and D. Baker, Improved side-chain modeling for protein-protein docking, *Protein Sci*, **14**, 1328–1339 (2005).

63. O. Schueler-Furman, C. Wang and D. Baker, Progress in protein-protein docking: atomic resolution predictions in the CAPRI experiment using RosettaDock with an improved treatment of side-chain flexibility, *Proteins, Structure Function and Bioinformatics*, **60**, 187–194 (2005).

64. M. D. Daily, D. Masica, A. Sivasubramanian, S. Somarouthu and J. J. Gray, CAPRI Rounds 3-5 reveal promising successes and future challenges for RosettaDock, *Proteins, Structure Function and Bioinformatics*, **60**, 181–186 (2005).

65. A. Heifetz, S. Pal and G. R. Smith, Protein-protein docking: Progress in CAPRI rounds 6-12 using a combination of methods: The introduction of steered solvated molecular dynamics, *Proteins* (2007).

66. E. Ben-Zeev, N. Kowalsman, A. Ben-Shimon, *et al.*, Docking to single-domain and multiple-domain proteins: old and new challenges, *Proteins, Structure Function and Bioinformatics*, **60**, 195–201 (2005).

67. S. R. Comeau, S. Vajda and C. J. Camacho, Performance of the first protein docking server ClusPro in CAPRI rounds 3-5, *Proteins*, **60**, 239–244 (2005).

68. R. A. Laskowski, J. A. Rullmannn, M. W. MacArthur, R. Kaptein and J. M. Thornton, AQUA and PROCHECK-NMR: programs for checking the quality of protein structures solved by NMR, *J Biomol NMR*, **8**, 477–486 (1996).

69. G. Vriend, WHAT IF: a molecular modeling and drug design program, *J Mol Graph*, **8**, 52–56, 29 (1990).

70. I. W. Davis, L. W. Murray, J. S. Richardson and D. C. Richardson, MolProbity: structure validation and all-atom contact analysis for nucleic acids and their complexes, *Nucleic Acids Research*, **32**, W615–W619 (2004).

71. A. Koglin, M. R. Mofid, F. Lohr, *et al.*, Conformational switches modulate protein interactions in peptide antibiotic synthetases, *Science*, **312**, 273–276 (2006).

72. Z. Lin, S. Sriskanthadevan, H. Huang, C. H. Siu and D. Yang, Solution structures of the adhesion molecule DdCAD-1 reveal new insights into Ca(2+)-dependent cell-cell adhesion, *Nat Struct Mol Biol*, **13**, 1016–1022 (2006).

73. A. P. Rideau, C. Gooding, P. J. Simpson, *et al.*, A peptide motif in Raver1 mediates splicing repression by interaction with the PTB RRM2 domain, *Nat Struct Mol Biol*, **13**, 839–848 (2006).

74. M. L. Reese, S. Dakoji, D. S. Bredt and V. Dotsch, The guanylate kinase domain of the MAGUK PSD-95 binds dynamically to a conserved motif in MAP1a, *Nat Struct Mol Biol*, **14**, 155–163 (2007).

75. M. B. Kamphuis, A. M. Bonvin, M. C. Monti, *et al.*, Model for RNA binding and the catalytic site of the RNase Kid of the bacterial parD toxin-antitoxin system, *J Mol Biol*, **357**, 115–126 (2006).

76. S. Cai, S. Zhu, Z. Zhang and Y. Chen, Determination of the three-dimensional structure of the Mrf2-DNA complex using paramagnetic spin labeling, *Biochemistry*, **46**, 4943–4950 (2007).

77. A. M. Wu, T. Singh, J. H. Liu, *et al.*, Activity-structure correlations in divergent lectin evolution: fine specificity of chicken galectin CG-14 and computational analysis of flexible ligand docking for CG-14 and the closely related CG-16, *Glycobiology*, **17**, 165–184 (2007).

78. L. Rutten, J. Geurtsen, W. Lambert, *et al.*, Crystal structure and catalytic mechanism of the LPS 3-O-deacylase PagL from Pseudomonas aeruginosa, *Proc Natl Acad Sci U S A*, **103**, 7071–7076 (2006).

79. S. Tomaselli, L. Ragona, L. Zetta, *et al.*, NMR-based modeling and binding studies of a ternary complex between chicken liver bile acid binding protein and bile acids, *Proteins*, **69**, 177–191 (2007).
80. A. W. Schuttelkopf and D. M. van Aalten, PRODRG: a tool for high-throughput crystallography of protein-ligand complexes, *Acta Crystallogr D Biol Crystallogr*, **60**, 1355–1363 (2004).
81. G. J. Kleywegt, K. Henrick, E. J. Dodson and D. M. van Aalten, Pound-wise but penny-foolish: How well do micromolecules fare in macromolecular refinement?, *Structure*, **11**, 1051–1059 (2003).
82. S. R. Comeau, D. Kozakov, R. Brenke, Y. Shen, D. Beglov and S. Vajda, ClusPro: Performance in CAPRI rounds 6-11 and the new server, *Proteins*, (2007).
83. S. R. Comeau, D. W. Gatchell, S. Vajda and C. J. Camacho, ClusPro: an automated docking and discrimination method for the prediction of protein complexes, *Bioinformatics*, **20**, 45–50 (2004).
84. A. Tovchigrechko and I. A. Vakser, GRAMM-X public web server for protein-protein docking, *Nucleic Acids Res*, **34**, W310–314 (2006).
85. D. Schneidman-Duhovny, Y. Inbar, R. Nussinov and H. J. Wolfson, PatchDock and SymmDock: servers for rigid and symmetric docking, *Nucleic Acids Res*, **33**, W363–367 (2005).
86. T. Man-Kuang Cheng, T. L. Blundell and J. Fernandez-Recio, pyDock: electrostatics and desolvation for effective scoring of rigid-body protein-protein docking, *Proteins*, **68**, 503–515 (2007).
87. S. Grosdidier, C. Pons, A. Solernou and J. Fernandez-Recio, Prediction and scoring of docking poses with pyDock, *Proteins* (2007).
88. S. Chaudhury, A. Sircar, A. Sivasubramanian, M. Berrondo and J. J. Gray, Incorporating biochemical information and backbone flexibility in RosettaDock for CAPRI rounds 6–12, *Proteins* (2007).
89. C. Wang, O. Schueler-Furman, I. Andre, *et al.*, RosettaDock in CAPRI rounds 6–12, *Proteins* (2007).
90. G. Terashi, M. Takeda-Shitaka, K. Kanou, M. Iwadate, D. Takaya and H. Umeyama, The SKE-DOCK server and human teams based on a combined method of shape complementarity and free energy estimation, *Proteins*, (2007).
91. C. J. Camacho and D. W. Gatchell, Successful discrimination of protein interactions, *Proteins*, **52**, 92–97 (2003).
92. J. R. Walker, G.V. Avvakumov, S. Xue, *et al.*, A novel and unexpected complex between the SUMO-1-conjugating enzyme UBC9 and the Ubiquitin-conjugating enzyme E2-25 kDa. PDB entry 2o25; DOI 10.2210/pdb2o25/pdb
93. Bonsor D.A., Grishkovskaya, I., Dodson, E.J., Kleanthous, C. Molecular mimicry enables competitive recruitment by a natively disordered protein. *J Am Chem Soc.* (2007) Apr 18; **129** (15): 4800–7. [Epub 2007 Mar 22].
94. M. Graille, E. A. Stura, M. Bossus, *et al.*, Crystal structure of the complex between the monomeric form of Toxoplasma gondii surface antigen 1 (SAG1) and a monoclonal antibody that mimics the human immune response, *J Mol Biol*, **354**, 447–458 (2005).

10

Protein Function Prediction via Analysis of Interactomes

Elena Nabieva and Mona Singh

10.1 Introduction

Genome sequencing efforts have resulted in an explosion of organisms whose entire protein complements have been determined. Nevertheless, for many proteins, little is known beyond their sequences, and for the typical proteome, between one-third and one-half of its proteins remain uncharacterized. As a result, a major challenge in modern biology is to develop methods for determining protein function at the genomic scale.

Computational methods to assign protein function have traditionally relied on identifying sequence similarity to proteins of known function. In recent years, however, other computational methods for predicting protein function have been developed (review: [1]). Many of these non-homology based methods still utilize sequence information, but can predict that two proteins share a function even when they have no sequence similarity. For example, in gene fusion methods,[2, 3] two proteins are believed to be related functionally if they appear as parts of a single protein in some other organism. Phylogenetic profiles[4, 5] predict proteins to be functionally related if they have similar patterns of occurrences across multiple genomes. Genomic context methods[6, 7] predict functional coupling between proteins if they tend to be contiguous in several genomes.

Increasingly, computational techniques for predicting protein function have analyzed data resulting from new high-throughput technologies. While there is a fascinating array of new functional genomics technologies that have enabled prediction of protein function, in this chapter we examine a family of methods that are based on analyzing large-scale protein–protein interaction data. Currently, several types of protein interactions have been

Prediction of Protein Structures, Functions, and Interactions   Edited by Janusz Bujnicki
© 2009 John Wiley & Sons, Ltd

determined via high-throughput experimental technologies. These include interactions between proteins that interact physically, that participate in a synthetic lethal or epistatic relationship, that are coexpressed, or where one phosphorylates or regulates another (review: [8]). Together, these interactions comprise the interactome and can be represented as networks or graphs, where interactions are undirected in the case of symmetric interactions, and directed otherwise.

Here, we focus primarily on predicting protein function via analysis of networks comprised of physical interactions. Most of these methods are based on the principle of *guilt-by-association*, where proteins are annotated by transferring the functions of the proteins with which they interact. The methods differ in whether they use local or global properties of the interactome in annotating proteins, in which particular topological features of the interactome they utilize, in whether they rely on first identifying tight clusters of proteins within the interactome before transferring annotations, and in whether they use guilt-by-association explicitly or employ some other similarity measure. While the focus of this chapter is on protein–protein physical interaction networks, it is often straightforward to apply the same methods to other types of networks. However, as the underlying topological features of these networks may differ, the methods may perform quite differently on them. We refer the reader to other reviews[9, 10] for alternative viewpoints that additionally consider function prediction methods that integrate physical interaction networks with network data derived from other experimental sources.

10.2   Further Background

10.2.1   Physical Interaction Networks

Large-scale physical interaction networks for several organisms have been obtained via two-hybrid experiments, where an interaction between a pair of proteins is determined via transcriptional activation in yeast.[11] An alternative high-throughput technology determines interactions of proteins via affinity purification of the target protein followed by mass spectrometry identification of the associated proteins (review: [12]). These two types of experiments are the most commonly used approaches for large-scale determination of physical interactions and have uncovered tens of thousands of interactions (see Tables 10.1 and 10.2 for data resources). However, they do impose certain features on the data that may be less than ideal. The yeast two-hybrid method may discover interactions that do not take place under physiological conditions and may miss interactions that do. The pull-down methods do not specify if the interactions inferred for a target protein are direct or are instead mediated through other associated proteins. Moreover, as with all experiments, especially high-throughput ones, a certain amount of noise is present in the results; this amount may differ between different experiments and between subsets of interactions found by the same experiment. To some extent, this noise can be handled computationally by incorporating an assessment of interaction reliability into the computational approach (see Section 10.3). It is worth noting as well that the interactomes determined to date are incomplete, and that comparisons between existing data sets for the same organism reveal only partial overlap; the latter is due both to noise in the data as well as different sets of proteins under consideration.

Table 10.1   *General (multi-organism) interaction databases*

BIND/BOND	http://bond.unleashedinformatics.com/
BioGRID[13]	http://www.thebiogrid.org/
DIP[14]	http://dip.doe-mbi.ucla.edu/
IntAct[15]	http://www.ebi.ac.uk/intact/
STRING[16]	http://string.embl.de

10.2.2   Protein Function

Protein function is a broad concept that has different meanings depending on context. In computational settings, protein function is typically described via terms from one of several controlled vocabularies. Because of the differing degrees of specificity with which protein function can be described, these controlled vocabularies are usually arranged as hierarchies or directed acyclic graphs that relate the different terms to each other. The Gene Ontology (GO)[17] is the most prevalent of such controlled vocabulary systems (see also Table 10.3). GO classifies protein function into three separate categories, each of which consists of a set of terms that may be related to each other via is-a or part-of relations; these relations can be represented as a directed acyclic graph. Protein function in the usual sense is described by two of the categories, molecular function and biological process. The molecular function of a protein describes its biochemical activity, whereas its biological process specifies the role it plays in the cell, or the pathway in which it participates. Additionally, GO has a cellular component category which describes the places where the protein is found. These views of protein function are largely orthogonal: for example, proteins with the same molecular function can play a role in different pathways, and a pathway is built of proteins of various molecular functions. This distinction affects which methods are the most applicable for computational prediction of protein function of each type. Because molecular function corresponds to the intrinsic features of the protein (e.g. its catalytic activity), it is often predicted based on sequence or structural similarity to proteins of known function. Biological processes, on the other hand, are fundamentally collaborative; therefore, it is natural to predict them based on a protein's interaction partners. In this chapter, when we refer to a protein's function, we will typically mean its biological process, though network analysis of interactomes can also be useful for predicting a

Table 10.2   *Network visualization software*

Osprey[18]	http://biodata.mshri.on.ca/osprey/	Software platform for visualization of complex biological networks
Cytoscape[19, 20]	http://www.cytoscape.org/	Open source bioinformatics platform for visualizing molecular interaction networks and integrating them with other state data
Pajek[21]	http://pajek.imfm.si/doku.php	General software for large network analysis
GraphViz[22]	http://www.graphviz.org/	Open source general graph visualization software

Table 10.3   *Functional ontologies and pathways*

The Gene Ontology[17]	http://geneontology.org/
MIPS Functional Catalogue[23]	http://mips.gsf.de/projects/funcat/
KEGG[24]	http://www.genome.jp/kegg/

protein's cellular component; for example, several of the clustering methods reviewed here focus as much on predicting membership within protein complexes (which are described by cellular component annotations) as on predicting biological processes.

10.2.3   Mathematical Formulation

It is natural to represent the collection of protein physical interactions discovered for an organism as an undirected graph or network, where the vertices represent proteins and the edges connect vertices whose corresponding proteins interact. Each vertex is then labeled with zero or more controlled vocabulary terms corresponding to the protein's function(s). The terms used as labels may furthermore participate in a relation described by a system like the Gene Ontology. The function prediction problem then becomes the task of assigning labels to all vertices in a network. This labeled graph representation makes the function prediction problem amenable to the wealth of techniques developed in the graph theory and network analysis communities. For example, the idea of guilt-by-association, which is used by most approaches, turns the problem of function prediction into the problem of identifying (possibly overlapping) regions in the network that participate in the same biological process (i.e. should be assigned the same vertex label). Broadly speaking, most of the methods used for the network-based functional annotation utilize and extend well-understood concepts from graph theory, graphical models and/or clustering.

10.2.4   Notation

More formally, a protein–protein interaction network is represented as a graph $G = (V, E)$, where there is a vertex $v \in V$ for each protein, and an edge $(u, v) \in E$ between two vertices u and v if the corresponding proteins interact. Since we are considering physical interactions between proteins, these edges are undirected. Throughout the chapter, we ignore self-interactions. Let N denote the number of proteins in the network. The network can also be represented by its $N \times N$ adjacency matrix A, where $A_{uv} = 1$ if $(u, v) \in E$ and 0 otherwise. Let $\mathcal{F}$ be the set of possible protein functional annotations. Each protein may be annotated with one or more annotations from $\mathcal{F}$. That is, each vertex $v \in V$ may have a set of labels associated with it. The edges in the network may be weighted; typically the weight $w_{u,v}$ on the edge between u and v reflects how confident we are of the interaction between u and v. If each interaction given in the network is considered equally trustworthy, the network may be considered unweighted or with unit-weighted edges.

Many approaches discussed below utilize the 'neighborhood' of a protein. Let $\mathcal{N}_r(u)$ denote the neighborhood of protein u within radius r; that is, $\mathcal{N}_r(u)$ is the set of proteins where each protein has some path in the network to u that is made up of at most r edges. Then $\mathcal{N}_0(u)$ consists of protein u, $\mathcal{N}_1(u)$ consists of protein u and all proteins that interact with u, $\mathcal{N}_2(u)$ consists of the proteins in $\mathcal{N}_1(u)$ along with all proteins that interact with any of the proteins in $\mathcal{N}_1(u)$, and so on. Note that the number of interactions of a protein u is given by $|\mathcal{N}_1(u)| - 1$, since self-interactions are not considered.

10.3   Incorporating Interaction Reliability

All methods for predicting protein function based on interaction networks face the issue of data quality, as it is well known that high-throughput physical interaction data are noisy, and that different experimental data sets have varying reliability, even if they are based on the same underlying technology.[25-27] A common practice to address the issue of noise is to include edge weights that are chosen to reflect the reliability of interactions. Here, we review a simple scheme for assessing physical interaction reliability,[28] that is essentially the same as the ones used in several approaches for the more general problem of data integration.[29, 30]

For each experimental source i (e.g. each high-throughput experiment may be considered one source, and the collection of all small-scale experiments may be considered as a single different source), let r_i denote the probability that an interaction observed in this experiment is a true physical interaction. Assuming that the observations and sources of error are independent for each experimental source, one can estimate the probability of a physical interaction between proteins u and v as:

$$1 - \Pi_i(1 - r_i), \tag{10.1}$$

where the product is taken over all experiments i which observe an interaction between u and v. This estimate can then be used as the weight $w_{u,v}$ of the edge between u and v. If r_i is chosen to be identical for all experimental sources, this approach simply gives higher reliability to physical interactions that have been observed multiple times. A more meaningful approach is to estimate r_i for each experimental source i by, for example, computing the fraction of interactions coming from that source that connect proteins with a known shared function. It has been shown that a wide range of network analysis algorithms perform better in predicting protein function when utilizing this scheme for assessing interaction reliability than when considering all interactions as equally likely.[28, 31] There are other alternatives for estimating data set reliability. For example, it is common for high-throughput experimental publications to report, along with data, some measure of reliability for each reported interaction; this measure may be as simple as the number of times an interaction has been observed or may be based on more sophisticated schemes.[32] Regardless of the specific method used to assess the reliability of an interaction, the importance of treating different data sources separately has been demonstrated.[33]

For well-studied organisms, the reliability of a physical interaction may also be estimated utilizing data integration schemes that attempt to combine many different types of data (e.g. expression, localization and physical and genetic interaction) in order to functionally link proteins.[30, 34–37] Each link is associated with a weight that represents the probability, or some other confidence measure, that the two corresponding proteins are functionally related. Physical interaction reliabilities may be justifiably estimated using functional linkage scores since a higher functional similarity between two proteins suggests that the observed interaction is more likely to be true. More generally, weighted networks derived via data integration techniques can themselves be used for protein function prediction. Note, however, that though the problems are closely related, predicting functional linkages is not the same as predicting the function of a protein, as a protein can be linked with varying levels of confidence to several proteins with multiple biological process annotations; some method or rule, such as one of those reviewed here, is still necessary to decide which annotations are transferred.

10.4   Algorithms

A wide range of methods have been developed for analysing protein–protein interaction networks in order to predict protein function. In the discussion below, we review some of these and categorize them based upon their underlying algorithmic ideas as well as upon the extent to which they utilize network structure (see also Tables 10.4 and 10.5 and Figure 10.3).

10.4.1   Neighborhood Approaches

The assumption of guilt-by-association naturally gives rise to a prediction method based on majority vote that assigns to each protein the biological process that is most frequent among its direct interactions.[38] In this case, the score for assigning to a protein u a particular annotation a could be the number of proteins that u interacts with that are annotated with a; alternatively, the score may be computed as the fraction of u's interactions that have annotation a. In the case of weighted interaction networks, a weighted sum can be used instead. This majority or neighborhood-counting method is limited in that it only uses local neighborhood information and takes no advantage of more global properties of the network; it also has limited efficiency for poorly annotated proteomes. Subsequent graph-theoretic approaches have attempted to generalize this principle to consider linkages beyond the immediate neighbors in the interaction graph, both to provide a systematic framework for analyzing the entire interactome as well as to make predictions for proteins with no annotated interaction partners.

A simple way to extend the majority approach is to look at all proteins within a neighborhood of specified radius and use the most over-represented functional annotation[39] as the prediction for the protein of interest. That is, for each protein u and a fixed radius r, this neighborhood approach considers all proteins in $\mathcal{N}_r(u)$ and then for each function, computes a score based on the χ^2 test. In particular, the score is computed as $\frac{(f-e)^2}{e}$, where f is the number of proteins within the neighborhood having the function under consideration and e is the number of proteins expected to have that function within the neighborhood, given the frequency of the function in the entire network. The function with the highest χ^2 score is assigned to the protein. With radius one, this approach is similar to the simpler majority approach; note, however, that if two functions annotate the same number of a protein's direct neighbors, the neighborhood approach favors the one that annotates fewer proteins in the entire interactome. While this approach moves beyond direct neighbors, it does not consider the network topology within the local neighborhood. For example, Figure 10.1 shows an interaction network where proteins u and v have the same count for each annotation within radius two; thus the neighborhood approach treats these proteins equivalently when considering a radius of two, despite the fact that the evidence for protein

Table 10.4 *Summary of methods for predicting protein function via network analysis*

Neighborhood approaches	Majority: consider how often a function is seen as annotation of a proteins' immediate interactors[38] Neighborhood: consider neighborhood of radius 1, 2, or 3 and compute over-representation of a function in that neighborhood, as judged by the χ^2 test[39] Weighted neighborhood: consider neighborhood of radius 2, and assign function based on weighted paths from the target protein to neighborhood proteins[31]
Cut-based	Multiway cut: consider all functions simultaneously[40] (NP-hard). Solve approximately via Monte Carlo approach[40] or exactly via ILP[28] Mincut: consider one function at a time.[41] Solve approximately via heuristic[41] or exactly via flow[42]
Flow-based	Assign functions via simulation of 'functional flow' from annotated nodes[28]
Markov network	Use pairwise potential over interacting proteins[43] or assume that the number of neighbors of a protein that have a particular function is binomially distributed according to whether the protein has the function in question or not.[44] One function is modeled at a time
Local graph clustering	Find high-density subgraphs of specified size via Monte Carlo methods[45] Starting from a locally dense node as seed, greedily add vertices according to their local neighborhood density (k-core clustering coefficient),[46] or according to their connectedness to the cluster while maintaining cluster density and vertex cluster property above a cutoff[47] Spectral analysis: build clusters consisting of nodes corresponding to the larger components of eigenvectors for positive eigenvalues of adjacency matrix[48]
Seeded module discovery	Add proteins to cluster that have sufficiently reliable paths to any seed protein[49] Add proteins to cluster that are grouped together with the seed proteins in sufficiently many random networks[50]
Network-based hierarchical clustering	Apply Girvan-Newman (GN) algorithm, building a hierarchical clustering by removing edges with highest edge-betweenness;[51] extend the GN algorithm to weighted graphs and modify to consider non-redundant paths;[52] extend the GN algorithm to additionally consider local measure (edge commonality);[53] perform agglomerative clustering in the reverse order of the GN edge removal[54]
Distance-based hierarchical clustering	Cluster proteins according to the overlap between their common interactors using hypergeometric distribution;[55] or Czekanowski-Dice distance[56] Cluster proteins according to a distance based on their shortest path distance and using randomization to break ties[57] Cluster proteins according to the similarity of their all-pairs shortest-path profiles;[58] combine this global measure with local measure based on direct interactors[59]

(continued overleaf)

Table 10.4   (continued)

Other graph clustering	Starting with a random initial clustering, apply moves to improve the clustering cost, which favors few missing edges within clusters and few present edges between clusters[60] Cluster proteins that belong to a path of adjacent *k*-cliques[61] Stochastic-flow clustering: alternate random-walk steps with steps that amplify the inter-cluster distance.[62] Has been applied to line graph transformation of network[66]
Network alignments	Identify conserved pathways[64] and complexes[65] by network alignment; use models of interactome evolution;[66] permit arbitrary multiple-species network alignments[67]
Supervised learning	Train SVM utilizing an appropriate kernel that captures the distance between two proteins in the network. Linear, diffusion, and locally-constrained diffusion kernels have been applied[68, 69]

u having the annotation depicted by the color black is much stronger than it is for protein *v*. Perhaps because the method completely ignores network topology within neighborhoods, its biological process predictions are best when considering neighborhoods of radius one.[39] Moreover, even the radius-one predictions perform worse than majority vote,[28] suggesting that the decision to penalize more frequent candidate functions may not be optimal; in fact, some of the methods we consider later in the chapter, such as those based on Markov network techniques, use a function's *a priori* frequency in the opposite way. A recent extension of the neighborhood approach attempts to include proteins at radius two while additionally utilizing some information about network topology by assigning weights to each protein in the neighborhood by favoring the number of shared interactors it has with the protein being annotated, and then scoring each function based on its weighted frequency in the neighborhood.[31]

10.4.2   Graph Cuts

One systematic approach to consider the entire network and its annotations in a way that uses information about network connectivity is to utilize the concept of graph cuts. A

Table 10.5   *Available websites and software for discussed methods*

MRF[44]	http://genomics10.bu.edu/netmark/
MCODE[46]	http://cbio.mskcc.org/~bader/software/mcode/
Complexpander[50]	http://llama.med.harvard.edu/cgi/Complexpander/Complexpander.pl
Cfinder[61]	http://www.cfinder.org/
MCL[62]	http://micans.org/mcl/
PRODISTIN[56]	http://crfb.univ-mrs.fr/webdistin/
UVCLUSTER[57]	http://www.uv.es/genomica/UVCLUSTER/
PathBlast[64, 65]	http://www.pathblast.org/
MaWish[66]	http://www.cs.purdue.edu/homes/koyuturk/mawish/
Graemlin[67]	http://graemlin.stanford.edu/

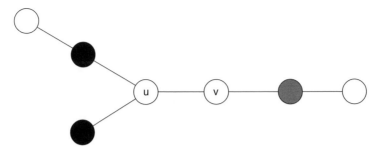

Figure 10.1 *A protein interaction graph annotated with two functions, depicted using black and grey. White nodes correspond to proteins that do not have biological process annotations. When annotating proteins u and v, a neighborhood approach[39] with radius two would make the same prediction, even though the evidence in favor of predicting the function depicted by black is much stronger for protein u than for protein v, and vice versa for the function depicted by grey.*

k-cut is defined as a partition of the vertices of a graph into k sets, and the cost of the cut is sum of the weights of the edges between vertices in different sets. This framework provides a natural application of the guilt-by-association assumption at the full-network scale, as the cut problem can be formulated so as to annotate proteins in a way that minimizes the weighted number of the edges that violate this assumption (i.e. connect proteins having different function). Several cut-based methods for function prediction have been developed;[28, 40, 41] they can either consider functions simultaneously[28, 40] or just one at a time.[41]

If all functions are considered at the same time, the function prediction problem is a generalization of the computationally difficult minimum multiway k-cut problem,[70] where the goal is to partition a graph in such a way that each of the k terminal nodes belongs to a different subset of the partition and such that the weighted number of edges that are 'cut' in the process is minimized. In the more general version of the multiway-cut problem relevant to the protein functional annotation problem, the goal is to assign a function to all unannotated nodes so as to minimize the sum of the weights of the edges joining nodes that have no (assigned or previously known) function in common (i.e. these edges define the cuts). Formally, the problem in the case of function prediction can be stated as minimizing

$$-\sum_{u,v} J_{uv}\delta(\sigma_u, \sigma_v) - \sum_u h_u(\sigma_u), \qquad (10.2)$$

Figure 10.2 *A protein interaction graph with two annotated functions, represented as black and grey nodes. White nodes do not have biological process annotations. There are four ways to annotate proteins so that only one edge is 'cut'. However, the second protein from the left is more likely to have the function depicted by the color black than the second protein from the right. A single cut of the graph does not take into account such distance effects.*

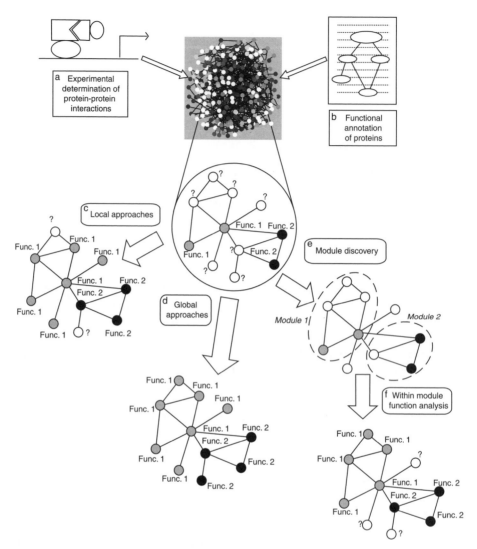

Figure 10.3 *Protein function prediction via network analysis. Experimentally determined interactions (a) are combined into an interaction network, and proteins are labeled with their functional annotations (b), when known. The problem then becomes assigning function to unannotated proteins. A range of methods have been developed which vary in the extent to which they exploit local or global network features. Local methods (c) only annotate proteins that are interacting with other proteins with known annotations; this may be generalized to include larger neighborhoods as well. Global methods (d) utilize the entire network structure to perform function annotation; cut, flow, and Markov random field methods typify this class of approaches. Module-based methods (e) perform functional annotation by first finding modules in the network and then typically finding enriched functions within modules (f).*

where σ_u is functional assignment to node u, $\delta(x, y) = 1$ if $x = y$ and 0 otherwise, J_{uv} is the adjacency matrix for unlabeled vertices, and $h_u(\sigma_u)$ is the number of classified neighbors of vertex u labeled with σ_u.[40] For the weighted version, J_{uv} and h_u can be easily modified to reflect edge weights.

In the case where one function is considered at a time, each protein that is known to have that function is labeled as a 'positive' and each protein that is known to have some function but not the one being considered is labeled as a 'negative'. The optimization problem in that case can be stated as minimizing

$$-\sum_u \sum_{v \neq u} w_{u,v} s_u s_v, \tag{10.3}$$

where $w_{u,v}$ is the weight of edge (u, v), and s_u is 1 if the vertex is labeled with the function being evaluated and -1 otherwise.[41] If the graph is unweighted, $w_{u,v}$ can be set uniformly to 1. It is straightforward to see that this is a basic minimum cut/maximum flow problem, and thus exact solutions are obtainable in polynomial time.[71] Several techniques have been applied to solve these cut problems for interactomes. In the case where one function at a time is considered, a deterministic approximation algorithm has been applied to obtain a single solution per function.[41] In this application, a version is also considered where edges are assigned (positive) weights based on the correlation of the corresponding proteins' expression profiles. In subsequent work, this formulation has been solved exactly using a minimum cut algorithm.[42] In the case where multiple functions are considered at once, simulated annealing has been applied and solutions from several runs have been aggregated.[40] That is, the score of a function for a particular protein is given by the number of runs in which the simulated annealing solution annotates the protein with the function. The simulated annealing approach is a heuristic and thus does not guarantee an optimal solution to the underlying optimization problem. However, an integer linear programming (ILP) formulation for the generalized multiway-cut problem has also been proposed.[28] While ILP is computationally difficult from a theoretical point of view, in practice optimal solutions to this ILP, and thus the original optimization problem, can be readily obtained for existing physical interactomes using AMPL[72] and the commercial solver CPLEX.[73]

An important shortcoming of the basic cut formulation is that it ignores distance in the network. For example, the network in Figure 10.2 has four minimum cuts of value one, and the cut criterion does not favor any one cut over the other. However, we expect proteins that are closer together in the network to have more similar biological process annotations than those that are further apart. Thus, in the network in Figure 10.2, we would want the proteins closer to the black node to be annotated with its function, and the proteins closer to the grey node to be annotated with its function. As suggested by [28], one may begin to address this problem in the cut-based framework by considering the multiplicity of optimal solutions. If we find all optimal cuts for the graph in Figure 10.2, we observe that proteins closer to the black node are found more frequently in the same cut as the black node than in the same cut as the grey node. Thus, the set of all optimal solutions contains a sense of distance to annotated nodes. In the earlier simulated annealing approach proposed for this problem, information from multiple solutions is utilized.[40] If each run does indeed converge to an optimal solution, considering multiple runs amounts to sampling from the space of optimal solutions. The ILP can also be modified to find multiple solutions.[28] The score for a function for a protein is then the number of obtained solutions in which this function is assigned to the protein.

10.4.3   Flow-based Methods

One attempt to overcome the cut-based methods' ignorance of distances in the network has been proposed based on another concept from computer science, namely, network flow.[28] Intuitively, network flow problems treat the graph as a collection of pipes having limited capacity (represented as weights), and pose the question of the maximum amount of liquid that can be sent from a specified source node to a specified sink node using those pipes. The network flow problem is dual to the notion of graph cut,[71] as the size of the minimum cut between the source and the sink turns out to be the limiting factor to maximum flow, and vice versa.

Network flow has been used as the inspiration for a simulation method for function prediction.[28] Informally, each protein of known functional annotation is an infinite 'source' of 'functional flow' that can be propagated to unannotated nodes, using the edges in the interaction graph as a conduit. Each protein has a 'reservoir' which represents the amount of flow that the node can pass on to its neighbors at the next iteration, and each edge has a capacity (its weight) limiting the amount of flow that can pass through the edge in one iteration. Each iteration of the algorithm updates the reservoirs using simple local rules, whereby flow only spreads from proteins with more filled reservoirs to those with less filled reservoirs, and a node pushes its flow to its neighbors proportionally to the capacities of the respective edges. The simulation is run for a fixed number of steps, and a functional score for each protein is obtained by summing the total amount of flow for that function that the protein has received over the course of the simulation. This method exploits network connectivity as multiple disjoint paths between functional sources and a protein result in more flow to the protein. It also incorporates a notion of distance in the network as the effect of each annotated protein on any other protein decreases with increasing distance between them: if the algorithm is run for d iterations, then a source's immediate neighbor in the graph receives d iterations worth of flow from the source, while a node that is two links away from the source receives $d - 1$ iterations worth of flow, and so on. Similarly, the number of iterations for which the algorithm is run determines the maximum number of interactions that can separate a recipient node from a source in order for the flow to propagate from the source to the recipient. For the protein interaction context, a relatively small number of iterations has worked well in practice (e.g. less than half the diameter of the network). The reader is referred to [28] for the exact formulation of the functional flow algorithm.

In subsequent work, a similar deterministic flow-based simulation approach has also been applied for finding clusters in protein interaction networks.[74]

10.4.4   Markov Network-based Methods

Cut-based methods for functional annotation have a more general probabilistic counterpart in methods based on Markov networks,[43, 44, 75] and these formulations can more fully address some of the weaknesses of the cut-based methods. A Markov network, also known as a Markov random field, is an undirected graphical model that represents the joint probability distribution of a set of random variables. It is specified by an undirected graph where each vertex represents a random variable and each edge represents a dependency between two random variables, such that the state of any random variable is independent of all others given the states of its neighbors. The joint distribution represented by a Markov random field is computed by considering a potential function over each of its cliques. For N random variables X_i, the probability of an assignment of the states is given by:

$$\Pr(X_1 = x_1, \ldots, X_N = x_N) = \frac{1}{Z} e^{-\sum_k \Phi_k(X_{\{k\}})},\tag{10.4}$$

where k enumerates all cliques, Φ_k is the potential function associated with the k-th clique, $X_{\{k\}}$ gives the states of the k-th clique's random variables, and Z is a normalizing constant.

In applications to network-based function annotation, one function has been considered at a time.[43, 75] Each protein has a random variable associated with it, and its state corresponds to whether the function under consideration is assigned to the protein or not. It is assumed that the joint distribution can be expressed in terms only of cliques of size at most two (i.e. edges). This means that the potential function evaluating the network is a linear expression composed of terms over the vertices and edges. So,

$$\Pr(X_1 = x_1, \ldots, X_N) = \frac{1}{Z} e^{-\left(\sum_{u \in V} \phi_1(X_{\{u\}}) + \sum_{(u,v) \in E} \phi_2(X_{\{u,v\}})\right)},\tag{10.5}$$

where φ_1 computes the vertex 'self-term' and the φ_2 computes the pairwise edge term. The self-term potential is chosen to correspond to the prior probability for annotating a protein with a particular function; it takes into account the frequency of the function in the network. Note that this is the opposite of what is done by the neighborhood method,[39] which prefers less frequent terms to those that are more frequent. The pairwise edge potential is chosen to have different values corresponding to the three cases where either the interacting proteins both have the function under consideration, or they both do not have that function, or one has that function and the other does not; these values are determined using a quasi-likelihood method. Note that these values are not necessarily the same for each function. As noted earlier,[76] this model is a generalization of the per-function cut-based method,[41] and is similar to that of the multiple function cut formulation.[40] In particular, the cut-based models assume the same fixed value for interactions between proteins of the same function (or for interactions between a protein of one function and any other), regardless of function; this may not be the best assumption, as the guilt-by-association assumption may be true to different degrees for different functions. To make a functional prediction for a protein, the posterior probability that a protein has the function of interest is computed using Gibbs sampling, and then if this value is above a chosen threshold, the function is predicted. Importantly, an exact computation of the posterior probability considers the probability of all assignments of the random variables, and thus implicitly incorporates a distance effect, where the impact of a protein's function on unannotated proteins decreases with distance.

An alternate Markov network approach for protein function annotation[44] assumes that the number of neighbors of a protein that have a particular functional annotation is binomially distributed according to a parameter that differs depending on whether the protein has that function or not. The posterior probabilities for each protein are computed via a heuristic modification of belief propagation (review: [77]).

10.4.5 Module Discovery

Another broad family of methods begins by first identifying components in the interaction network that are likely to correspond to functional units, and then assigning functions to proteins based on their membership in the functional unit. The underlying philosophy for most of these methods is that cellular networks are organized in a modular fashion,[78] and that these modules correspond to sets of proteins that take part in the same cellular process or together comprise a protein complex. Identification of functional modules is thus a somewhat stronger goal than simple functional assignment. Most of the methods for identifying modules operate on the underlying assumption that proteins within modules are more tightly connected than proteins in different modules; one may think of this as the module-discovery problem's analog of the guilt-by-association assumption.

Once functional modules, or clusters, are identified, they can be used for annotating uncharacterized proteins, as the most common functional annotation within a cluster can be transferred to its uncharacterized proteins. Alternatively, one can look at overrepresentation instead of frequency and transfer the functions that are enriched in a cluster according to the hypergeometric distribution. Such an approach computes a *p*-value for a particular function in a cluster as:

$$p = 1 - \sum_{i=0}^{i=f-1} \frac{\binom{F}{i}\binom{N-F}{n-i}}{\binom{N}{n}}, \tag{10.6}$$

where N is the number of proteins in the network, F is the number of proteins in the network annotated with the function under consideration, n is the size of the cluster, and f is the number of proteins within the cluster annotated with that function. Like the neighborhood overrepresentation method,[39] if two functions annotate the same number of proteins within a cluster, this method favors the function that annotates fewer proteins in the interactome. We also note that one feature of cluster-based function prediction methods is that it is possible and indeed not uncommon for certain modules not to contain any annotated proteins, in which case functional assignment to such a cluster cannot be made in a straightforward fashion.

Cluster analysis is a rich area with applications in many diverse fields. A large number of clustering methods have been developed, both for the more familiar problem of clustering general data that comes with some natural measure of similarity, and, to a lesser extent, for the more specific problem of graph clustering. Many of these methods have been applied to interactome data. Broadly speaking, the clustering methods we consider are either specific to the network domain, or are based on standard distance- or similarity-based clustering techniques; in the latter case, the key issue is typically in deciding on a suitable measure of distance or similarity between two proteins in an interaction network. Additionally, the methods differ in the extent to which the network features they exploit are local. In this regard, we note that some methods use only local neighborhood information when clustering whereas others use more global features of the network; nevertheless, even when using local features to cluster proteins, clustering can be performed on the entire interactome, and thus in some sense, such clustering approaches incorporate the global organization of the interactome as well.

10.4.5.1   *Network-based Clustering*

Of the clustering approaches, those based on network clustering are perhaps the closest in spirit to the cut-or flow-based annotation schemes: they explicitly attempt to partition the network into contiguous components in such a way that there are more connections between proteins within a component than between proteins belonging to different components. However, unlike the former group of methods, cluster-based approaches typically do not begin with the prior information about the partial assignment of function to neighbors; moreover, several graph clustering-based methods focus on the more specific problem of identifying protein complexes.

Local clustering. A number of local clustering approaches attempt to isolate highly connected or dense components within the larger protein interaction network. The density of a set of vertices may be defined in many ways. The density of a set of vertices V' is sometimes computed as the total number of edges among the vertices in V' divided by the total number of possible edges within V' (i.e. $\binom{|V'|}{2}$). Finding the densest subgraph of a particular size is a computationally hard problem, and thus a number of heuristic approaches have been developed. In one approach, a Monte Carlo procedure is developed that attempts to find a set of k nodes with maximum density.[45] A special case of the density measure that has also been exploited to uncover dense components is the clustering coefficient. It is computed for a vertex v as the density of the neighbors of v (i.e. $\mathcal{N}_1(v)$ with v excluded). Each vertex can be weighted using a measure similar to its clustering coefficient, but that instead tries to exclude the effects of low-degree vertices.[46] Low degree vertices are frequent in protein interaction networks, and may artificially lower the clustering coefficients of highly connected vertices in dense regions of the network that are also connected to several vertices of low degree. The clustering coefficient is thus computed instead over a k-core of the neighbors of each vertex, where k-cores are maximal subgraphs of degree $\geq k$. The vertex with the highest weight seeds the search process, and clusters are greedily expanded from it, with vertices being included in the cluster if their weights are above a given threshold. Once no more vertices can be added, this process is repeated for the next highest weighted unseen vertex in the network.

A greedy graph clustering approach is also taken where a cluster is grown so as to maintain the density of the cluster above a particular threshold, and to ensure that each vertex that is added to the cluster is connected to a large enough number of vertices already in the cluster.[47] The process is initialized by finding the vertex that takes part in the largest number of triangles (i.e. has the largest number of common neighbors with its neighbors).

Dense substructures within protein networks have also been uncovered via spectral analysis.[48] Here, eigenvalues and eigenvectors of the adjacency matrix of the network are computed. For each positive eigenvalue, its corresponding eigenvector is used to group together proteins. In particular, the proteins corresponding to the larger components of the eigenvector tend to form dense subgraphs. Groupings are further filtered to be of sufficient size and to have large enough interconnectivity.

Seeded module discovery. Rather than finding clusters in protein–protein physical interaction networks without any functional annotations, a few approaches start with a set of proteins in the interaction network and attempt to identify modules around these 'seed' proteins.[49, 50] In the context of protein function prediction, the seeds are proteins that are known to share some biological process or take part in the same complex. In one approach,[49] each interaction is labeled with confidence or reliability value in the range of 0 and 1, and a protein is added to the cluster if there exists a path from any seed protein to it such that the product of the reliabilities of the edges in the path is greater than a preselected threshold; for each protein, this corresponds to computing its shortest path to any seed protein when mapping each edge reliability to its negative logarithm. This approach thus scores the membership of a protein to the initial seed set using the probability of its connection via the single-most probable path. In another approach,[50] random networks are used to compute the probability that protein u is a member of the same group as the seed set of proteins. This probability is estimated as the fraction of random networks in which a path exists from u to any protein in the seed set. Each random network is generated by taking every edge in the original network and adding it into the network with probability proportional to its reliability in the original network. This approach thus attempts to compute the probability of a connection to the initial seed set using any path in the network.

Divisive hierarchical network clustering. Girvan and Newman have proposed a divisive hierarchical clustering procedure that is based on edge betweenness.[79] For any edge, its betweenness is defined as the number of shortest paths between all pairs of vertices that run through that edge. This technique, thus, uses global information about the protein network. Edges between modules are expected to have more shortest paths through them than those within modules, and therefore should have higher betweenness values. The overall hierarchical procedure partitions the network by successively deleting edges with highest betweenness values. It has been applied to yeast and human interaction data.[51] The Girvan-Newman algorithm has also been modified so that shortest paths are computed on weighted networks. In one approach, instead of counting the total number of shortest paths through an edge, the total number of 'non-redundant' shortest paths through an edge are counted by considering paths that do not share an endpoint.[52] Edge weights are also considered by this method; in this case, weights correspond to dissimilarities between endpoints, rather than similarities or edge reliabilities.

The Girvan-Newman algorithm has also been modified so that the edge with lowest edge clustering coefficient is iteratively deleted.[80] The edge clustering coefficient is a generalization of the usual clustering coefficient, and measures the number of triangles to which a given edge belongs, normalized by the number of triangles that might potentially include it. To deal with the special case where the edge is found in no triangles, the edge clustering coefficient for edge (u, v) is actually defined as:

$$ECC(u, v) = \frac{z_{u,v} + 1}{\min\{|\mathcal{N}_1(u)| - 2, |\mathcal{N}_1(v)| - 2\}}, \tag{10.7}$$

where $z_{u,v}$ gives the number of triangles that edge (u, v) participates in. Unlike the edge betweenness measure, the edge clustering coefficient is a local measure; however, in principle, this definition can be extended to handle higher-order cycles as well. The edge clustering coefficient has been used to uncover modules in yeast.[53] A related algorithm that combines both the global edge betweenness measure with a local measure similar to the edge clustering coefficient has also been proposed.[53] This algorithm computes a local measure called the commonality index for each edge as

$$C(u, v) = \frac{z_{u,v} + 1}{\sqrt{|\mathcal{N}_1(u) - 1| \cdot |\mathcal{N}_1(v) - 1|}}. \tag{10.8}$$

The edge evaluation measure is then based on the observation that an edge connecting different modules should have a low commonality and high edge betweenness. Therefore, the algorithm removes edges (u, v) in the order of decreasing $B(u, v)/C(u, v)$ ratio, where $B(u, v)$ is the Girvan-Newman betweenness, and $C(u, v)$ is the commonality index.

Divisive methods do not necessarily specify how to get modules or clusters from the hierarchical grouping process. One proposed approach is to consider a set of vertices $V' \subset V$ as a module if, for each of its vertices, the number of interactions it has within V' (its indegree) is greater than the number of interactions it has with vertices in $V - V'$ (its outdegree).[80] This condition can be weakened so that a module only requires that the sum of the indegrees for the vertices in the module be greater than the sum of their outdegrees. The partitioning of the network can now be performed so that an edge with highest edge betweenness or lowest edge clustering coefficient is only removed if it results in two modules.[80] A modified definition considers a set V' a module if the ratio of the number of edges within V' to the number of edges from vertices in V' to vertices outside of this set is greater than one;[54] this is almost the same criterion as that for a weak module,[80] except that edges within V' are not counted twice. This definition has been used to uncover modules in an agglomerative procedure, where singleton vertices are considered initially and edges are added back in, using the reverse Girvan-Newman ordering, only if the edge is not between two modules. Modules have also been defined in terms of the structure of the hierarchical cluster subtrees.[53] Here, a module consists of the nodes of a maximal subtree where all non-leaf nodes have at least one child being a leaf, and two modules that have the same parent are merged when the maximal commonality of edges between them is larger than a pre-defined cut off.

Other network clustering approaches. In one approach,[60] an initial random partitioning of the network is modified by iteratively moving one protein from one cluster to another in order to improve the clustering's cost. For each protein, the cost measure considers the number of proteins within its assigned cluster with which it does not interact, as well as the number of interactions from it to proteins not assigned to its cluster; both should be small in ideal clusterings. In order to avoid local minima, the local search is modified so as to occasionally disperse the contents of a cluster at random. Additionally, a list of forbidden moves is kept to prevent cycling back to a previous partition. Resulting clusters are then filtered for size, density, and functional homogeneity.

Another approach for clustering is based on uncovering so-called k-clique percolation clusters.[61] A k-clique is a complete subgraph over k nodes, and two k-cliques are considered adjacent if they share exactly $k - 1$ nodes. A k-clique percolation cluster consists of nodes that can be reached via chains of adjacent k-cliques from each other. An advantage of such an approach is that each protein can belong to several clusters. Since a protein can have different roles in the cell, membership in several clusters is biologically meaningful, and it may be useful to identify a strategy that can recover multiple functions.

A clustering approach based on (modified) random walks within a network has also been developed.[62, 81] The interaction network is transformed into a Markov process, where transition probabilities from u to v and v to u are associated with each edge (u, v); that is, the adjacency matrix is converted to a stochastic matrix. The stochastic-flow algorithm alternates between an expansion step, which causes flow to dissipate within clusters, and an inflation step, which eliminates flow between different clusters. In the expansion step, the probability transition matrix is squared; this corresponds to taking one more step in a random walk. In the inflation step, each entry in the stochastic matrix is raised to the r-th power and then normalized to ensure that the resulting matrix is stochastic again; for $r \geq 1$, the inflation step tends to favor higher probability transitions, and thus tends to boost the probabilities of intra-cluster walks and demote those of inter-cluster walks. This process is repeated until convergence, at which point the connected directed components are evident. Note that in this algorithm, the inflation step distinguishes it from simply taking (traditional) random walks on a graph. This stochastic flow-based clustering procedure has been applied to a protein interaction network that has been transformed into a line graph.[63] Here, each vertex in the new graph represents an interaction in the original network, and any two vertices are adjacent if the corresponding interactions in original network involve a common protein. Note that the line graph formulation allows the stochastic flow-based clustering to place each protein into several clusters.

10.4.5.2   General Distance-based Clustering

Rather than use the guilt-by-association assumption directly and explicitly attempt to keep connected nodes in the same cluster, many approaches to clustering interactomes rely instead on assumptions about the similarity of cluster co-members' patterns of connections to other vertices in the graph. This makes it possible to use standard distance-based clustering techniques, such as hierarchical clustering, on the resulting similarity or distance matrix. Various similarity measures have been proposed for clustering interaction networks. In one approach,[55] the similarity between two proteins is determined by considering each protein's interactions, and computing the significance of their number of shared interactions via the hypergeometric distribution. An alternate approach that also measures the overlap between the sets of interactions for each pair of proteins uses the Czekanowski-Dice distance.[56] For proteins u and v, this is given by:

$$CD(u, v) = \frac{|\mathcal{N}_1(u) \Delta \mathcal{N}_1(v)|}{|\mathcal{N}_1(u) \cup \mathcal{N}_1(v)| + |\mathcal{N}_1(u) \cap \mathcal{N}_1(v)|}, \quad (10.9)$$

where Δ computes the symmetric difference between two sets. In addition to these two measures,[55, 56] there are a number of other ways of computing the similarity or distance between two proteins by considering only the overlap among their direct interactions.[82, 83] In contrast to these purely local measures, a more global measure can be used where the distance between two proteins is calculated as the shortest path distance between them in the network.[57] In a related earlier approach,[58] each protein is associated with a vector that contains its shortest path distance to all other proteins in the network. A similarity between two proteins is computed as the correlation coefficient between their corresponding shortest-path vectors. Since global and local similarity measures may be quite different, this global shortest-path based similarity measure has also been used in conjunction with a local connectivity coefficient based on the common interactors of two proteins.[59] For any of these measures, agglomerative hierarchical clustering is then performed by progressively merging groups of proteins that are closest or most similar to each other. Neighbor-joining[84] has also been used in the context of clustering interactomes;[56] it favors merging items that are close to each other but also considers distances from the remaining items. As discussed earlier, hierarchical clustering methods do not automatically give the final partitioning of the network. In one approach,[56] the separation into clusters is performed using existing biological process annotations, whereby each cluster must have at least half of its proteins annotated by the same term. This function is then transferred to the other proteins in the cluster. In some applications of distance-based hierarchical clustering, there can be a problem where distances among several items are identical. This is the case, for example, when setting the distance between two proteins as their shortest path distance in the network. One possible solution to this problem is a two phase approach.[57] In the first phase, hierarchical clustering is performed multiple times, and each time there is a 'tie in proximity', a random pair is chosen for merging. Each clustering run is stopped according to a threshold that considers the distances between all proteins in a cluster. In the second phase, the fraction of solutions in which each protein pair is clustered together is then used as a similarity measure for a final round of clustering.

10.4.5.3   Network Alignment Approaches

Network alignments are a relatively new approach to uncover modules within biological networks.[64] As implied from their name, network alignments are a natural counterpart to sequence alignments: they align subgraphs of interaction networks which consist of homologous proteins having a similar pattern of interactions among them. The resulting subnetworks highlight conserved functional modules in the network(s), and thus enriched biological process annotations may be transferred to proteins within the modules. Network alignments can be performed between different interactomes, or with one interactome aligned against itself (to search, e.g. for duplicated pathways), or with a query subnetwork (e.g. proteins making up a known pathway) that is aligned against an interaction network to search for similar pathways.[64] Network alignment is a research area worthy of its own review;[85] here we briefly mention a subset of early methods. The first method for network alignment aligned up to two interactomes;[64] this was later extended to align multiple interactomes.[65] Evolutionary models of interactome change have been incorporated,[66] and efforts to scale the methods to align arbitrary numbers of interactomes have recentlybeen made.[67] Approaches to globally align interactomes have also developed.[86]

10.4.6   Kernel-based Learning Methods

Discriminative learning methods are another broad area in computer science that has been applied to the problem of predicting protein function using interaction networks. The methods discussed here use support vector machines (SVMs), machine learning methods which embed positive and negative examples in a feature space and then find a maximal separating hyperplane in this space between the positive and negative examples.[87, 88] In the context of function prediction via network analysis, SVMs have been applied by considering each function in turn and labeling each protein as either positive or negative based upon whether it is annotated with the function of interest.[68, 69] The key technical difficulty is how each protein u in the network is mapped to a point x_u in the feature space. If proteins are 'close' in the network, then they should also be close in the feature space. The mapping to the feature space can be given implicitly via a positive definite kernel matrix K specifying the inner product (i.e. $K_{uv} = x_u^T x_v$); since the discriminant function for SVMs is specified via inner products, explicit representations of the points are not necessary.

In the first paper describing kernel-based methods,[68] two kernels are considered. First, a linear kernel is created where each entry K_{uv} is the dot product of the N-dimensional vectors representing the interactions of proteins u and v. The more similar the interaction patterns for the proteins, the larger this value is in the kernel matrix; this kernel is similar in spirit to local clustering methods based on the similarity of immediate interactors. It does not capture more global properties of the network. Second, a diffusion kernel[89] is created where the kernel value K_{uv} can be interpreted as the probability that a random walk starting from u will be at v after infinite time steps; the transition probabilities between nodes are dependent on a parameter specifying the rate of diffusion. The diffusion kernel accounts for all possible paths connecting two proteins, and nodes that are connected with shorter paths or by several paths are considered more similar. Thus this kernel utilizes some of the same network features as the flow-based function prediction method and the stochastic-flow clustering approach. It has been shown that the diffusion kernel captures the global constraint that the sum of the Euclidean distances between connected samples is bounded, but that this can lead to large variances in the pairwise distances.[69] This observation has led to the development of a locally constrained diffusion kernel, which captures additional local constraints requiring that the Euclidean distance between connected samples be more tightly bound. SVMs using the locally constrained diffusion kernel are found to better predict protein function than those using the original diffusion kernel.

10.5   Assessment of Prediction Quality

It is natural to ask how different network-based methods for the function prediction problem perform in comparison to each other. Unfortunately, a comprehensive comparative evaluation of these methods has not been done. Therefore, we briefly outline a couple of evaluation frameworks that have been proposed and showcase the performance of some of the reviewed methods in these frameworks. Overall, it is difficult to judge the comparative performance of different methods by surveying the literature. This is due in part to differences in the evaluation frameworks; such differences include the measures used to assess performance quality, the treatment of multiple annotations and predictions, the selection of a gold standard for functional annotation, the treatment of the functional hierarchy, and the precise (and always changing) interaction networks under consideration.

Some common features of evaluation frameworks are that most of the existing testing has been performed in the baker's yeast *Saccharomyces cerevisiae*, because of the quality and quantity of data available for that organism, and that all frameworks use cross-validation testing. In this type of testing, the annotations of one (or more) protein are considered as unknown, and the annotations of the remaining proteins, along with the network, are used to predict its annotations.

10.5.1   Evaluation frameworks

One way to treat the issue of multiple predictions and multiple annotations is by using each prediction in the calculation of performance measurements. This is the approach taken by,[43] using annotations from YPD functional categories[90] and considering all predictions with score above a cutoff. In this work, for each annotated protein u with at least one annotated interaction partner, it is assumed to be unannotated and its function is predicted. Then, performance measurements are computed in terms of: k_u, the number of known functions for protein u; p_u, the number of predicted functions for protein u; and o_u, the amount of overlap between the set of known and predicted functions. The precision (or positive predictive value) is defined as:

$$\text{Precision} = \frac{\sum_u o_u}{\sum_u p_u}. \tag{10.10}$$

The recall (or sensitivity) is defined as:

$$\text{Recall} = \frac{\sum_u o_u}{\sum_u k_u}. \tag{10.11}$$

In follow up work,[76] 134 GO biological process terms are chosen for consideration if they annotate more than 50 proteins and if none of their child biological process terms annotates the same set of proteins. Since GO is a directed acyclic graph and functional terms can be related to each other via is-a or part-of relationships, the authors suggest modifications to this basic scheme to accommodate this hierarchical structure. A possible weakness in this per-prediction framework is that proteins that have more annotations will have a larger effect on performance measurements.

An alternative approach[28] is to treat each *protein* as a data point when measuring performance. In particular, for each protein, if the top scoring functional annotation is above some threshold, it is the prediction for the protein. If a prediction is a known functional annotation, it is considered a true positive, and otherwise, it is a false positive. Measuring performance per-protein avoids the problem of proteins with many multiple annotations or predictions from dominating the results, and makes the performance measures easily interpretable in terms of the number of proteins that can be annotated at a certain confidence. This criterion still permits ties between top-scoring predictions; in this case, a protein's predicted annotation is counted as a true positive if more than half of its top-scoring predictions are correct, and a false positive otherwise. This approach is taken as a compromise between two extreme cases. In the first case, a prediction for a protein can be counted as a true positive if at least one of the predictions made for it is correct; however, in this case, a method that predicts every protein to participate in every function would only have true positives in this framework. At the other extreme, a protein can be counted as a true positive if every prediction made for it is correct. This, however, would count as false positives those proteins that get many correct predictions and only one incorrect one. An alternative and perhaps better approach would be to compute the precision and recall per-protein, and then average the results over proteins. Here, a flat set of functional terms coming from the MIPS[23] functional hierarchy was used for evaluation, with 72 biological process terms chosen from the second level of hierarchy.

A number of clustering approaches have been evaluated,[9] based on how well they recapitulate known yeast protein complexes. While this is not the same as assessing the performance of function prediction, there is likely to be a relationship between the two; moreover, this study is likely to be useful in designing a similar evaluation of clustering approaches in the context of function prediction. The clustering algorithms are run both on simulated networks where complexes are embedded into the graph, and edges are added and removed at various proportions, as well as on data sets obtained in high-throughput experiments. Performance is measured by computing recall values (i.e. for each complex, find the cluster which has the highest fraction of its proteins) and precision values (i.e. for each cluster, find the maximal fraction of its proteins found in the same annotated complex). In theory, at least, it is also possible to use either of the above approaches[28, 43] to evaluate how well the enriched biological processes in each cluster predict protein function. We expect the evaluation of clustering for prediction of complexes to give different results than clustering for function prediction, as, on the one hand, complex prediction may be a more specific problem than function prediction, but, on the other hand, the dense network components that are readily identified by clustering methods may be 'easy cases' for function prediction, while more ambiguous proteins in sparse regions may be left out of the clusters identified by some of the methods.

10.5.2   Comparative Performance

Nabieva *et al.*[28] test the majority, neighborhood, multiway-cut and flow formulations in two-fold cross-validation on the yeast proteome using Receiver Operating Characteristic (ROC) analysis. They find that the flow-based method generally outperforms other methods. They also find, perhaps surprisingly, that the next best method is majority, which outperforms neighborhood and multiway-cut formulations and performs as well as the flow-based method for proteins with at least three neighbors annotated with the same function.

The multiway-cut formulation was previously found to outperform the majority method.[40] However, the measure of success used to judge performance there was the fraction of times the top prediction for each protein is correct, and the score of the top prediction was not considered. ROC analysis, with a varying threshold gives a more complete picture of performance, particularly with respect to high-confidence predictions, and shows that majority outperforms the cut-based method over a large false positive range, but the cut method is able to make predictions when majority cannot.[28] A subsequent paper[42] also finds that a cut-based approach does not outperform a strictly local approach which predicts function based on the fraction (instead of number) of neighbors with a particular function. In their case, the cut-based approach considered is the pairwise min-cut problem.[41]

Deng *et al.* find in leave-one-out testing that the Markov network approach [43] outperforms the majority[38] and neighborhood approaches[39] on the yeast interactome. The significant added generality of the Markov network approach over the cut-based approach and its implicit use of distance in the network may potentially explain why it performs better than majority whereas the related cut-based methods do not; however, a weakness with the testing as performed[43] is that the functional prediction for a protein is scored according to its rank when using the majority and neighborhood methods. This means that the strength of the evidence for a functional prediction from the protein's neighborhood is not considered; for example, for the majority method, it does not matter in this testing framework if the top-scoring function for a protein appears nine times or one time among its direct interactions – both are treated equivalently. It remains to be seen whether the

Markov network approach will outperform the local method when scores – not ranks – are considered.

Other findings revealed in cross-validation testing include the necessity of multiple solutions for the cut-based method in order to get higher confidence predictions, and a deteriorating performance of the neighborhood method with increasing radius, reflecting the peril of using more distant nodes without considering their distance or connectivity to the target node.[28] It is also observed that all methods, including majority, multiway-cut, and functional flow improve when incorporating interaction reliability.[28, 31]

These evaluations show that the strength of the functional signal from the local neighborhood is the best indicator of whether or not a high-confidence prediction can be made: if a protein is interacting with many proteins with known annotation, a majority scheme performs well, as do other methods. Also, the results suggest that the information from immediate neighbors can be used directly, and statistical information, such as that used in the χ^2 criterion, is not necessarily helpful. On the other hand, when a protein is known to interact with only unannotated proteins, local approaches such as majority cannot make any predictions, whereas the cut, flow, Markov network, and clustering methods can. More broadly, for proteins with few interactions or few interactions with annotated proteins, which is likely to be the case for more recently characterized proteomes, more global methods are necessary for functional predictions. Thus, global methods are likely to be an important tool in characterizing proteins in unusual or less-studied proteomes.

As mentioned earlier, clustering methods have largely not been evaluated with respect to function prediction. However, a recent study[9] finds that the stochastic flow-based clustering procedure[62] is robust to alterations in the simulated data and clearly outperforms the other methods tested[46, 60, 92] in extracting complexes from high-throughput physical interaction datasets.

10.6   Conclusions

The emergence of high-throughput techniques for determining protein physical interactions at the genomic scale has provided large amounts of data that can be used for answering the challenge of predicting protein function. Here, we have reviewed a number of methods that have been developed for this problem. There are several promising directions for further research in this area.

First of all, it is clear that the area of function prediction via network analysis is in need of a comprehensive and systematic evaluation framework. Such an evaluation will ideally attempt to answer not only which methods perform better but also why. We expect different methods to perform well in different circumstances, and ideally an evaluation would bring to light which method should be used in which situation. In particular, it should be possible to relate topological features and annotation density of the network to performance. For example, local methods may be expected to perform well on dense or well-annotated networks. Since the experimentally determined interactomes of various organisms in their present state differ with respect to their coverage, network density, and known annotations, such an evaluation will be vital for guiding researchers towards an appropriate prediction method for their particular needs.

In terms of methodological directions, a potentially fruitful area that is in need of principled exploration is a closer study of protein annotations, and in particular, of the relationships between functions. One promising line of research is to accommodate functional relationships between interacting proteins that go beyond guilt-by-association, which forms the basis of most methods currently used for network-based function prediction. Simply stated, if guilt-by-association were completely true, all proteins in an organism would be engaged in the same non-trivial biological process. This is clearly not the case; moreover, biological 'cross-talk' is evident in interactomes,[38] as there are many pairs of different biological processes recurring as annotations for interacting proteins. Understanding and leveraging the interplay between biological processes should benefit future methods for predicting protein function. A related research direction involves developing network analysis function prediction methods that incorporate the hierarchical nature of protein function ontologies. Currently, many methods address the issue only at the evaluation step, and often use a flat set of terms which are then treated as unrelated labels; the flat set may include the leaf terms of the functional hierarchy (i.e. the most specific descriptors), or a hand-picked or heuristically selected set of terms. In the latter case, more specific functional annotations may be 'upcast' to their ancestor term(s) in the flat set, and less specific annotations are ignored. The development of methods that more directly exploit the functional hierarchy as part of their network analysis is likely to be fruitful. Promising research along both of these lines has been initiated.[93, 94] Lastly, an intriguing possibility is to relate modularity to the distance in the network along which functional connections hold. The assumption of modularity suggests that guilt-by-association may hold on mezoscale, along the size of a functional module, but at larger network distances understanding of the hierarchical and cross-talk relationships between processes may become more relevant for function prediction.

The area of function prediction via network analysis is based on recently available data and is thus relatively new, yet its graph-theoretic formulation enables it to tap into decades of algorithmic and methodological advances in computer science and applied mathematics. In the coming years, we expect to see further methodological developments in this area, as well as the establishment of more uniform testing frameworks. Together with the growth of interaction data and the improvement of the accuracy of experimental techniques for interaction determination, these developments promise to give network analysis methods a position of increasing prominence in computational function prediction.

Acknowledgments

This work has been supported by NSF CCF-0542187, NSF IIS-0612231, NIH GM076275, and the NIH Center of Excellence grant P50 GM071508.

References

1. T. Murali, C.-J. Wu, and S. Kasif. The art of gene function prediction. *Nat. Biotechnol.*, **24**: 1474–1475, 2006.
2. V. Batagelj and A. Mrvar. Analysis and visualization of large networks. In M. Junger and P. Mutzel, editors, *Graph drawing software*, pp. 77–103. Springer, 2003.

3. M. Kanehisa, S. Goto, S. Kawashima, Y. Okuno, and M. Hattori. The KEGG resource for deciphering the genome. *Nucleic Acids Res.*, **32**: D277–D280, 2004.

4. S. Letovsky and S. Kasif. Predicting protein function from protein/protein interaction data: a probabilistic approach. *Bioinformatics*, **19** Suppl. 1: i197–i204, 2003.

5. S. Carroll and V. Pavlovic. Protein classification using probabilistic chain graphs and the Gene Ontology structure. *Bioinformatics*, **22**: 1871–1878, 2006.

6. S. Kerrien, Y. Alam-Faruque, B. Aranda, *et al*. IntAct-open source resource for molecular interaction data. *Nucl. Acids Res.*, **35**(Suppl. 1): D561–565, 2007.

7. T. Joshi, Y. Chen, J. Becker, N. Alexandrov, and D. Xu. Genome-scale gene function prediction using multiple sources of high-throughput data in yeast. *OMICS*, **8**: 322–333, 2004.

8. B. Kelley, R. Sharan, R. Karp, *et al*. Conserved pathways within bacteria and yeast as revealed by global protein network alignment. *Proc. Natl. Acad. Sci. USA*, **100**: 11394–11399, 2003.

9. M. Cline, M. Smoot, E. Cerami, *et al*. Integration of biological networks and gene expression data using Cytoscape. *Nat. Protoc.*, **2**(10): 2366–2382, 2007.

10. B. Breitkreutz, C. Stark, and M. Tyers. Osprey: a network visualization system. *Genome Biol.*, **4**: R22, 2003.

11. C. Wang, C. Ding, Q. Yang, and S. R. Holbrook. Consistent dissection of the protein interaction network by combining global and local metrics. *Genome Biol.*, **8**: R271, 2007.

12. J. Chen and B. Yuan. Detecting functional modules in the yeast protein-protein interaction network. *Bioinformatics*, **22**: 2283–2290, 2006.

13. C. von Mering, R. Krause, B. Snel, M. Cornell, S. Oliver, S. Fields, and P. Bork. Comparative assessment of large-scale data sets of protein-protein interactions. *Nature*, **417**:399–403, 2002.

14. I. Lee, S. Date, A. Adai, and E. Marcotte. A probabilistic functional network of yeast genes. *Science*, **306**(5701): 1555–1558, 2004.

15. U. Karaoz, T. M. Murali, S. Levotsky, *et al*. Whole-genome annotation by using evidence integration in functional-linkage networks. *Proc. Natl. Acad. Sci. USA*, **101**: 2888–2893, 2004.

16. V Vapnik. *Statistical Learning Theory*. John Wiley & Sons, Inc. 1998.

17. J. Yedidia, W. Freeman, and Y. Weiss. Understanding belief propagation and its generalizations. In *Exploring Artificial Intelligence in the New Millennium*, pp. 239–269. Morgan Kaufmann Publishers Inc., San Francisco, CA, USA, 2003.

18. M. Kirac, G. Ozsoyoglu, and J. Yang. Annotating proteins by mining protein interaction networks. *Bioinformatics*, **22**: e260–e270, 2006.

19. B. Schwikowski, P. Uetz, and S. Fields. A network of protein-protein interactions in yeast. *Nat. Biotechnol.*, **18**: 1257–1261, 2000.

20. A. Enright, S. Van Dongen, and C. Ouzounis. An efficient algorithm for large-scale detection of protein families. *Nucleic Acids Res*, **30**: 1575–1584, 2002.

21. Y.-R. Cho, W. Hwang, M. Ramanathan, and Aidong Zhang. Semantic integration to identify overlapping functional modules in protein interaction networks. *BMC Bioinformatics*, **8**: 265, 2007.

22. N. Saitou and M. Nei. The neighbor-joining method: a new method for reconstructing phylogenetic trees. *Mol. Biol. Evol.*, **4**: 406–425, 1987.

23. C. von Mering, L. J. Jensen, M. Kuhn, S. Chaffron, T. Doerks, B. Kruger, B. Snel, and P. Bork. STRING7 - recent developments in the integration and prediction of protein interactions. *Nucl. Acids Res.*, **35**(Suppl. 1): D358–362, 2007.

24. A. Ruepp, A. Zollner, D. Maier, *et al*. The FunCat, a functional annotation scheme for systematic classification of proteins from whole genomes. *Nucleic Acids Res.*, **32**: 5539–5545, 2004.

25. S. Asthana, O. King, F. Gibbons, and F. Roth. Predicting protein complex membership using probabilistic network reliability. *Genome Res.*, **14**: 1170–1175, 2004.

26. P. Shannon, A. Markiel, O. Ozier, *et al*. Cytoscape: a software environment for integrated models of biomolecular interaction networks. *Genome Research*, **13**(11): 2498–2504, 2003.

27. T. Dandekar, B. Snel, M. Huynen, and P. Bork. Conservation of gene order: a fingerprint of proteins that physically interact. *Trends Biochem. Sci.*, **23**(9): 324–328, 1998.

28. R. Sharan and T. Ideker. Modeling cellular machinery through biological network comparison. *Nat. Biotech.*, **24**: 427–433, 2006.

29. H. Hishigaki, K. Nakai, T. Ono, A. Tanigami, and T. Takagi. Assessment of prediction accuracy of protein function from protein-protein interaction data. *Yeast*, **18**: 523–531, 2001.
30. A. King, N. Przulj, and I. Jurisica. Protein complex prediction via cost-based clustering. *Bioinformatics*, **20**: 3013–3020, 2004.
31. E. Sprinzak, S. Sattath, and H. Margalit. How reliable are experimental protein-protein interaction data? *J. Mol. Biol.*, **327**(5): 919–923, 2003.
32. T. Aittokallio and B. Schwikowski. Graph-based methods for analysing networks in cell biology. *Briefings in Bioinformatics*, **7**: 243–255, 2006.
33. M. Blatt, S. Wiseman, and E. Domany. Superparamagnetic clustering of data. *Phys. Rev. Lett.*, **76**: 3251–3254, 1996.
34. J. Poyatos and L. Hurst. How biologically relevant are interaction-based modules in protein networks? *Genome Biol.*, **5**: R93, 2004.
35. L. Salwinski, C. S. Miller, A. J. Smith, F. K. Pettit, J. U. Bowie, and D. Eisenberg. The Database of Interacting Proteins: 2004 update. *Nucl. Acids Res.*, **32**(Suppl. 1): D449–451, 2004.
36. V. Arnau, S. Mars, and I. Marin. Iterative cluster analysis of protein interaction data. *Bioinformatics*, **21**: 364–378, 2005.
37. B. Adamcsek, G. Palla, I. Farkas, I. Derenyi, and T. Vicsek. Cfinder: locating cliques and overlapping modules in biological networks. *Bioinformatics*, **22**: 1021–1023, 2006.
38. Thomas H. Cormen, Charles E. Leiserson, and Ronald L. Rivest. *Introduction to Algorithms*. MIT Press/McGraw-Hill, 1990.
39. A. J. Enright, I. Iliopoulos, N. C. Kyrpides, and C. A. Ouzounis. Protein interaction maps for complete genomes based on gene fusion events. *Nature*, **402**: 86–90, 1999.
40. R. Singh, J. Xu, and B. Berger. Pairwise global alignment of protein interaction networks by matching neighborhood topology. *Proceedings of the 11th International Conference on Research in Comp. Mol. Biol(RECOMB)*, **4453**: 16–31, 2007.
41. C. Stark, B. Breitkreutz, T. Reguly, L. Boucher, A. Breitkreutz, and M. Tyers. BioGRID: A general repository for interaction datasets. *Nucleic Acids Res.*, **34**: D535–D539, 2006.
42. R. Sharan, I. Ulitsky, and R. Shamir. Network-based prediction of protein function. *Molecular Systems Biology*, **3**: 88, 2007.
43. L. Giot, J. Bader, C. Brouwer, *et al.* A protein interaction map of Drosophila melanogaster. *Science*, **302**: 1727–1736, 2003.
44. M. Samanta and S. Liang. Predicting protein functions from redundancies in large-scale protein interaction networks. *Proc. Natl. Acad. Sci. USA.*, **100**: 12579–12583, 2003.
45. J. Flannick, A. Novak, B. Srinivasan, H. McAdams, and S. Batzoglou. Graemlin: general and robust alignment of multiple large interaction networks. *Genome Res.*, **16**(9): 1169–81, 2006.
46. M. Deng, K. Zhang, S. Mehta, T. Chen, and F. Sun. Prediction of protein function using protein-protein interaction data. *J. Computational Biol.*, **10**: 947–960, 2003.
47. E. Dalhaus, D. S. Johnson, C. Papadimitriou, P. Seymour, and M. Yannakakis. The complexity of multiway cuts. In *Proc. 24th Annual STOC*, pp. 241–251. ACM, 1992.
48. D. Goldberg and F. Roth. Assessing experimentally derived interactions in a small world. *Proc. Natl. Acad. Sci. USA*, **100**: 4372–4376, 2003.
49. R. Overbeek, M. Fonstein, M. D'Souza, G. Pusch, and N. Maltsev. The use of gene clusters to infer functional coupling. *Proc. Natl. Acad. Sci. USA*, **96**(6): 2896–2901, 1999.
50. D. Bu, Y. Zhao, L. Cai, *et al.* Topological structure analysis of the protein-protein interaction network in budding yeast. *Nucl. Acids. Res.*, **31**: 2443–2450, 2003.
51. E. Marcotte, M. Pellegrini, H. Ng, D. Rice, T. Yeates, and D. Eisenberg. Detecting protein function and protein-protein interactions from genome sequences. *Science*, **285**: 751–753, 1999.
52. X. Zhu, M. Gerstein, and M. Snyder. Getting connected: analysis and principles of biological networks. *Genes and Development*, **21**: 1010–1024, 2007.
53. C. von Mering, M. Huynen, D. Jaeggi, S. Schmidt, P. Bork, and B. Snel. STRING: a database of predicted functional associations between proteins. *Nucleic Acids Res.*, **31**: 258–261, 2003.
54. S. Suthram, T. Shlomi, E. Ruppin, R. Sharan, and T. Ideker. A direct comparison of protein interaction confidence assignment schemes. *BMC Bioinformatics*, **7**: 360, 2006.
55. F. Radicchi, C. Castellano, F. Cecconi, V. Loreto, and D. Parisi. Defining and identifying communities in networks. *Proc. Natl. Acad. Sci. USA*, **101**(9): 2658–2663, 2004.

56. R. Fourer, D. M. Gay, and B. W. Kernighan. AMPL: A Modeling Language for Mathematical Programming. Brooks/Cole Publishing Company, Pacific Grove, CA, 2002.

57. ILOG CPLEX 7.1, 2000. http://www.ilog.com/products/cplex/.

58. R. Krause, C. von Mering, and P. Bork. A comprehensive set of protein complexes in yeast: mining large-scale protein-protein interaction screens. *Bioinformatics*, **19**: 1901–1908, 2003.

59. J. Pereira-Leal, A. Enright, and C. Ouzounis. Detection of functional modules from protein interaction networks. *Proteins*, **54**: 49–57, 2004.

60. M. Altaf-Ul-Amin, Y. Shinbo, K. Mihara, K. Kurokawa, and S. Kanaya. Development and implementation of an algorithm for detection of protein complexes in large interaction networks. BMC *Bioinformatics*, **7**: 207, 2006.

61. L. Hartwell, J. Hopfield, S. Leibler, and A. Murray. From molecular to modular cell biology. *Nature*, **402**: C47–52, 1999.

62. T. Gaasterland and M. Ragan. Microbial genescapes: phyletic and functional patterns of ORF distributions among prokaryotes. *Microb. Comp. Genomics*, **3**: 199–217, 1998.

63. M. Deng, Z. Tu, F. Sun, and T. Chen. Mapping gene ontology to proteins based on protein-protein interaction data. *Bioinformatics*, **20**: 895–902, 2004.

64. G. Bader and C. Hogue. An automated method for finding molecular complexes in large protein interaction networks. *BMC Bioinformatics*, **4**: 2, 2003.

65. M. Ashburner, C. Ball, J. Blake, D. Botstein, H. Butler, J. Cherry, *et al.* Gene Ontology: tool for the unification of biology. The Gene Ontology consortium. *Nat. Genet.*, **25**(1): 25–29, 2000.

66. E. Nabieva, K. Jim, A. Agarwal, B. Chazelle, and M. Singh. Whole-proteome prediction of protein function via graph-theoretic analysis of interaction maps. *Bioinformatics*, **21** Suppl. 1: i302–i310, 2005.

67. Koyuturk, M., A. Grama, and W. Szpankowski. Pairwise local alignment of protein interaction networks guided by models of evolution. In *Lecture Notes in Bioinformatics*, volume **3500**, pp. 48–65, 2005.

68. V. Spirin and L. A. Mirny. Protein complexes and functional modules in molecular networks. *Proc. Natl. Acad. Sci. USA.*, **100**: 12123–12128, 2003.

69. F. Luo, Y. Yang, C. Chen, R. Chang, J. Zhou, and R. Scheuermann. Modular organization of protein interaction networks. *Bioinformatics*, **23**: 207–214, 2007.

70. M. C. Costanzo, M. E. Crawford, J. E. Hirshman, *et al.* YPD™, PombePD™, and WormPD™: Model organism volumes of the BioKnowledge library, an integrated resource for protein information. *Nucl. Acids Res.*, **29**: 75–79, 2001.

71. O. Troyanskaya, K. Dolinski, A. Owen, R. Altman, and D. Botstein. A Bayesian framework for combining heterogeneous data sources for gene function prediction (in S. cerevisiae). *Proc. Natl. Acad. Sci. USA*, **100**: 8348–8353, 2003.

72. C. Brun, F. Chevenet, D. Martin, J. Wojcik, A. Guenoche, and B. Jacq. Functional classification of proteins for the prediction of cellular function from a protein-protein interaction network. *Genome Biol.*, **5**: R6, 2003.

73. G. Lanckriet, T. Bie, N. Cristianini, M. Jordan, and W. Noble. A statistical framework for genomic data fusion. *Bioinformatics*, **20**: 2626–2635, 2004.

74. E. Gansner and S. North. An open graph visualization system and its applications to software engineering. *Softw. Pract. Exper.*, **30**(11): 1203–1233, 2000.

75. H. Chua, W.-K. Sung, and L. Wong. Exploiting indirect neighbors and topological weight to predict protein function from protein-protein interactions. *Bioinformatics*, **22**: 1623–1630, 2006.

76. K. Tsuda and W. Noble. Learning kernels from biological networks by maximizing entropy. *Bioinformatics*, **20** Suppl. 1: i326–i333, 2004.

77. M. Pellegrini, E. Marcotte, M. Thompson, D. Eisenberg, and T. Yeates. Assigning protein functions by comparative genome analysis: protein phylogenetic profiles. *Proc. Natl. Acad. Sci. USA*, **96**(8): 4285–4288, 1999.

78. A. Bauer and B. Kuster. Affinity purification-mass spectrometry. *Eur. J. Biochem.*, **270**: 570–578, 2003.

79. S. van Dongen. Graph clustering by flow simulation. PhD thesis, University of Utrecht, 2000.

80. R. H. Jansen, H. Yu, D. Greenbaum, *et al.* A Bayesian networks approach for predicting protein-protein interactions from genomic data. *Science*, **302**: 449–453, 2003.

81. M. Deng, T. Chen, and F. Sun. An integrated probabilistic model for functional prediction of proteins. In *Proc. 7th Annual RECOMB*, pp. 95–103. ACM, 2003.
82. S. Brohee and J. van Helden. Evaluation of clustering algorithms for protein-protein interaction networks. *BMC Bioinformatics*, **7**: 488, 2006.
83. S. Fields and O.-K. Song. A novel genetic system to detect protein-protein interactions. *Nature*, **340**: 245–246, 1989.
84. M. Galperin and E. Koonin. Who's your neighbor? New computational approaches for functional genomics. *Nat. Biotechnol.*, **18**: 609–613, 2000.
85. A. Rives and T. Galitski. Modular organization of cellular networks. *Proc. Natl. Acad. Sci. USA*, **100**(3): 1128–1133, 2003.
86. C. Myers, D. Robson, A. Wible, *et al.* Discovery of biological networks from diverse functional genomics data. *Genome Biol.*, **6**: R114, 2005.
87. A. Vazquez, A. Flammini, A. Maritan, and A. Vespignani. Global protein function prediction from protein-protein interaction networks. *Nat Biotechnol.*, **21**: 697–700, 2003.
88. M. Girvan and M. Newman. Community structure in social and biological networks. *Proc. Natl. Acad. Sci. USA*, **99**: 7821–7826, 2002.
89. R. Dunn, F. Dudbridge, and C. Sanderson. The use of edge-betweenness clustering to investigate biological function in protein interaction networks. *BMC Bioinformatics*, **6**: 39, 2005.
90. C. Burges. A tutorial on support vector machines for pattern recognition. *Data Mining and Knowledge Discovery*, **2**(2): 121–167, 1998.
91. M. Deng, F. Sun, and T. Chen. Assessment of the reliability of protein-protein interactions and protein function prediction. In *Pac. Symp. Biocomput.*, pp. 140–151, 2003.
92. R. Kondor and J. Lafferty. Diffusion kernels on graphs and other discrete input spaces. In *Proc. Intl. Conf. on Machine Learning*, pp. 315–322, 2002.
93. R. Sharan, T. Ideker, B. Kelley, R. Shamir, and R. M. Karp. Identification of protein complexes by comparative analysis of yeast and bacterial protein interaction data. *Journal of Computational Biology*, **12**(6): 835–846, 2005.
94. J. Bader. Greedily building protein networks with confidence. *Bioinformatics*, **19**: 1869–1874, 2003.

11

Integrating Prediction of Structure, Function, and Interactions

Michael Tress, Janusz M. Bujnicki, Gonzalo Lopez and Alfonso Valencia

11.1 Introduction: A Flood of Sequence Data and the Limits of Homology-based Function Prediction

Dramatic improvements in high throughput sequencing technologies have lead to a substantial increase in whole-genome sequencing projects. According to Entrez[1] there were 24 completed eukaryotic and 599 completed microbial genomes by mid-2008, with another 1,281 genomes in progress or already in the assembly stage. The rapid growth in sequenced genomes is leading to radical changes in our understanding of genomics and provides unparalleled opportunities for research. In addition to the genome sequencing projects, environmental sequencing projects[2] are producing colossal numbers of new sequences, though the fact that most sequences from environmental sequencing projects are fragments of whole sequences means that they should be handled with care.[3]

However, while genome-sequencing projects are generating almost unimaginable numbers of protein sequences, these sequences are not annotated with functional information. The spectacular increase in unannotated sequences is widening the gap between sequenced genes and known protein functions. Experimental procedures for characterizing protein function are expensive, time consuming and difficult to automate, so researchers will have to turn increasingly to computational annotation to close the gap. Providing functional annotations for the torrent of new sequence information is one of the greatest challenges facing computational biology today and it is clear that function prediction is becoming an increasingly important field.[4–7]

Prediction of Protein Structures, Functions, and Interactions   Edited by Janusz Bujnicki
© 2009 John Wiley & Sons, Ltd

The sheer range of sequenced prokaryotic species could allow a systematic and complete categorization of genes and their biological functions. Similarly the sequencing of multiple related eukaryotic species is a powerful tool to help understand genomes and gene products. The recent sequencing of 12 related *Drosophila* genomes[8] leads to the prediction of several thousand new functional elements, including protein-coding genes and exons.

There are two main problems facing researchers when it comes to functional annotation. The first is that function is often multi-faceted and hard to define. A protein's function may be defined by its role and location in the cell, the metabolic pathway or regulatory network that it forms part of or by the physico-chemical effects that it brings about. In addition many proteins have more than one cellular role[9, 10] and some are even known to perform different functions in different tissues.[11] Functional ontologies such as the Gene Ontology (GO) terms developed by the GO consortium[12] or the nomenclature used by the Enzyme Commission (EC number, [13]) are an attempt to standardize, systematize and compartmentalize protein functions (for a detailed review see the chapter by Chitale *et al.* in this volume). Beyond EC numbers and GO terms protein function can also be annotated with functional keywords that describe biochemical function and the interaction with cofactors, substrates, regulators and other cellular components. It is also possible to be more specific about the function of a protein at the level of individual residues. Some amino acids are directly implicated in molecular function by having catalytic activity or binding a substrate. Pinpointing residues of functional interest is especially important for studying biochemical function at the cellular level and for designing experiments. In addition, a full description of biological function has to take into account both its spatial and temporal aspects.[14]

The second problem is that protein function annotation is still predominantly transferred by homology, which is typically established from the detection of sequence similarity. Transference works on the assumption that proteins that are evolutionarily related will have similar structures and functions. The transfer of general functional information (in the form of GO terms or keywords) typically requires little more than a simple BLAST[15] homology search, but more powerful methods such as profile-based methods[16, 17] or hidden Markov models[18] can also be used (see the chapter by Chitale *et al.* for a detailed review of sequence-based function prediction methods). Homology-based prediction of protein function is widely used, in part because of successes with structure prediction and in part because it is so simple. However, the identification of a functional relationship by sequence comparison is nowhere near as reliable as it is with structure (see Chapter 1 by Kaminska *et al.* and Chapter 7 by Kinoshita *et al.* in this volume). While there is a direct relationship between sequence similarity and conservation of protein structure, the hypothesis holds less true in relation to protein function. For example, while orthologous genes, related by speciation, are likely to share a common function, paralogous genes, related by duplication events, are more likely to have divergent functions. However, orthologs do not necessarily exhibit stronger sequence similarity than paralogs and disambiguating the two types of relationship by pairwise sequence comparisons is very difficult.

Common evolutionary origin does not guarantee functional similarity to the same extent it guarantees structural similarity, and the more distant the evolutionary relationship, the less reliable the transfer. Large-scale studies have shown that transfer of annotation is only accurate for highly similar pairs of proteins.[19–21] Substrate specificity (approximating to the fourth EC number) can be reliably transferred between two proteins that have more than 60–70% pairwise identity while the 3rd EC number (relating to the broader catalytic mechanism) tends to be conserved in pairs of proteins that have sequence identities of 35–40% or above (Figure 11.1). However, there are many exceptions to this rule, as even single residue changes have been shown to cause changes in substrate specificity.[22] If two proteins have less than 30% sequence identity or different domain compositions they are very likely to have significant functional differences. The difficulties of reliable sequence-based annotation and the intrinsic problem of defining biological function mean that sequence-based annotation pipelines will introduce many errors during the process of function annotation.[23]

Another difficulty is that despite the availability of an increasing number of protein sequences from a wide range of organisms, it may not always be possible to find homologues for a protein. Although a large fraction of gene products from both genomic and metagenomics sequencing projects can be assigned to known protein families by homology, the number of novel small families of sequences with no detectable similarity to other sequences continues to grow linearly over time.[24] This suggests that we are nowhere near the limits of the protein function space.[25] Because of this, and because of the difficulty of obtaining completely reliable annotations of protein function through homology-based transfer, techniques that are capable of assigning function in a more sophisticated manner have been developed.

11.2   Predicting Function with 1D Sequence Features

If no sequence homologue exists, function must be predicted *de novo*. A number of tools have been developed that can predict certain aspects of function purely from short motifs present in the amino acid sequence. Even when two proteins do not appear to be homologous, their sequences may contain short conserved sequence motifs characteristic of a particular functional features. Features that can be predicted are usually collectively described as 1D features[26] and can be associated to a single residue and include features such as secondary structure, solvent accessibility, post-translational modifications and the sub-cellular localization of a protein. These methods are particularly suitable for orphan sequences – those proteins that have no detectable homologues or for protein families that show no detectable relationship to other families of known function.

The identification of sub-cellular localization is an important first step in identifying a protein's functional role in the cell. Experimental high-throughput procedures to determine the subcellular location have been carried out with yeast,[27] but in the absence of experimental results and reliable homology-based annotation, computational prediction is still important. Methods that predict sub-cellular localization use signal sequences, overall amino acid composition and the evolutionary information contained in profiles.[28] Although these methods are reliable, they tend to provide only limited functional information. For instance there are many methods that predict trans-membrane helices and likely membrane polarity with high levels of accuracy (reviewed in Chapter 2 by Majorek *et al.* in this volume), but they cannot predict where the protein will insert into the membrane. Post-translational modification prediction methods are also relatively reliable, at least for those post-translational modifications that have well-known highly conserved

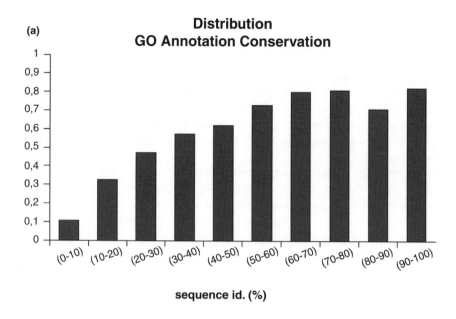

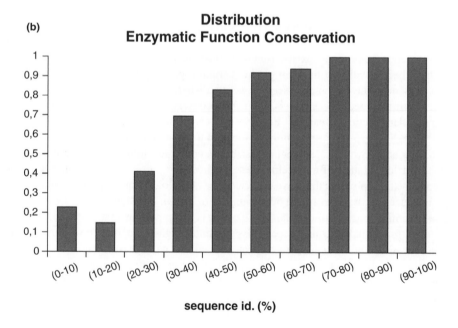

Figure 11.1 *Conservation of function and sequence similarity: (a) conservation of GO terms plotted against pairwise sequence identity (metric); (b) conservation of the four EC numbers plotted against pairwise sequence identity in pairs of proteins sharing a given sequence similarity, where a score of 1 indicates that all four numbers are conserved. All four EC numbers tend to be conserved above 70% pairwise identity*

sequence motifs. Predicted post-translational modifications include phosphorylation sites, and N- and O-glycosylations.[29] However, the existing methods can only predict whether a given modification is introduced if the substrate is presented to the respective modification enzyme under appropriate conditions, but are not able to predict whether the same modifications are introduced or removed dynamically within biological processes. Nonetheless, the combination of sub-cellular localization and modification prediction can provide useful hints as to the cellular function and process. An example of a computational tool that integrates these two features is the ProtFun server,[30] which uses a neural network to predict functional classes for proteins with no known homology to functionally characterized proteins. Functional assignment is by enzyme class and Gene Ontology category and is made directly from the amino acid sequence. The neural network combines predicted sub-cellular location and post-translational modifications, as well as the length, isoelectric point and amino acid composition of the polypeptide chain.

Methods focused on 1D features can not predict function with the same level of precision as homology-based transfer methods, but for those proteins without homologues of known function these methods can provide predictions with useful levels of accuracy and detail.

11.3   Predicting Function with 3D Structures

While whole-genome sequencing projects are contributing to the growth of sequence databases, structural genomics projects are helping fuel a similar, but more modest, growth in the number of proteins with known structures.[31] As of July 4, 2008 the Protein Data Bank (PDB, [32]) repository of protein structures contained a total of 63,409 unique chains of at least 30 residues in length (around 20,000 when non-redundant at 97%). On the whole the proteins in the PDB are better characterized in terms of function than those in the sequence databases. However, many of the proteins deposited by structural genomics initiatives have little associated functional information and indeed the driving force for the selection of candidates for structure determination is often the lack of homologues with known function.[33] The fact that structural genomics projects preferentially select proteins of unknown function for structure resolution means that the PDB contains an increasing pool of proteins with determined structure but no known biological role. This makes structure-based prediction of protein function both relevant and necessary.

Protein 3D structure can be of use in predicting function, in particular because structure is more conserved than sequence between homologous proteins. Homology detection is easier and more reliable with structure than with sequence (see Chapter 4 by Kosinski *et al.* in this volume), so structural data can be used to detect proteins with similar functions for those proteins whose sequences have diverged beyond a level of similarity that can be reliably detected using sequence comparison methods. However, besides identification of homology, there are no simple rules for structure-based functional annotation. Knowledge of the specific three-dimensional fold adopted by a given protein does not directly imply a function. For example, there are 27 different homologous superfamilies that adopt the TIM barrel fold and exhibit as many as 60 different enzymatic functions according to the EC classification.[4] Moreover, while structure may be conserved within a superfamily of proteins, it is not always true that function is conserved to the same extent.[19]

Although the resolution of protein three-dimensional structure rarely leads to direct function prediction, it can often reveal important specific information. From the structure it is possible to see which residues are buried in the core of the protein and which residues are on the surface. The shape and molecular composition of the surface may indicate the binding of particular ligands (for a review see Chapter 7 by Kinoshita *et al.* in this volume) and the surface may have also clefts that might be indicative of catalytic sites ([34]; see also Chapter 8 by Torrance and Thornton in this volume).

A further way to use structure to help predict function is to identify short structural motifs formed by residues that are not necessarily adjacent in sequence. Methods employing this approach can predict function for known structures with unknown function and no known functional homologues. For instance the PINTS server[35] can suggest possible function by finding local structural similarities between protein structures. The server compares protein structures that do not have to have the same fold. Instead the server determines similarity by evaluating whether the two structures have similar patterns of amino acids clustered in structural space (for example, catalytic triads).

Another possibility is to use the structure as a target for virtual screening, i.e. computational docking of a large collection of candidate ligands, in which the best-fitting molecules are most likely similar to the true ligands that bind to the target molecule in the cell. This approach has been found to be successful both for small molecules[36] and peptides.[37] However, such methods are extremely costly in terms of computing power and require expert knowledge and preparation of compounds to be docked, thus they are very difficult to automate and in the near future probably will not become generally available as automated servers to lay users. At the other end of the spectrum of structure-based methods there are those that analyze very general features and that are computationally inexpensive, despite requiring expert knowledge to interpret the results. For example SURF'S UP![38] provides clues as to the function of proteins of known structure by performing comparative analysis of surface features such as charge and hydrophobicity. An advantage of this particular method is its applicability to theoretical models, as long as they correctly reproduce the most important general features of the structure. Thus, for proteins without known structure, models may be generated first, e.g. by comparative or *de novo* modelling approaches (see Chapter 4 by Kosinski *et al.*, and Chapter 5 by Gront *et al.* in this volume), and then used to make general functional inferences: something that would not be possible directly from sequence.

Structural motifs can be also combined with sequence motifs to form hybrid motifs and these can be used to predict conserved function for hypothetical proteins with known structure. For example ProKnow[39] integrates structures, sequence information, motifs and a database of protein features and functional annotations. The server associates sequence and structure features of the query structure with the functional annotations in the database and predicts function in the form of GO terms. As is almost always the case, the best approach is to use as many structure-based and sequence-based methods as possible, as this increases the chance of finding similarities. In addition, predictions are more likely to be reliable if more than one method makes a similar prediction. ProFunc[40] is a 'meta-server' that was developed as an aid to the prediction of function for proteins with known structure. It employs a range of separate sequence and structure-based methods. The range of methods used means that more often than not one of the methods will provide interesting results and taken together the results

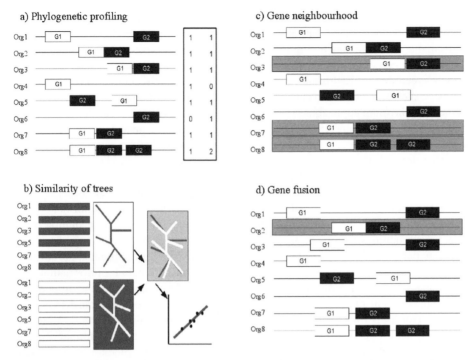

Figure 11.2 *Prediction of protein interactions based on genomic and sequence features. Information from homologues of proteins P1 and P2 in one species can be used to predict an interaction between these proteins in another organism: (a) phylogenetic profiling: the presence/absence of a homologue of the two genes in other species is shown in the matrix on the right – interaction is predicted if the profiles are very similar; (b) similarity of phylogenetic trees: phylogenetic trees are derived from multiple sequence alignments built for both proteins – proteins with highly similar trees are predicted to interact; (c) gene neighbourhood – genome closeness – interaction is predicted if the genes coding for the two homologues are close to each other in many organisms; (d) gene fusion – if the genes for the two proteins are fused in one species, they are predicted to interact in another*

from these procedures are often able to provide insight as to the likely function of the protein.

11.4   Predicting Function from Whole Genome Information

The growing number of completely sequenced genomes allows us to look at protein function in the context of entire genomes and proteomes across different species. This class of methods focuses on protein function defined by cellular pathways or the complexes the protein participates in, rather than by suggesting a specific biochemical activity (Figure 11.2). They exploit the well-known tendency of organisms to physically link genes, proteins and domains that code for activities that must be spatially brought together in order to carry out some higher-order function. In Prokaryota, genes encoding functionally related proteins usually cluster together in the genome and often form operons, in which transcription is coordinated. Furthermore, genes that appear as independently expressed ORFs in one organism may be 'fused' as part of the same polypeptide sequence in another organism. Fused composite genes in one genome are strong indicators potential interactions between the unfused orthologous or homologous proteins in other genomes. It has also been shown that fusion events are particularly common in metabolic proteins.[41] This method may be more applicable to Eukaryota, which possess more multi-domain proteins, but exhibit very little functional clustering on the level of the genome. Obviously, genes related by neighbourhood or fusion are even more relevant if they are conserved across different species,[42] so the significance of gene proximity can be reinforced by its conservation in evolutionary distant species.

Another class of genomic methods explicitly considers the evolutionary history of genes. Phylogenetic analyses can be used to predict protein function by using information from multiple sequences in two complementary ways: by studying divergent evolution within homologous families (to discriminate between proteins that preserve common function and those that could have evolved new functions) and by studying correlated evolution between families. The first approach, termed 'phylogenomics', combines information about whole genomes with homology-based reconstruction of evolutionary trees, in particular with respect to gene duplications, losses within genomes and horizontal transfers between genomes (for a review see [43]). The general protocol requires the collection of homologous sequences related to the query sequence, the constructing of a multiple sequence alignment, the calculation of a phylogenetic tree, the identifying of sub-trees with statistical support, overlaying the phylogenetic tree with the available experimental data, the identifying of tree nodes corresponding to gene duplication or speciation, and finally the inferring of molecular function. Identification of gene duplication or speciation may be greatly enhanced by including some aspects of context analysis, for instance discrimination of subfamilies with different domain architectures. An example of a computational tool for phylogenomic analysis is the recently developed PhyloBuilder web server,[44] which provides an integrated pipeline starting with a user-supplied protein sequence, proceeding to homologue identification, multiple alignment, phylogenetic tree construction, subfamily identification and structure prediction.

Phylogenetic analysis can be also used to look for relationships that do not necessarily rely on homology. The simplest methods compare the patterns of presence and absence of genes from different families across a set of genomes. A complementary distribution of two families (a member of one family tends to be present while the corresponding member of the other family is absent and vice versa) may indicate non-orthologous gene displacement, an exchange of one gene encoding a particular function for a non-orthologous (unrelated or distantly related) but functionally analogous gene.[45] Thus, protein families with complementary phylogenetic distribution may exhibit the same function, regardless of their homology. On the other hand, genes that have the same pattern of presence and absence over a wide range of genomes are likely to interact. The rationale is that reductive evolution will remove proteins that form part of pathways that are no longer important, so genes that are always present or absent together should be in the same pathway. Although similar patterns may suggest a related functional role, it does not necessarily imply a direct physical interaction between the proteins. The approach is powerful because it enables the exploration of interdependencies between proteins. The main drawbacks of this approach are that it can only be applied to complete genomes (as only then is it possible to rule out the absence of a given gene) and that it cannot be used to characterize essential genes that tend to be universally conserved.[46, 47]

Similarity in the topology of phylogenetic trees of interacting proteins has been qualitatively observed in a number of closely studied cases. For example it has been possible to show that insulin and its receptors,[48] and dockerins and cohesins[49] co-evolve because the corresponding phylogenetic trees show a greater degree of similarity than would be expected. If these observations are to be extended to a quantitative method for predicting protein interactions, the correlation between the similarity matrices of the families of the two proteins must be quantified. Goh *et al.*[50] measured the similarity of the phylogenetic trees of the two domains of phosphoglycerate kinase using the correlation between the distance matrices that had been used to construct the trees. This approach has been extended to predict interactions for large sets of proteins and protein domains.[51] Another related approach looks at correlation at the residue level. Correlated mutations are residues in pairs of proteins that undergo compensatory mutations that stabilize the interaction between the two proteins. It has been shown that correlated mutations in interacting surfaces can help select the correct structural arrangement of two known protein structures.[52] This idea has been extended to predicting pairs of interacting proteins (*in silico* two-hybrid). Predictions can be made based on correlated mutations between the interacting proteins and within individual proteins.[53]

While these genomic methods can often provide quite reliable predictions, the degree of coverage provided by each of these methods is relatively small. For example, a study of the *Mycoplasma genitalium* genome[54] showed that, the gene order method could be applied to 37% of the genes, but just 11% of the genes could be predicted by phylogenetic profiles and only 6% by gene fusion. The main use of these methods is likely to be as a complement to other methods of function prediction and in combination with experimental methods or homology-based prediction. The combination of computational and experimental approaches will broaden our view of protein interaction networks and allow studies of key nodes in the networks, the distribution and number of interactions and differences in network organization from one organism to another. For example, recent network analyses have revealed interesting new properties of biological interaction networks that may have consequences for drug design.[55, 56]

11.5   Function Prediction from Large-scale Experimental Methods and Networks

While individual proteins are often regarded as having distinct and independent functions, the molecular bases of cellular operations are largely sustained by macromolecular interactions, in particular those involving proteins. Protein interactions and networks that are ultimately responsible for much of cellular function and the great majority of biochemical processes in the cell are regulated by physically interacting protein entities that can come together either in the form of temporary interactions or in relatively long-lived complexes. Indeed many of the properties of complex systems seem to be more closely determined by their interactions than by the characteristics of their individual components.

It has only recently become possible to combine the traditional study of proteins as isolated entities with the analysis of large protein interaction networks. As with genome sequencing projects and structural genome initiatives, recent technological advances are generating a flood of molecular interaction data. High-throughput experimental techniques such as yeast two-hybrid,[57–59] large-scale mass spectroscopy of protein complexes,[60–62] and genome-wide chromatin immunoprecipitation[63, 64] are generating protein-protein interaction data on a genome-wide scale, allowing researchers to study protein–protein interactions in a number of model species. The available interaction data goes beyond individual proteins and small complexes to encompass the protein–protein interaction network for entire organisms and opens up new opportunities for investigating cellular biology and disease. Researchers have sought to analyse this interaction data, leading to the development of computational approaches that can unravel and complete the large-scale protein–protein interaction maps for entire proteomes.[65, 66] Many recent studies have concentrated on comparing and contrasting networks from different species and under varying conditions and there have also been a plethora of methods that predict protein function based on the data that comes from large scale interaction experiments.[67, 68]

The aim of network comparison is to determine which proteins, protein interactions and complexes have equivalent functions in distinct species and to use these equivalences to predict function for poorly characterized proteins and interactions. Pairwise alignment of networks involves identifying analogous (or sometimes also homologous) pairs of interactions from two different molecular interaction networks. Methods to align networks are computationally hard and their solution usually requires heuristics, such as network alignment graphs.[69] As new information about complete genomes becomes available, these methodologies are extensively used for comparative genomics. Comparisons of protein-protein interaction networks across species[70, 71] have been used to identify pairs of interactions from the best sequence matches in other species. Network alignments can also be used to predict protein properties; if many proteins in a sub-network have the same function the other proteins are also likely to have that function and the network alignment can also be used to describe protein–protein interactions and links between cellular processes.

Since proteins that interact will be part of the same biochemical pathway, information in the form of interacting partners and pathways can help in the prediction of broad protein function. The principle applied here is 'guilt by association' – the closer proteins are to one another in the interaction network the more likely they are to have similar function. This is more true of certain definitions of function than others, while proximity in interaction networks is likely to be indicative of similarity in biological process and cellular localization, it remains unclear to what extent this information can be used to predict molecular function.

The recent availability of large-scale networks of molecular interactions for model species have made it possible to develop computational methods that can predict protein function from interaction network data (see Chapter 10 by Nabieva and Singh in this volume). In addition methods that can predict function from high-throughput experimental data are especially attractive because of reduced costs. New interactions, modules, motifs and functional properties (phenotypes, diseases and others) can be predicted simply by integrating complementary information from functional and structural interactions. More recent methods are based on network topology, with the network or multiple networks represented as a graph with nodes representing proteins and edges representing the interactions. The protein interaction data is often combined with sequence similarity, protein structural information and gene expression data.

There are two broad types of computational methods, those that predict function directly from functional information in the network and those that annotate function in terms of functional modules. Methods for function prediction from large-scale interaction data include basic neighbourhood counting (where protein function is predicted from the functions of nearby proteins,[72] graph theoretic methods that treat interaction networks as graphs,[68] and integrating data from multiple sources.[73] For example one recent method[74] integrated gene expression data, protein motif information, mutant phenotype data, and protein localization data with protein-protein interaction data to predict function (in Gene Ontology terms) for *Saccharomyces cerevisiae* genes. Integrating external information with the interaction data increased recall by 18% and allowed GO terms to be assigned to 463 proteins that had no annotated function.

Even though there are a myriad of different methods for comparing networks and predicting function, very few of the algorithms are publicly accessible and even fewer have graphical interfaces for easy viewing.[75–77] There are limitations to methods that predict function from interaction information. The false positive rate from interaction data is estimated to be as much as or as greater than 50%.[78] In addition to being noisy[79] protein interaction data is also partial.[80] It is therefore difficult to know which interactions are real binding events and whether certain interactions only occur under certain conditions. A further problem is that interaction data from large-scale high throughput techniques comes from a single tissue type under fixed conditions – interaction data does not come with spatial or temporal information.

Despite these problems, interaction network analysis is very active research area and still developing. Network-based function prediction techniques will continue to improve, in particular because there is still much room for improvement. Some of these problems will be addressed by the growth in high throughput data, for example false positive interactions are less likely to be reproduced across many interaction networks.

11.6   Information from Text Mining

Databases and repositories of experimental information are further growth areas. The challenge here is to retrieve content from text repositories, to filter for biologically relevant information, and to cross-link the information with data mined from databases that contain experimental data such as molecular interactions, metabolic pathways, and signalling cascades. The development of text-mining and information-extraction methods over the last decade is in direct response to this challenge and the development of text-mining tools has accelerated the process of database standardization. The prediction of functional associations between proteins by text mining techniques is driven by the assumption that protein names that appear together within a sentence, an abstract or even a whole text are more likely to have a biological relationship. In this way it ought to be possible to predict whole biological pathways based on the extraction of associations between proteins. However, reconstructing pathways in this way has turned out to be more difficult than expected.[81]

It is also possible to characterize the biological significance of the interactions. Blaschke *et al.*[82] proposed sets of expressions to classify the biological relations, such as 'complex of protein x and protein y,' or 'phosphorylation of protein x by protein y'. Other approaches[83] classify protein interactions using natural language processing techniques. However, the major difficulty for all these systems is the problem of organizing and presenting the information in a biologically relevant form. The same techniques can also be applied to the inference of indirect relations. The detection process is best illustrated by the discovery[84] of the potential use of magnesium to treat migraine ('magnesium loss can have an effect on stress' + 'stress is associated with migraines'). This approach has had some successes.[85]

Text mining methods for function prediction have many problems. For example, predicting relations between enzymes in metabolic pathways is problematic because enzymes acting in metabolic pathways tend not to be mentioned together. A much bigger problem is the fact that the use of standard gene and protein nomenclature is not enforced. A further factor complicates prediction, common functions can be described in many different ways in many different articles and relating these definitions to a standard definition of function (for example, the Gene Ontology) is a problem, as the results of the protein annotation extraction task of BioCreAtIvE (Critical Assessment of Information Extraction systems in Biology[86]) challenge shows.

While current text-mining methods are not yet be able to predict function directly and reliably, they come into their own when integrating information extracted from a range of sources. With the help of text mining methods it is possible to connect proteins and genes in ordered pathways and extend them to form model protein networks. The Valencia group has developed the iHOP (Information Hyperlinked over Proteins[87]) system, which covers genes and proteins that co-occur in PubMed.[88] Users can navigate (or hop) from one related sentence to another in a form that is closer to human intuition than conventional keyword searches. iHOP allows for stepwise and controlled acquisition of information and provides direct links to the IntAct protein interaction database,[89] thus allowing users to superimpose experimental information onto the network. In this way, novel and existing knowledge can be explored simultaneously.

11.7 Recent Development and Future Directions

Much of the current work is centered around macromolecular complexes. At any one time cells may contain hundreds of distinct multi-protein complexes and these functional modules form a crucial part in almost all biological processes. Complexes have keys roles in metabolism, transcription and translation, cell signalling and cellular transport, and complete understanding of the structure and function of these complexes in the context of the cell would go a long way towards deriving working models of the whole-cell.

Much of the impetus for the work on macromolecular complexes comes from the recent flood of protein interaction data from high-throughput experimental techniques such as yeast two-hybrid,[57–59] and tandem affinity-purification/mass spectroscopy.[60–62] While these experiments have been able to provide detailed lists of the proteins and their cellular interactions, one drawback is that the interaction data from these studies does not yet have high levels of coverage or accuracy, and there is little overlap between the interactions found with yeast two-hybrid and those detailed by tandem affinity-purification. Two promising recent analysis of *Saccharomyces cerevisiae* have attempted to put a value on the reliability of the interactions.[90, 91] Although the data from these studies can give clues as to which proteins may be involved in complexes, they cannot demonstrate how the protein sub-units interact physically. Knowledge of the 3D structural arrangement of the proteins in complexes is vital to deduce cellular function, but up to now relatively few protein complexes have been crystallized.

Despite recent improvements in crystallization procedures, solving the structures of the hundreds of unresolved protein complexes by experimental means alone will take many years. However, improvements in experimental methodologies, such as the mass spectrometry of complexes[92] and, in particular, cryo-electron microscopy[93] have been a huge step forward in the prediction of large 3D complexes. Cryo-electron microscopy can take two-dimensional electron-microscopy images of three-dimensional cellular structures from many different angles, allowing researchers to reconstruct low-resolution images of large complexes. Because the images are recorded *in situ* in the cell, cryo-electron microscopy can even construct images of transient complexes, those functional modules that come together only fleetingly and that cannot currently be treated by crystallization methods. The low-resolution images generated by these new methods – one combination of mass spectrometry and electron microscopy[94] is particularly promising – will allow scientists to build up 3D models of complexes based on their constitutive protein subunits.

The first step in reconstructing the large complexes detected by electron microscopy is to determine the proteins that make up the complex (e.g. with mass spectrometry). Then the 3D structures of the constituent proteins (obtained by experimental methods or by modelling) must be fitted into the cellular tomograms. Here the assembly of the individual subunits is guided by information from the known homologous complexes, contact information that comes from mass spectrometry[95] and by the tomographic image of the complex. The final models of the complexes have low resolution, and at present only the larger complexes (such as spliceosomes, or nuclear pore complexes) can be reconstructed, but even low resolution models can provide atomic level detail sufficient for making hypotheses about the functioning of the complex, and it will be possible to study smaller complexes as the imaging capabilities improve.

Aloy *et al.*[96] first used electron microscopy to help model a large set of yeast complexes. They were able to show that it was possible to model over a hundred complexes in this way, though many of the complexes were incomplete because the component protein structures were lacking.

More recently a model for the nuclear pore complex has been described.[97] Nuclear pore complexes are large protein complexes that span the nuclear envelope. They regulate transport between the cytoplasm and the nucleoplasm in eukaryotes.[98] The electron-microscopy images of the complex showed that the nuclear pore complex was formed of two stacked rings. Within each ring there were eight repeat units and each repeated unit was made up of approximately 30 different nucleoporins protein.

Structures had only been solved for 5% of the domains in the nuclear pore complex,[99] so the remaining domains were assigned folds through a combination of sequence search and structure prediction methods,[99] demonstrating strong synergy between predictive and experimental methods. The spatial arrangement of the nucleoporin proteins was determined from restraints calculated from the protein sequences, from ultracentrifugation, from immuno-electron microscopy, from the results of affinity purification and from cryo-electron microscopy. The final model shows that the nucleoporins form a structural scaffold analogous to vesicle-coating complexes.[97, 99]

Multi-protein complexes have also been predicted using many other hybrid techniques. For example, models have been constructed for the synaptic vesicle based on the protein and lipid composition, vesicle size, density, and mass, and protein copy number.[100] Researchers are also attempting to make predictions of spatial arrangement on a much smaller scale. Wollacott *et al.*[100] had some success with predicting the orientation of domains from the same protein by first predicting the form and orientation of the linker regions that join the two domains and a theoretical model of apolipoprotein B100, a 4,500 residue protein that stabilizes low density lipoprotein particles has been predicted based on restraints from physico-chemical properties and a low resolution electron microscopy map.[102] Here, combination of high-accuracy structure prediction, macromolecular docking, and low-resolution experimental analyses is very likely to help in making functional inferences for the whole, based on knowledge of activities of individual parts (Figure 11.3).

11.8   Assessment of Function Prediction

Computational prediction of protein function from sequence, structure, and any other source of information can only become truly useful if the accuracy of predictions can be reliably estimated. While such estimators of prediction quality have been recently developed in the field of protein structure prediction (see Chapter 6 by Wallner and Elofsson in this volume), the field of protein function prediction has still a long way to go. However, the first steps in this direction have already been made. The Critical Assessment of Structure Prediction (CASP)[103, 104] is an experiment that evaluates the state of the art of structure prediction. In the CASP experiment known structures are hidden from the predictors, so the structure predictions are blind. The prediction of function has been included as a category in the last two editions of the experiment;[105, 106] in theory the predictors could use structural information in the form of models to make their predictions.

Methods for the prediction of function from structure, whether *ab initio* or by homology, are not as well developed as those for the prediction of structure from sequence and the assessment of function prediction from structure suffered from three problems. The first difficulty (as already mentioned in this chapter) was that the definition of protein function was difficult to pin down. Function, unlike structure, cannot be defined by atomic co-ordinates. Secondly, the predictors were actually predicting the function from sequence – if predictors were to use the target structure in the prediction they would first have to build a model or predict facets of the structure in some way. Finally, targets are chosen for CASP because their structures, not their functions, are close to being resolved, so there was little new functional information available to the assessors. With the exception of bound ligands, the CASP7 assessors had no more functional information at the end of the experiment than was available to the predictors during the experiment.[106] Even a year after the CASP7 evaluation only a few of the targets had newly annotated GO terms based on experimental data. The fact that the target proteins had little new functional information complicated the way that the function prediction category was assessed.

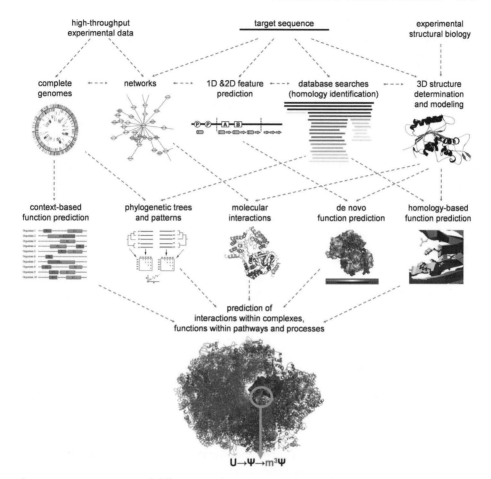

high-throughput
experimental data

target sequence

experimental
structural biology

complete
genomes

networks

1D &2D feature
prediction

database searches
(homology identification)

3D structure
determination
and modeling

context-based
function prediction

phylogenetic trees
and patterns

molecular
interactions

de novo
function prediction

homology-based
function prediction

prediction of
interactions within complexes,
functions within pathways and processes

U→Ψ→m³Ψ

Figure 11.3 *Integration of different prediction approaches and experimental data. Genomic context and network analyses can be used to identify various types of interactions and suggesting cellular pathways the protein participates in, rather than by suggesting a specific biochemical activity. On the other hand, homology-based predictions can be used to make complementary predictions about biochemical details, starting from identification of putative active sites of enzymes and suggesting potential reaction types to be catalyzed. De novo methods are capable of identifying structural and functional analogies between unknown systems and previously studied systems, regardless of the presence or absence of evolutionary relationships. Combination of context-based methods with homology-based methods and de novo predictions (e.g. macromolecular docking of models guided by restraints) can be used to predict not only which proteins interact, but also how, e.g. what is the structure of the complex, allowing further inferences to be made about the spatial and temporal aspects of biological processes* (See insert for color representation of figure)

In CASP7, function prediction was evaluated by three different measures, EC numbers, GO molecular function terms and the prediction of ligand binding residues. The number of groups making predictions in this section was disappointingly poor given the importance of function prediction and the number of new prediction servers that had recently become available. Many groups that had published methods in this area did not participate in the experiment. It is almost certainly true that the slow release of functional information that hampered the assessment was also the cause of this low turnout. However, binding site predictions were something that could have readily been evaluated, so it was surprising that only two groups made consistent predictions for this category.

There were a number of difficulties in running a function prediction assessment in CASP, but nevertheless the function prediction category is important and should be maintained in some form. At present the main problem is the fact that it may take several years for the functional annotations of known structures to become public. This is not ideal for a rapidly developing field where predictors need to make use of the evaluation in order to refine their methods. Thus, it is imperative for the prediction community to coordinate its efforts with the experimental community to obtain data that could be readily used to develop and test new methods. This may be difficult, as the experimental community shows little coordination towards systematic experimental characterization of proteins that could be used (in advance) as targets for computational methods. Nonetheless, in recent years there have been several calls for community action to systematize efforts to functionally characterize so far uncharacterized protein families. The call to work together towards deciphering the role of the 'hypothetical proteins' encoded in microbial genomes by Richard J. Roberts was addressed both to bioinformaticians and experimentalists.[107] It was suggested that

> *a consortium of bioinformaticians [should] produce a list of all of the conserved hypothetical proteins that are found in multiple genomes, to carry out the best possible bioinformatics analysis, and then to offer those proteins to the biochemical community as potential targets for research into their function.*

However, one problem to solve would be how to avoid biasing the experimental function determination by bioinformatics predictions, which can possibly lead to artefacts. In fact, there are known cases of apparently erroneous computational predictions that seemed to have been supported by experimental evidence, whose value subsequently become questioned when the original computational predictions were refuted and replaced by alternative strongly supported predictions.[108]

Another related call for community action has been published by Peter D. Karp, who proposed obtaining at least one protein sequence for each previously biochemically characterized enzyme (with an assigned EC number) that has no sequence in the public databases.[109] This has interesting applications for the computational community, as it requires 'reverse prediction' of sequences for already known functions, and would require a process analogous to high-throughput 'virtual screening'. In order to identify sequences (or protein families) that have the highest likelihood to exhibit the target enzymatic activity, function prediction (in terms of EC numbers) would have to be carried out for representatives of all families. The results of such predictions can be more objectively evaluated, as they can be more easily combined with high-throughput enzymatic screens that do not necessarily take any particular predictions into account. In any case, progress in protein function prediction will be achieved only if theoreticians and experimentalists strengthen their collaborations, analogously to the very successful co-operations in the framework of the CASP experiment.

Acknowledgements

The authors would like to thank David Juan and Angela del Pozo for their invaluable contributions.

References

1. D. Maglott, J. Ostell, K. D. Pruitt and T. Tatusova, Entrez Gene: gene-centered information at NCBI, *Nucleic Acids Res*, **35**, D26–31 (2007).
2. M. Y. Galperin, Metagenomics: from acid mine to shining sea, *Environ Microbiol*, **6**, 543–545 (2004).
3. M. L. Tress, D. Cozzetto, A. Tramontano and A. Valencia, An analysis of the Sargasso Sea resource and the consequences for database composition, *BMC Bioinformatics*, **7**, 213 (2006).
4. D. Lee, O. Redfern and C. Orengo, Predicting protein function from sequence and structure. *Nat Rev Mol Cell Biol*, **8**, 995–1005 (2007).
5. D. Frishman, Protein annotation at genomic scale: the current status, *Chem Rev*, **107**, 3448–3466 (2007).
6. Godzik, M. Jambon and I. Friedberg, Computational protein function prediction: are we making progress? *Cell Mol Life Sci*, **64**, 2505–2511 (2007).
7. R. Sharan, I. Ulitsky and R. Shamir, Network-based prediction of protein function, *Mol Syst Biol*, **3**, 88 (2007).
8. Drosophila 12 Genomes Consortium, Evolution of genes and genomes on the Drosophila phylogeny, *Nature*, **450**, 203–218 (2007).
9. P. Mhawech, 14-3-3 proteins–an update, *Cell Res*, **15**, 228–236 (2005).
10. S. G. Park, K. L. Ewalt, S. Kim, Functional expansion of aminoacyl-tRNA synthetases and their interacting factors: new perspectives on housekeepers, *Trends Biochem Sci*, **30**, 569–574 (2005).
11. J. Piatigorsky and G. J. Wistow, Enzyme/crystallins: gene sharing as an evolutionary strategy, *Cell*, **57**, 197–199 (1989).
12. M. Ashburner, C. A. Ball, J. A. Blake, *et al.* Gene ontology: tool for the unification of biology. *Nat Genet*, **25**, 25–29 (2000).
13. Enzyme Nomenclature 1992 of IUBMB. Academic Press, New York (1992).
14. P. Bork and L. Serrano, Towards cellular systems in 4D. *Cell*, **121**, 507–509 (2005).
15. S. F. Altschul, W. Gish, W. Miller, E. W. Myers and D. J. Lipman, Basic local alignment search tool, *J Mol Biol*, **215**, 403–410 (1990).
16. S. F. Altschul, T. L. Madden, A. A. Schaffer, *et al.* Gapped BLAST and PSI-BLAST: a new generation of protein database search programs, *Nucleic Acids Res*, **25**, 3389–3402 (1997).
17. L. Jaroszewski, L. Rychlewski, Z. Li, W. Li and A. Godzik, FFAS03: a server for profile-profile sequence alignments. *Nucleic Acids Res*, **33**, W284–288 (2005).
18. S. R. Eddy, Hidden Markov models, *Curr Opin Struct Biol*, **6**, 361–365 (1996).
19. D. Devos and A. Valencia, Practical limits of function prediction, *Proteins*, **41**, 98–107 (2000).
20. E. Todd, C. A. Orengo and J. M. Thornton, Evolution of function in protein superfamilies, from a structural perspective, *J Mol Biol*, **307**, 1113–1143 (2001).
21. W. Tian and J. Skolnick, How well is enzyme function conserved as a function of pairwise sequence identity? *J Mol Biol*, **333**, 863–882 (2005).
22. L. Wang, A. Brock, B. Herberich and P. G. Schultz, Expanding the genetic code of Escherichia coli, *Science*, **292**, 498–500 (2001).
23. Valencia, Automatic annotation of protein function, *Curr Opin Struct Biol*, **15**, 267–274 (2005).
24. R. L. Marsden, D. Lee, M. Maibaum, C. Yeats and C. A. Orengo, Comprehensive genome analysis of 203 genomes provides structural genomics with new insights into protein family space, *Nucleic Acids Res*, **34**, 1066–1080 (2006).

25. J. Raes, E. D. Harrington, A. H. Singh and P. Bork, Protein function space: viewing the limits or limited by our view? *Curr Opin Struct Biol*, **17**, 362–369 (2007).
26. B. Rost, Prediction in 1D: secondary structure, membrane helices, and accessibility, *Methods Biochem Anal*, **44**, 559–587 (2003).
27. W. K. Huh, J. V. Falvo, L. C. Gerke, *et al.* Global analysis of protein localization in budding yeast, *Nature*, **425**, 686–691 (2003)
28. Pierleoni, P. L. Martelli, P. Fariselli and R. Casadio, BaCelLo: a balanced subcellular localization predictor, *Bioinformatics*, **22**, e408–e416 (2006).
29. K. Nakai, Prediction of in vivo fates of proteins in the era of genomics and proteomics, *J Struct. Biol*, **134**, 103–116 (2001).
30. L. Juhl Jensen, R. Gupta, N. Blom, et al.. *Ab initio* prediction of human orphan protein function from post-translational modifications and localization features, *J Mol Biol*, **319**, 1257–1265 (2002).
31. M. Levitt, Growth of novel protein structural data, *Proc Natl Acad Sci USA*, **104**, 3183–3188 (2007).
32. H. M. Berman, J. Westbrook, Z. Feng, *et al.* The Protein Data Bank, *Nucleic Acids Res*, **28**, 235–242 (2000).
33. Friedberg and A. Godzik, Functional differentiation of proteins: implications for structural genomics, *Structure*, **15**, 405–415 (2007).
34. D. Baker and A. Sali, Protein structure prediction and structural genomics, *Science*, **294**, 93–96 (2001).
35. Stark, S. Sunyaev and R.B. Russell, A model for statistical significance of local similarities in structure, *J Mol Biol*, **326**, 1307–1316 (2003).
36. C. Hermann, R. Marti-Arbona, A. A. Fedorov, *et al.* Structure-based activity prediction for an enzyme of unknown function, *Nature*, **448**, 775–779 (2007).
37. E. Sánchez, P. Beltrao, F. Stricher, *et al.* Genome-wide prediction of SH2 domain targets using structural information and the FoldX algorithm, *PLoS Comp Biol*, **4**, 4 (2008).
38. M. Sasin, A. Godzik and J. M. Bujnicki, SURF'S UP! – protein classification by surface comparisons, *J Biosci*, **32**, 97–100 (2007).
39. R. A. Laskowski, J. D. Watson, J. M. Thornton, ProFunc: a server for predicting protein function from 3D structure, *Nucleic Acids Res*, **33**, W89–93 (2005).
40. D. Pal and D. Eisenberg, Inference of protein function from protein structure, *Structure*, **13**, 121–130 (2005).
41. S. Tsoka and C.A. Ouzounis, Prediction of protein interactions: metabolic enzymes are frequently involved in gene fusion, *Nat Genet*, **26**, 141–142 (2000).
42. Tamames, G. Casari, C. Ouzounis and A. Valencia, Conserved clusters of functionally related genes in two bacterial genomes, *J Mol Evol*, **44**, 66–73 (1997).
43. Sjolander, Phylogenomic inference of protein molecular function: advances and challenges, *Bioinformatics*, **20**, 170–179 (2004).
44. G. Glanville, D. Kirshner, N. Krishnamurthy and K. Sjölander, Berkeley Phylogenomics Group web servers: resources for structural phylogenomic analysis. *Nucleic Acids Res*, **35**, W27–W32 (2007).
45. E. V. Koonin, A. R. Mushegian and P. Bork, Non-orthologous gene displacement. *Trends Genet*, **12**, 334–336 (1996).
46. Pellegrini, E.M. Marcotte, M.J. Thompson, D. Eisenberg and T.O. Yeates, Assigning protein functions by comparative genome analysis: protein phylogenetic profiles, *Proc Natl Acad Sci USA*, **96**, 4285–4288 (1999).
47. T. Gaasterland and M.A. Ragan, Microbial genescapes: phyletic and functional patterns of ORF distribution among prokaryotes, *Microb Comp Genomics*, **3**, 199–217 (1998).
48. K.J. Fryxell, The coevolution of gene family trees, *Trends Genet*, **12**, 364–369 (1996).
49. S. Pages, A. Belaich, J.P. Belaich, *et al.* Species-specificity of the cohesin-dockerin interaction between *Clostridium thermocellum* and *Clostridium cellulolyticum*: prediction of specificity determinants of the dockerin domain, *Proteins*, **29**, 517–527 (1997).
50. C.-S. Goh, A.A. Bogan, M. Joachimiak, D. Walther and F.E. Cohen Co-evolution of proteins with their interaction partners, *J Mol Biol*, **299**, 283–293 (2000).

51. F. Pazos and A. Valencia, Similarity of phylogenetic trees as indicator of protein-protein interaction, *Protein Eng*, **14**, 609–614 (2001).
52. F. Pazos, M. Helmer-Citterich, G. Ausiello and A. Valencia, Correlated mutations contain information about protein-protein interaction, *J Mol Biol*, **271**, 511–523 (1997).
53. F. Pazos and A. Valencia, In silico two-hybrid system for the selection of physically interacting protein pairs, *Proteins*, **47**, 219–227 (2002).
54. J. Huynen, B. Snel, W. Lathe and P. Bork, Predicting protein function by genomic context: quantitative evaluation and qualitative inferences, *Genome Res*, **10**, 1204–1210 (2000).
55. H. Jeong, S.P. Mason, A.L. Barabasi and Z.N. Oltvai, Lethality and centrality in protein networks, *Nature*, **411**, 41–42 (2001).
56. T. Ideker, V. Thorsson, J.A. Ranish, *et al.* Integrated genomic and proteomic analyses of a systematically perturbed metabolic network, *Science*, **292**, 929–934 (2001).
57. T. Ito, T. Chiba, R. Ozawa, M. Yoshid, M. Hattori and Y. Sakaki. A comprehensive two-hybrid analysis to explore the yeast protein interactome, *Proc Natl Acad Sci USA*, **98**, 4569–4574 (2001).
58. P. Uetz, L. Giot, G. Cagney, *et al.*, A comprehensive analysis of protein-protein interactions in *Saccharomyces cerevisiae*, *Nature*, **403**, 623–627 (2000).
59. S. Fields, High-throughput two-hybrid analysis. The promise and the peril, *FEBS J*, **272**, 5391–5399 (2005).
60. Y. Ho, A. Gruhler, A. Heilbut, *et al.*, Systematic identification of protein complexes in Saccharomyces cerevisiae by mass spectrometry, *Nature*, **415**, 180–183 (2002).
61. C. Gavin, M. Bösche, R. Krause, *et al.*, Functional organization of the yeast proteome by systematic analysis of protein complexes, *Nature*, **415**, 141–147 (2002).
62. R. Aebersold and M. Mann, Mass spectrometry-based proteomics, *Nature*, **422**, 198–207 (2003).
63. Simon, J. Barnett, N. Hannett, *et al.*, Serial regulation of transcriptional regulators in the yeast cell cycle, *Cell*, **106**, 697–708 (2001).
64. T. I. Lee, N. J. Rinaldi, F. Robert, *et al.*, Transcriptional regulatory networks in *Saccharomyces cerevisiae*, *Science*, **298**, 799–804 (2002).
65. G. D. Bader, D. Betel and C. W. Hogue, BIND: the Biomolecular Interaction Network Database, *Nucleic Acids Res*, **31**, 248–250 (2003).
66. C. von Mering, L. J. Jensen, B. Snel, *et al.*, STRING: known and predicted protein-protein associations, integrated and transferred across organisms, *Nucleic Acids Res*, **33**, 433–437 (2005).
67. G. D. Bader and C. W. Hogue, Analyzing yeast protein-protein interaction data obtained from different sources, *Nat Biotechnol*, **20**, 991–997 (2002).
68. Vazquez, A. Flammini, A. Maritan and A. Vespignani, Global protein function prediction from protein-protein interaction networks. *Nat Biotechnol*, **21**, 697–700 (2003).
69. G. G. Loots and I. Ovcharenko, Mulan: multiple-sequence alignment to predict functional elements in genomic sequences, *Methods Mol Biol*, **395**, 237–254 (2007).
70. R. Matthews, P. Vaglio, J. Reboul, *et al.*, Identification of potential interaction networks using sequence-based searches for conserved protein-protein interactions or 'interologs', *Genome Res*, **11**, 2120–2126 (2001).
71. H. Yu, N. M. Luscombe, H. X. Lu, *et al.*, Annotation transfer between genomes: protein-protein interologs and protein-DNA regulogs, *Genome Res*, **14**, 1107–1118 (2004).
72. B. Schwikowski, P. Uetz and S. Fields, A network of protein-protein interactions in yeast, *Nat Biotechnol*, **18**, 1257–1261 (2000).
73. T. Joshi, Y. Chen, J. M. Becker, N. Alexandrov and D. Xu, Genome-scale gene function prediction using multiple sources of high-throughput data in yeast *Saccharomyces cerevisiae*, *OMICS*, **8**, 322–333 (2004).
74. Nariai, E. D. Kolaczyk and S. Kasif, Probabilistic protein function prediction from heterogeneous genome-wide data, *PLoS ONE*, **2**, e337 (2007).
75. G. D. Bader and C. W. Hogue, An automated method for finding molecular complexes in large protein interaction networks, *BMC Bioinformatics*, **4**, 2 (2003).

76. Baudot, D. Martin, P. Mouren, *et al.*, PRODISTIN Web Site: a tool for the functional classification of proteins from interaction networks, *Bioinformatics*, **22**, 248–250 (2006).

77. Adamcsek, G. Palla, I. J. Farkas, I. Derényi and T. Vicsek, CFinder: locating cliques and overlapping modules in biological networks, *Bioinformatics*, **22**, 1021–1023 (2006).

78. M. Deane, L. Salwiński, I. Xenarios and D. Eisenberg, Protein interactions: two methods for assessment of the reliability of high throughput observations, *Mol Cell Proteomics*, **1**, 349–356 (2002).

79. von Mering, R. Krause, B. Snel, *et al.*, Comparative assessment of large-scale data sets of protein-protein interactions, *Nature*, **417**, 399–403 (2002).

80. G. T. Hart, A. K. Ramani and E. M. Marcotte, How complete are current yeast and human protein-interaction networks? *Genome Biol*, **7**, 120 (2006).

81. C. Blaschke, L. Hirschman and A. Valencia, Information extraction in molecular biology, *Brief Bioinf*, **3**, 154–165 (2002).

82. C. Blaschke and A. Valencia, The frame-based module of the Suiseki information extraction system, *IEEE Intell. Syst*, **17**, 14–20 (2002).

83. C. Friedman, P. Kra, H. Yu, M. Krauthammer and A. Rzhetsky, GENIES: A natural-language processing system for the extraction of molecular pathways from journal articles, *Bioinformatics*, **17**, S74–82 (2001).

84. D. R. Swanson, Migraine and magnesium: Eleven neglected connections, *Perspect Biol Med*, **31**, 526–557 (1988).

85. C. Blaschke and A. Valencia, The potential use of SUISEKI as a protein interaction discovery tool, *Genome Inf Ser Workshop Genome Inf*, **12**, 123–134 (2001).

86. C. Blaschke, E. Andres, M. Krallinger and A. Valencia, Evaluation of BioCreative assessment of task 2. *BMC Bioinformatics*, **6**, S16 (2005).

87. R. Hoffmann and A. Valencia, A gene network for navigating the literature, *Nature Genet*, **36**, 664 (2004).

88. E. Sequeira, PubMed Central – three years old and growing stronger, *ARL*, **228**, 5–9 (2003).

89. S. Kerrien, Y. Alam-Faruque, B. Aranda, *et al.*, IntAct – open source resource for molecular interaction data, *Nucleic Acids Res*, **35**, D561–565 (2007).

90. C. Gavin, P. Aloy, P. Grandi, *et al.*, Proteome survey reveals modularity of the yeast cell machinery, *Nature*, **440**, 631–636 (2006).

91. J. Krogan, G. Cagney, H. Yu, *et al.*, Global landscape of protein complexes in the yeast *Saccharomyces cerevisiae*, *Nature*, **440**, 637–643 (2006).

92. F. Sobott, M. G. McCammon, H. Hernández and C. V. Robinson, The flight of macromolecular complexes in a mass spectrometer, *Philos Transact A Math Phys Eng Sci*, **363**, 379–391 (2005).

93. V. Lucić, F. Förster and W. Baumeister, Structural studies by electron tomography: from cells to molecules, *Annu Rev Biochem*, **74**, 833–865 (2005).

94. J. L. Benesch, B. T. Ruotolo, D. A. Simmons and C. V. Robinson, Protein complexes in the gas phase: technology for structural genomics and proteomics, *Chem Rev*, **107**, 3544–3567 (2007).

95. H. Hernández and C. V. Robinson, Determining the stoichiometry and interactions of macromolecular assemblies from mass spectrometry, *Nat Protoc*, **2**, 715–726. (2007).

96. Aloy, B. Böttcher, H. Ceulemans, *et al.*, Structure-based assembly of protein complexes in yeast, *Science*, **303**, 2026–2029 (2004)

97. F. Alber, S. Dokudovskaya, L. M. Veenhoff, *et al.*, The molecular architecture of the nuclear pore complex, *Nature*, **450**, 695–701 (2007).

98. Y. Lim and B. Fahrenkrog, The nuclear pore complex up close, *Curr Opin Cell Biol*, **18**, 342–347 (2006).

99. D. Devos, S. Dokudovskaya, R. Williams, *et al.*, Simple fold composition and modular architecture of the nuclear pore complex, *Proc Natl Acad Sci USA*, **103**, 2172–2177 (2006).

100. Takamori, M. Holt, K. Stenius, *et al.* Molecular anatomy of a trafficking organelle, *Cell*, **127**, 831–846 (2006).

101. M. Wollacott, A. Zanghellini, P. Murphy and D. Baker, Prediction of structures of multidomain proteins from structures of the individual domains, *Protein Sci*, **16**, 165–175 (2007).

102. Krisko and C. Etchebest, Theoretical model of human apolipoprotein B100 tertiary structure, *Proteins*, **66**, 342–358 (2007).

103. J. Moult, A decade of CASP: progress, bottlenecks and prognosis in protein structure prediction. *Curr Opin Struct Biol*, **15**, 285–289 (2005).

104. 'J. Moult, K. Fidelis, A. Kryshtafovych, B. Rost, T. Hubbard and A. Tramontano. Critical assessment of methods of protein structure prediction-Round VII, *Proteins*, **69**, 3–9 (2007).

105. Soro and A. Tramontano, The prediction of protein function at CASP6, *Proteins*, **61**, 201–213 (2005).

106. G. López, A. Rojas, M. L. Tress and A. Valencia, Assessment of predictions submitted for the CASP7 function prediction category, *Proteins*, **69**, 165–174 (2007).

107. R. J. Roberts, Identifying protein function – a call for community action, *PLoS Biol*, **2**, E42 (2004).

108. M. Iyer, L. Aravind, P. Bork, *et al.*, Quoderat demonstrandum? The mystery of experimental validation of apparently erroneous computational analyses of protein sequences. *Genome Biol*, **2**, research0051 (2001).

109. P. D. Karp, Call for an enzyme genomics initiative, *Genome Biol*, **5**, 401 (2004)..

Index

Prediction of Protein Structures, Functions, and Interactions Edited by Janusz Bujnicki
© 2009 John Wiley & Sons, Ltd